# ENZYME-CATALYSED REACTIONS

# ENZYME-CATALYSED REACTIONS

C. J. GRAY

*Department of Chemistry, University of Birmingham*

VAN NOSTRAND REINHOLD COMPANY
LONDON
NEW YORK CINCINNATI TORONTO MELBOURNE

VAN NOSTRAND REINHOLD COMPANY
Windsor House, 46 Victoria Street, London S.W.1

INTERNATIONAL OFFICES
New York Cincinnati Toronto Melbourne



Library of Congress Catalog Card No. 75-125951
SBN 442 02808 3

*First published 1971*

Printed in Great Britain by
Butler & Tanner Ltd, Frome and London

To Jennifer

# CONTENTS

# PREFACE

Recent advances in biology have meant that more attention is being paid to the study of events which take place, at the molecular level, in living systems. For this reason the distinctions between biology, biochemistry and chemistry are becoming increasingly difficult to define. Progressively more biological scientists are finding it necessary to become acquainted with various subjects which were once regarded as primarily the domain of the physical or organic chemist. Similarly, chemists are applying their ideas and methods to the study of biological processes. One field in particular which holds the attention of both these groups of scientists is that of the study of enzyme-catalysed reactions. One aim of this book is to help to bridge the gap between the chemical and biological disciplines even further in this field, by illustrating for the biologist that the enzymes with which he is familiar can be considered in terms of the fundamental principles of organic chemistry, and for the organic chemist that these principles, established for simple molecules, may be extended to the extremely complex molecules upon which life itself depends.

On the whole this book is directed mainly to the senior undergraduate and postgraduate students of both biochemistry and chemistry. In the first chapter some of the basic principles of the theory of reaction mechanisms, including some thermodynamic aspects, are discussed, the treatment being neither very detailed nor too rigorous. This, it is hoped, will provide a suitable introduction for those who are not familiar with the subject. Those who wish to study the subject in more depth should refer to the appropriate textbooks [1, 2, 3, 4]†. The chapter should be of value, too, to those who are familiar with the subject since it is treated from a slightly different point of view which emphasizes the rates of reactions and the ways in which organic compounds can influence those rates.

The purpose of Chapter 2 is to describe some of the methods (mainly chemical) which have been used to study the enzymes and obtain information about their modes of action, so providing a general background to the greater detail contained in the later chapters. These later chapters (Chapters 3–8) describe many of the enzymes individually. In Chapter 9 a brief summary is given of some of the factors which are thought at present to be important to the mechanisms of catalysis by enzymes.

In presenting this kind of material the writer is inevitably confronted with the choice of making his discussion of individual enzymes either

† See references after Chapter 1.

more comprehensive or more selective. Apart from the fact that he is limited by the physical size of the final book, the writer who chooses to be comprehensive in his approach and to present as much information as possible runs the real risk of having the important principles of the subject missed by his reader, because they are clouded by the overwhelming detail. On the other hand, if the writer chooses to be more selective in that he discusses only a very few examples emphasizing the main principles involved, then he is in danger of presenting to the reader only a very small fraction of the vast amount of information which may be available on the subject. In addition, selectivity is of necessity subjective on the part of the writer, and this subjectivity may be undesirable in certain circumstances. Although the field is too large for this book to include a discussion of every enzyme, I have tended towards the more comprehensive, and I hope more objective, approach in that a relatively large number of enzymes have been discussed individually. This allows the reader to see for himself as broad a picture as possible of the whole field. I believe that this more comprehensive treatment fills a gap in the published literature where there are already some good examples of the selective approach.

I am grateful to all those who have given me permission to use their photographs and line drawings in this book, especially those showing the three-dimensional structures of molecules based on X-ray crystallography. I wish to thank Professor S. A. Barker for his invaluable advice and encouragement, Mrs C. M. J. Law for her assistance with drawings and references and Mr. T. H. Yeo for helping with the reading of the proofs.

Thanks are also due to the Efficient Typing Service, Birmingham, for their typing of the manuscript, and especially to the staff of Van Nostrand Reinhold for their advice and help in the preparation of this book.

The jacket design is based on the model of ribonuclease as determined by G. Kartha and co-workers.

1970 C. J. GRAY

# CLASSIFICATION OF ENZYMES

On the whole, enzymes are considered in this book in the groups which correspond to the classification recommended by the Commission on Enzymes (1956–61) of the International Union of Biochemistry. The major advantage of this classification system is that the enzymes are considered according to the overall nature of the reactions catalysed. This means that an enzyme can be classified and assigned a systematic name and number as soon as it has been detected since an enzyme is of course detected by means of the reaction which it brings about. An alternative method of classifying the enzyme could be envisaged based on the mechanism of action of each enzyme. In many cases this would coincide with the Commission's classification but there would be certain differences. However, this alternative method would possess the great disadvantage that an enzyme could not be classified, and therefore given a systematic name, until its mechanism of action has been established. Since in only a few cases have we any clear idea at all of the mechanism by which an enzyme functions, an impossible situation would develop. Because we are considering the enzymes from the point of view of their possible reaction mechanisms, in order that as far as possible a unified approach is taken, certain enzymes must be considered out of the order recommended by the Commission. The most important change is that the hydrolases are considered in Chapters 4 and 5 before the transferases. In this way certain principles and concepts are introduced at an earlier stage. Some other exceptions to the classification order are that the thiamine pyrophosphate-dependent *transketolase* is considered in the same section of Chapter 7 as are the thiamine pyrophosphate-dependent *lyases*, and the enzyme *transaldolase* which functions by a mechanism involving an active lysine residue is discussed with the active lysine lyases in Chapter 7. Those *ligases* which contain biotin are also considered out of turn, being taken with the biotin *transferases*. However, it is not a general rule that all enzymes which require one particular coenzyme are considered together. For example the vitamin $B_{12}$ coenzyme transferases are discussed in Chapter 6 while other enzymes requiring this coenzyme are referred to in Chapter 8. Similarly, the pyridoxal-dependent enzymes are not all considered together in the same chapter.

One further difficulty arises from the Enzyme Commission Classification which should be noted; this is that in some cases enzymes which catalyse the same reactions are given the same nomenclature and number, even where these enzymes may be obtained from different sources. For example, *alcohol dehydrogenase* (alcohol: NAD oxidoreductase, 1.1.1.1) may be obtained from mammalian liver or from yeast. The enzymes from these two sources are not identical and there are certain differences in their mechanisms of action. On the whole, conclusions about the mechanism of action of an enzyme from one source may not be drawn from experiments on a similar enzyme from a different source. This statement does not merely apply to enzymes obtained from different types of organism (such as yeast, bacteria, mammals, etc.) but even to enzymes from different sources within the same organism. *ATPase* (ATP phosphohydrolase, 3.6.1.3) from muscle may differ from the ATPase from cell membranes in the same animal.

The following terms are used in this book without further definition:

Apo-enzyme—The protein-only component of a complex enzyme, from which coenzyme and metal-ion has been removed.

Holo-enzyme—A complete enzyme, comprising apo-enzyme and any required coenzyme or metal-ion.

# CHAPTER 1

# RATES AND REACTION MECHANISMS

## REACTION RATES

One of the most important properties of a chemical reaction is the rate at which it proceeds. Among the factors which determine that rate may be listed the type of reaction which is involved, the nature and intrinsic molecular properties of the chemical species which undergo the reaction, and various external factors such as the environment of the molecules (whether or not a solvent is used) and the temperature. In addition, the rates of many chemical reactions may be influenced by the presence of foreign substances, that is, by catalysts.

### Rates and Rate Constants

If a chemical reaction actually proceeds according to the equation

$$X + Y \rightarrow P + Q \qquad (1.1)$$

that is, at the molecular level a single stage is involved in which one molecule of species X reacts with one of Y to produce one molecule each of P and Q, then experimentally we would find that the rate of the reaction could be expressed by the equation

$$\text{Rate} = k[X][Y], \qquad (1.2)$$

where $k$ is a constant known as the *rate constant* for the reaction and the term [X] represents the *activity* of the species X, etc. On the whole we shall be concerned with reactions occurring in solution where the concentrations of the reacting species are low, and in these cases activities become equivalent to concentrations. Hence, [X] refers to the concentration of X, etc. This reaction is said to be first order in X because the concentration term [X] appears in the rate equation (1.2) raised to the power one. Similarly the reaction is first order in Y. This is in fact a reflection of the original statement that one molecule of X reacts with one of Y. If the process had involved two molecules of X reacting with one of Y in a single step,

$$2X + Y \rightarrow \text{Products}, \qquad (1.3)$$

then the rate equation would have been as follows:

$$\text{Rate} = k[X]^2[Y] \qquad (1.4)$$

in which the reaction would have been second order in X. Thus by studying experimentally the way in which the rate of a reaction depends on the concentrations of the reactants, one may determine the order of the reaction and hence obtain some information about its mechanism. If the reaction proceeds other than by a single step, the situation is more complicated.

One of the factors, therefore, which determine the rate of a chemical reaction is the concentration of each reactant according to the order. All the other rate-determining factors are included in the rate constant $k$.

Most reactions in organic chemistry, and certainly most of those with which we shall be dealing, are reversible. Thus, for reaction (1.1), the products P and Q can themselves react to re-form the starting materials X and Y. The reaction may be written

$$\mathrm{X} + \mathrm{Y} \underset{k_{-1}}{\overset{k_{+1}}{\rightleftharpoons}} \mathrm{P} + \mathrm{Q}, \tag{1.5}$$

where $k_{+1}$ and $k_{-1}$ are the rate constants for the forward and reverse reactions respectively. The reaction can proceed until the concentrations of the species present are such that the rate of the forward reaction is exactly equalled by that of the reverse process. At this stage,

$$k_{+1}[\mathrm{X}][\mathrm{Y}] = k_{-1}[\mathrm{P}][\mathrm{Q}].$$

Hence

$$\frac{k_{+1}}{k_{-1}} = \frac{[\mathrm{P}][\mathrm{Q}]}{[\mathrm{X}][\mathrm{Y}]} = K\,\mathrm{eq}. \tag{1.6}$$

The system is in a state of dynamic equilibrium, and the ratio of the two rate constants, $K$ eq, is known as the *equilibrium constant*. This constant is of great importance from the thermodynamic point of view because it can be related to the *free energy* change which accompanies the reaction. If $\Delta G^0$ is the standard free energy change, that is, the change in free energy when one mole of X and one mole of Y react together to give one mole each of P and Q under standard conditions, then

$$\Delta G^0 = -RT.\ln K\,\mathrm{eq}, \tag{1.7}$$

when $K$ eq is expressed in terms of the standard conditions.

The free energy change, $\Delta G^0$, may be expressed in terms of the changes in *heat content* and in *entropy*:

$$\Delta G^0 = \Delta H^0 - T.\Delta S^0, \tag{1.8}$$

where $\Delta H^0$ is the standard heat of reaction and $\Delta S^0$ is the standard entropy of reaction (for an introduction to thermodynamics and for definitions of the thermodynamic terms used here the reader is referred to Everett [1]).

From these equations, it is possible to derive the following relationships which show how the equilibrium constant varies with temperature.

$$\frac{d(\ln K\,\text{eq})}{dT} = \frac{\Delta H^0}{RT^2} \quad \text{or} \quad \frac{d(\ln K\,\text{eq})}{d\left(\frac{1}{T}\right)} = \frac{-\Delta H^0}{R}. \tag{1.9}$$

If the heat of the reaction is independent of temperature then (1.9) gives

$$\ln K\,\text{eq} = \frac{-\Delta H^0}{R}\left(\frac{1}{T}\right) + \text{a constant} \tag{1.10}$$

and thus, from a graph of ln $K$ eq against $1/T$ one may obtain by experiment a value for the heat of reaction. Since we may use (1.7) to obtain $\Delta G^0$ and (1.8) for $\Delta S^0$, it is possible to find values for all the thermodynamic parameters of the reaction.

## The Arrhenius Equation and the Collision Theory

Most chemical reactions proceed far more rapidly when the temperature is raised; very approximately an increase of ten degrees may cause the rate to be doubled [5]. Arrhenius [6] found that the relationship between the rate constant and the temperature could be expressed by the equation

$$\ln k = \ln A - \frac{E}{RT} \quad \text{or} \quad k = A.e^{-E/RT}, \tag{1.11}$$

where $A$ and $E$ are constants. This is known as the Arrhenius equation, and from a graph of ln $k$ against $1/T$ one may obtain values for $A$ and $E$.

A simple explanation for this relationship is afforded by the *collision theory*, which may be briefly described in the following way. In order that the reaction (1.1) may take place, a molecule of X must come into contact, or collide, with one of Y. The rate of reaction will therefore depend on the frequency of such collisions. If the reaction occurs in the gas phase the kinetic theory of gases may be used to calculate the collision frequency. Our discussion will concentrate on reactions in solution and the kinetic theory may be extended to apply to these conditions. Among other things the collision frequency depends upon the concentration of each species, according to the order of the collision.

$$\text{Frequency} = \text{constant}.[\text{X}][\text{Y}]. \tag{1.12}$$

The calculated collision frequency turns out to be very much greater than that expected from the reaction rate, and it is clear that only a small proportion of the collisions lead to a reaction. In fact, only collisions involving molecules which possess an exceptional amount of energy will give rise to products. This amount of energy which the molecules must have in order to undergo a fruitful collision is the *energy of activation*. The Maxwell–Boltzman law indicates that the fraction of molecules in a system which possess at least an amount of energy $E$ at temperature $T$ is approximately proportional to the term $e^{-E/RT}$. The frequency of fruitful collisions, which should be equal to the rate of the reaction, may therefore be expressed

$$\text{Frequency} = \text{rate} = A[X][Y].e^{-E/RT}. \qquad (1.13)$$

Comparison with (1.2) indicates that the rate constant $k$ is given by

$$k = A.e^{-E/RT}.$$

This is equation (1.11) and from this we infer that in the Arrhenius equation, $A$ must be a factor related to the frequency of collisions (the frequency factor) and $E$ is the energy of activation for the reaction. It has already been said that it is possible for values for $A$ and $E$ to be obtained experimentally.†

It became apparent to early workers that, in many cases, the experimentally-determined values for $A$ did not correspond with the calculated values based on collision frequencies and so a further modifying factor was introduced. For almost all chemical reactions, the molecules which collide must do so with the correct alignment or orientation to each other. The majority of organic reactions are controlled by severe steric requirements in this way. For example, in the case of the reaction between methyl bromide and the chloride ion, which is represented semipictorially below, only a collision involving relative movement in the direction of the arrow (a) will result in reaction. Relative movement in any other direction such as (b) will not be fruitful. In other words the methyl bromide must present to the chloride ion the face opposite the bromine atom. In more complicated reactions, both reacting species must present the correct faces at the moment of collision:

$$Cl^{\ominus} + CH_3Br \longrightarrow Cl.CH_3 + Br^{\ominus}$$

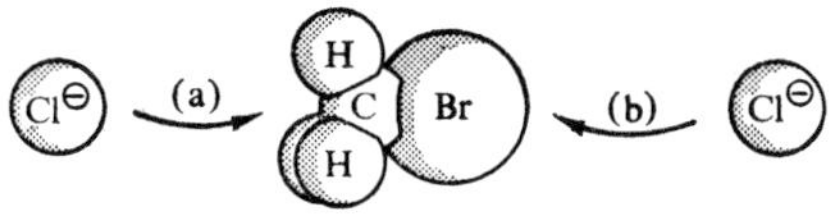


Because of this effect the number of collisions leading to reaction is further decreased, and the steric factor introduced to allow for it is usually less than one, although in some cases, especially reactions involving ions, the correction factor may be greater than unity. If we define this steric

† The graphical method used for the determination of $A$ and $E$ is based on the assumption that $A$ does not vary with temperature, and hence that the plot will be a straight line. In fact, the collision number does vary with temperature, being proportional to $T^{1/2}$. However, over small temperature ranges this variation is not great, especially in comparison with changes in the term $e^{-E/RT}$. For example, over the range 273–303 °K, $A$ will increase by a factor of about 1·05, whereas the term $e^{-E/RT}$ will increase by a factor of about 18 (assuming a value for $E$ of approximately 22 kcal/mole). Early experimentalists also failed to observe a very small change in $E$ with temperature.

factor $P$ as the probability that colliding molecules with sufficient energy will react, the expression for the rate constant becomes

$$k = A.P.e^{-E/RT}. \tag{1.14}$$

**The Transition-State Theory**

The steric factor $P$ mentioned above is purely empirical and can only be obtained by comparing theoretical with experimental results. For this reason, the collision theory leaves much to be desired. The *transition-state theory*, proposed by Eyring and his colleagues [7], approaches the problem from an entirely different point of view. If we consider in detail the processes which take place during the reaction just mentioned, that is, between methyl bromide and the chloride ion, we may envisage the intermediate situation where the bond between the carbon and chlorine atoms is only partially formed and that between the carbon and bromine atoms is only partially broken. At a certain stage the free energy of the system is at a maximum and this state is known as an activated *complex* or the

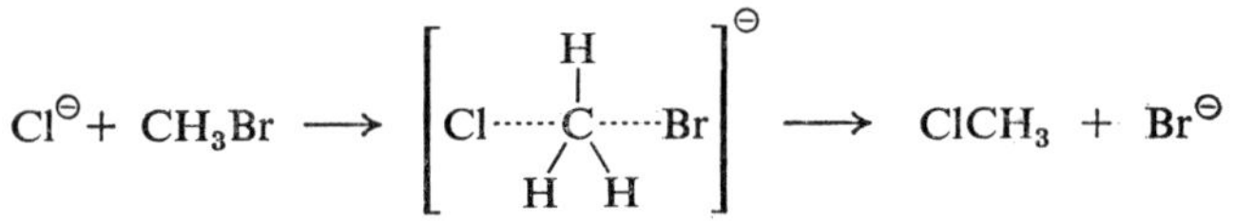

*transition state.* All reactions are pictured as passing through such a transition state, and the transition state is considered to be in equilibrium with the starting materials. If the equilibrium constant for this process is $K^*$ then the rate constant can be shown to be proportional to $K^*$, the constant of proportionality being almost exactly equal to $\kappa T/h$, where $\kappa$ is Boltzman's constant, $T$ is the temperature and $h$ is Planck's constant. Thus,

$$k = (\kappa T/h)K^*. \tag{1.15}$$

As before, (1.7), we may relate an equilibrium constant to a free energy change and if $\Delta G^*$ is the free energy change in passing from starting materials to activated complex, i.e. the *free energy of activation*, then

$$k = \left(\frac{\kappa T}{h}\right) \exp\left(\frac{-\Delta G^*}{RT}\right). \tag{1.16}$$

Also as before, $\Delta G^*$ may be considered in terms of the heat and entropy changes

$$\Delta G^* = \Delta H^* - T\,\Delta S^*. \tag{1.17}$$

From (1.17) we obtain

$$k = \left(\frac{\kappa T}{h}\right) \exp\left(\frac{\Delta S^*}{R}\right) \exp\left(\frac{-\Delta H^*}{RT}\right), \tag{1.18}$$

where $\Delta H^*$ is the heat of activation and $\Delta S^*$ the entropy of activation.

It may be shown that the following three relationships hold, and these may be used to obtain values for the three activation parameters.

$$\Delta G^* = -RT \ln \frac{kh}{\kappa T}$$

$$\Delta H^* = -R\left[\frac{(d\ln k)}{d(1/T)} + T\right] \tag{1.19}$$

$$\Delta S^* = \frac{\Delta H^* - \Delta G^*}{T} = R\left[\frac{Td\ln k}{dT} + \ln\frac{kh}{\kappa T} + T\right]. \tag{1.20}$$

Inspection of equation (1.19) and the equivalent form of (1.11) shows that the Arrhenius activation energy, obtained from a plot of $\ln k$ against $1/T$ differs from the heat of activation $\Delta H^*$ by an amount equal to $RT$. At normal temperatures this is equal to about 0·6 kcal/mole and this difference is often within the experimental error of the determination. Sometimes this difference is ignored and the terms energy of activation and heat or enthalpy of activation are used to refer to the same quantity. Comparison of equations (1.18) and (1.14) suggests that the steric factor $P$ might be interpreted in terms of the entropy of activation. We have already referred to $P$ as the probability, given that the energy (or better enthalpy) requirement is satisfied, that the collision will result in reaction. Entropy is a measure of the randomness of a system, or a measure of the freedom from restraint of that system. If the transition state for a reaction is more ordered or has fewer degrees of freedom than the starting materials then the entropy of activation will be negative, a decrease in entropy having occurred in the formation of the transition state. Such a situation exists when there are strict steric requirements to be met, i.e. when we might expect $P$ to be low.

At this stage it is worthwhile for us to examine some examples where the entropies of activation have easily discernible effects. These examples help to explain the nature of entropy effects. Reactions which occur by the association of molecules, that is when the total number of species decreases, are accompanied by a negative entropy of activation. Clearly when two molecules coalesce to form an activated complex their freedom of movement is severely restricted and so the transition state possesses a lower entropy than the unassociated starting materials. Therefore the overall order of a reaction, which for a single-step process is a measure of the number of molecular species which become associated with each other, can have a significant effect on the entropy of activation and therefore on the rate. This phenomenon is very important in connection with proximity effects which are discussed on page 52.

If two similar reactions proceed by the same mechanism and in one of them the transition state is highly crowded then this reaction will have the larger negative entropy of activation. This is clearly shown in several nucleophilic substitution reactions [8, 9 and 10]. For example, the entropy of activation for the bromide replacement

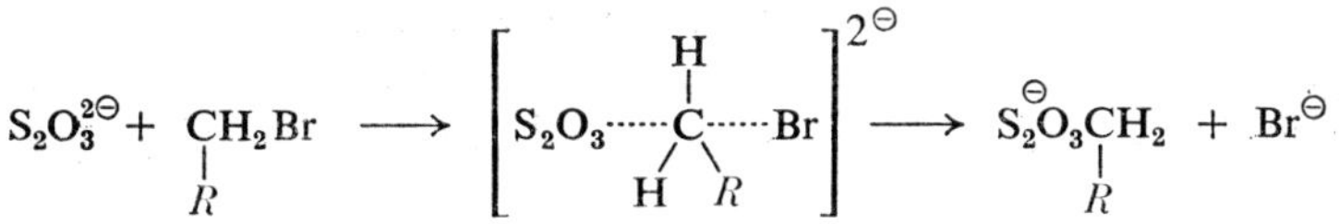

is −0·6 kcalories per degree per mole if *R* is hydrogen, but when *R* is the iso-propyl group the entropy of activation is found to be −3·4 kcalories per degree per mole [8]. The relatively large iso-propyl group clearly restricts the freedom of movement of all the groups in the transition state.

Just as $\Delta G^*$ represents the change in free energy of the *whole system* in progressing from starting materials to transition state, so $\Delta S^*$ represents the change in entropy of the whole system, and this must include the solvent in which the reaction is taking place. Solvent effects are especially important when ionic species are involved. Molecules of a solvent such as water arrange themselves regularly around each charged species and thus a considerable amount of 'orderliness' will be associated with the system as a whole. When the charged species react, the solvent molecules must rearrange themselves with the result that there is a change in the overall entropy. If the process involves the reaction of a positively-charged molecule with one that is negatively charged then in the transition state the charges are partially neutralized and the orderly arrangement of the solvent molecules around the charged species is destroyed. This will be reflected in an increase in entropy, that is, in a positive entropy of activation. Some reactions between oppositely-charged species do indeed show a value for *P* greater than unity.

Similarly if a reaction involves the production of charged species from neutral molecules then the increase in orientation of the solvent around the transition state leads to a decrease in entropy and hence to a reduction in the reaction rate.

From equation (1.16) it is clear that the most important factor upon which the rate of a reaction depends is the free energy of activation $\Delta G^*$. We may regard this as being a composite term consisting of two parts, the heat of activation $\Delta H^*$ and the entropy of activation–temperature product $T.\Delta S^*$. Just as a high negative entropy of activation may cause the reaction to be slow, a positive value for $T.\Delta S^*$ may increase a reaction rate even though the heat of activation may be high. For example, the denaturation of proteins can occur quite rapidly at relatively low temperatures despite the fact that the heats of activation have rather large values. This is because the reactions proceed through transition states which are more 'open' than the original proteins and which represent therefore increases in randomness. This leads to a positive value for the entropy of activation.

A convenient way of representing the course of a reaction as it proceeds through its transition state is afforded by the energy profile diagram such as that shown in Fig. 1.1. In this diagram the horizontal scale, representing

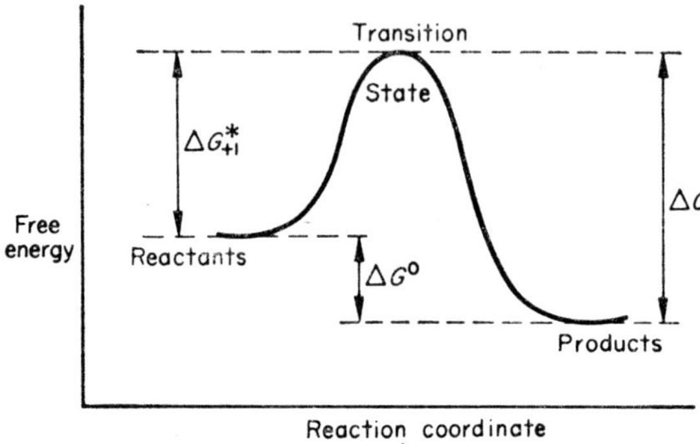


FIG. 1.1. Energy-profile diagram for a single-step process

the reaction coordinate, is a measure of the extent of reaction. $\Delta G^*_{+1}$ is the difference between the free energies of the starting materials and the transition state and is therefore the free energy of activation for the forward reaction. $\Delta G^*_{-1}$ is the free energy of activation of the reverse process. It can be clearly seen that the difference between these, $\Delta G^0$, is the standard free energy change for the whole reaction and so the *kinetic* parameters can be related to those associated with the equilibrium situation.

The energy profile diagram shown in Fig. 1.1 represents a situation where the products are at a lower free energy level than the reactants and in such a case $K$ eq will be greater than unity. An equilibrium mixture will result in which the products will predominate.

In the foregoing discussion it has been assumed that the energy level of the transition state for the forward reaction is the same as that for the reverse process. This is in fact obligatory and is a consequence of an important principle encountered in the field of reaction mechanisms. This is the principle of *microscopic reversibility* and it may be briefly expressed as follows: the mechanism of a reversible reaction is the same in microscopic detail for the reaction in one direction as in the other (given the same conditions). Application of this principle can be very valuable especially when reactions involving more than one step are under consideration.

The distinction between kinetic and thermodynamic factors becomes important when we consider the case of a reagent which can undergo two competing reactions. Thus if we consider a species E which may undergo

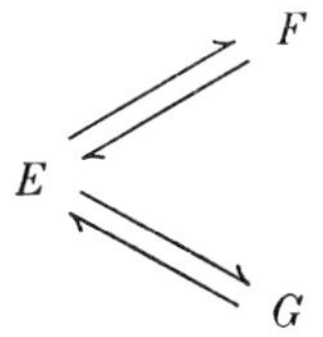


reversible reactions forming either F or G, with the energy profiles as shown (Fig. 1.2) the formation of the products is regarded as being under either kinetic or thermodynamic control. The reaction in which E is converted into F, having a lower energy of activation, will proceed more rapidly and hence F will accumulate at an early stage. However, because

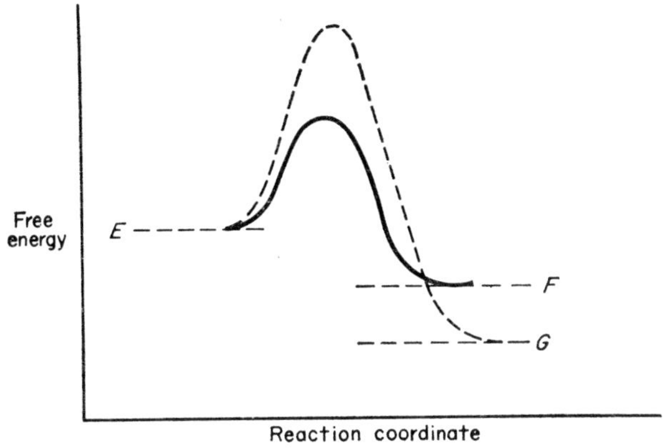


FIG. 1.2. Energy-profile diagram for reactions under kinetic and thermodynamic control

both the reactions are reversible and because the energy level of G is lower than that of F or of E then in an equilibrium mixture G will predominate. Therefore, although F would predominate after a relatively short time, eventually G would become the major product. If we were to carry out the reaction experimentally we could isolate high yields of either F or G according to the timing of the isolation. The formation of F is said to be under kinetic control and that of G under thermodynamic control.

An example of this phenomenon is afforded by the sulphonation of naphthalene, where two products are possible [11]. Reaction may occur at either the 1- or the 2-position giving naphthalene-1-sulphonic acid or naphthalene-2-sulphonic acid.

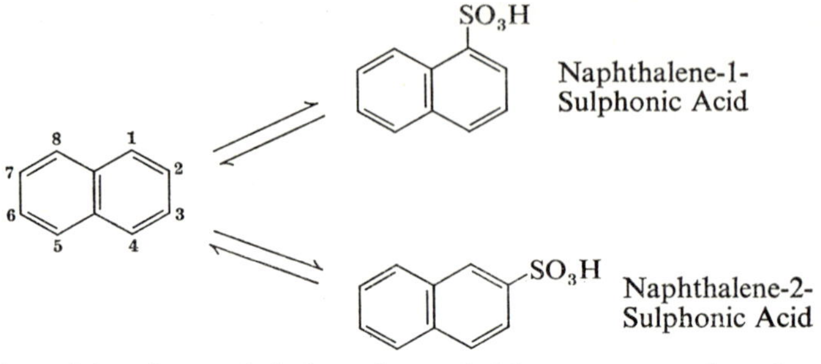


The 1-position in naphthalene is probably more reactive than the 2-position, which is another way of saying that the energy of activation for reaction at the 1-position is lower than that for the 2-position. The 1-sulphonic acid is therefore formed rapidly, in significant quantities. However, the 2-sulphonic acid is the more stable (because in the 1-isomer a great deal of strain is produced by interaction between the bulky sulphonic acid group and the hydrogen atom in position 8—the *peri*-position) and on prolonged reaction, the 2-isomer is obtained as by far the major product.

So far we have discussed only reactions which proceed in one step. Many chemical processes, however, require several stages and each stage is governed by the appropriate rate constant. For a complete description of such a process all the rate constants must be considered. For example, if a reaction in which a species A is converted into species C involves two stages with species B as an intermediate, thus

$$A \underset{k_{-1}}{\overset{k_{+1}}{\rightleftharpoons}} B \underset{k_{-2}}{\overset{k_{+2}}{\rightleftharpoons}} C$$

and the rate constants for the separate steps are as shown, then we may represent the process by one of several energy profile diagrams. Two of

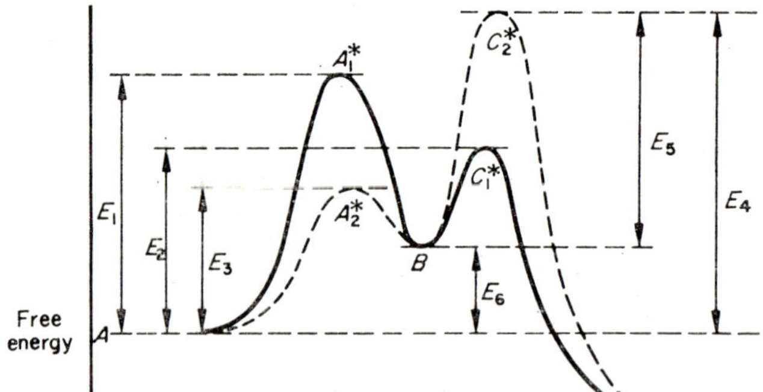


FIG. 1.3. Two possible energy-profile diagrams for a two-stage process. Pathway 1—continuous line; pathway 2—broken line

the possibilities are presented in Fig. 1.3. In the first of these, pathway 1 which is represented by the continuous line, $A_1^*$ is the transition state for the first stage. B, the intermediate, lies at a position of minimum energy and $C_1^*$ is the transition state for the second stage. It is also of course the transition state for the first step (C $\rightarrow$ B) of the reverse process. The transition states for the alternative pathway are $A_2^*$ and $C_2^*$, respectively. For pathway 1, $A_1^*$ is at a higher energy level than $C_1^*$ and therefore the rate of the first step would be lower than that of the second (other things being equal). In a sequence of reactions the overall rate is equal to that of the slowest step and so in this case we might say that the first step is the rate-determining step. We shall use the term *rate-determining step* to describe that step in a reaction sequence for which the transition state corresponds to the state of maximum energy over the whole sequence. We can express the same thing in another way. The rate of the first step depends on the number of molecules of A which have sufficient energy $E_1$, to reach the activated state $A_1^*$. Now provided that this energy is not dissipated in some other way during passage through B these molecules will clearly have sufficient energy $E_2$ to reach $C_1^*$ and hence proceed to C.

In the case of pathway 2, however, molecules which have the energy $E_3$ to reach $A_2^*$ still require more energy (up to $E_4$) before the second step can proceed. Hence, the rate at which the molecules can proceed through the maximum-energy transition state will determine the overall rate of reaction.

It is important to note that if the rate-determining step in a sequence is also the first step then the overall energy of activation (that is, the energy difference between the starting materials and the state of maximum energy) is equal to the energy of activation for that first step. In any other case the overall energy of activation is a complex term. For pathway 2 for example the overall energy of activation ($E_4$) is made up of the energy of activation for the second step ($E_5$) plus the free energy *change* for the first step ($E_6$).

Although almost all of the reaction mechanisms discussed later in this book are considered mainly in terms of the transition-state theory it must be pointed out that for theoretical chemists this theory may be no less unsatisfactory than the simple collision theory. Indeed, it would appear unreasonable to suggest that each individual molecule during its conversion into product passes through a rigidly-defined transition state whose free energy content is exactly the same for all molecules. Just as the temperature of a sample of matter represents an average value for the vibrational and translational energies of every molecule, so does the experimentally-determined energy of activation represent an average value for the free energy content of the range of transition states which may be involved. In addition it must be emphasized that no transition state exists for any finite length of time. Transition states are by their very nature of only temporary existence.

However, we may justify our continued use of the transition-state theory, at least as a basis for discussion of organic reaction mechanisms, by suggesting that probably the majority of molecules undergoing a reaction pass through a situation which is very similar to the hypothetical transition state. This hypothetical transition state is of course defined by the experimentally-determined values for the enthalpy and entropy of activation. On this basis we may modify our energy profile diagram in Fig. 1.1 so that it now appears as Fig. 1.4. In Fig. 1.4 the broken line

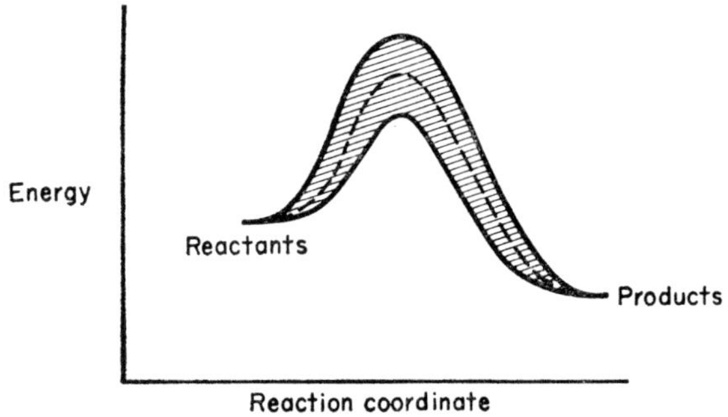


FIG. 1.4. Energy-profile diagram for a single-stage process; a certain arbitrary fraction of the molecules pass through a transition state which falls within the area bounded by the solid lines

represents the hypothetical reaction path where the transition state has the experimentally-determined free energy of activation. We may define an area such as that shaded so that a given fraction (say 90%) of the molecules pass through transition states which fall within this area. A similar device is used for representing the electron densities around atoms.

## ACIDS AND BASES

The development of chemistry as we know it has placed enormous emphasis on reactions which involve water, either as a reagent or as a solvent. This was inevitable for several reasons. Water is not only the most abundant of all the chemical compounds available on earth but also one of the most easily purified. It is a very powerful solvent. In addition, all living systems depend on it, and it was natural that the human beings who have studied chemistry should have given an important place to such an essential substance. One of the most common processes to occur in aqueous solutions is that of hydrogen-ion transfer and it is understandable that this process, along with the related concepts of acids and bases, should have received a great deal of attention.

We shall deal with these concepts only in so far as they help us in our examination of the catalytic process, and for the most part we shall confine ourselves to the Brønsted–Lowry definition of an acid. The more general definition is considered where it is relevant. According to the Brønsted–Lowry definition an *acid* is a chemical species which is able to donate a hydrogen ion (a proton, $H^{\oplus}$) to another species. A *base* is a species able to accept a hydrogen ion, and therefore a proton exchange process must involve both a base and an acid. For example,

$$\text{A—H} + \text{B} \rightleftharpoons \text{A}^{\ominus} + \text{BH}^{\oplus}. \tag{1.21}$$

In this reaction A—H is the acid and B the base. During the course of the reaction two new species are formed; $A^{\ominus}$ is the *conjugate base* of the acid A—H and $BH^{\oplus}$ is the *conjugate acid* of the base B. The process is reversible, proton exchange between $BH^{\oplus}$ and $A^{\ominus}$ being possible, and the equilibrium will be under the control of an equilibrium constant.

It is often convenient for us to refer to such a proton transfer in terms of two half-reactions:

$$\text{A—H} \rightleftharpoons \text{A}^{\ominus} + \text{H}^{\oplus}$$
$$\text{B} + \text{H}^{\oplus} \rightleftharpoons \text{BH}^{\oplus}.$$

Some chemical species are able to act both as acids and bases, the most common and most important of these being water itself, as shown by the following half-reactions:

$$\text{H}_2\text{O} + \text{H}^{\oplus} \rightleftharpoons \text{H}_3\text{O}^{\oplus}$$
$$\text{H}_2\text{O} \rightleftharpoons \text{H}^{\oplus} + \text{OH}^{\ominus}.$$

In any sample of water, proton exchanges are occurring continuously between water molecules and the equilibrium constant $K_w$ has a value of $10^{-14}$ under normal conditions.

$$\text{H}_2\text{O} + \text{H}_2\text{O} \rightleftharpoons \text{H}_3\text{O}^{\oplus} + \text{OH}^{\ominus}$$
$$K_w = [\text{H}_3\text{O}^{\oplus}][\text{OH}^{\ominus}] = 10^{-14}.$$

(Concentration terms involving solvent molecules, being regarded as constant, are generally not included in the equilibrium constants for acid–base reactions.) This low value for $K_w$ means that in pure water most of the molecules are present as neutral molecules, only a very small number being converted into the protonated ($H_3O^{\oplus}$) or hydroxyl ($OH^{\ominus}$) forms.

Water, as a base, can accept protons from any acid and the so-called dissociation of acetic acid involves this process.

$$\underset{\text{acetic acid}}{\text{CH}_3\text{COOH}} + \text{H}_2\text{O} \rightleftharpoons \text{CH}_3\text{COO}^{\ominus} + \text{H}_3\text{O}^{\oplus}$$

$$K_A = \frac{[\text{CH}_3\text{COO}^{\ominus}][\text{H}_3\text{O}^{\oplus}]}{[\text{CH}_3\text{COOH}]}. \tag{1.22}$$

The equilibrium constant $K_A$ for such a reaction is an important measure of the ability of an acid to donate a proton to water, that is a measure of the strength of the acid. The strengths of different acids are usually compared through their values for $K_A$. In general where numerical values are given, the expression $pK_A$ ($= -\log_{10} K_A$) is used rather than $K_A$ itself.

From equation (1.22) we can see that under conditions where the acid is half-dissociated, that is when $[CH_3COO^{\ominus}]$ is equal to $[CH_3COOH]$, the $pK_A$ becomes equivalent to the pH ($= -\log_{10} [H_3O^{\oplus}]$) of the solution. A strong acid is one for which the $pK_A$ is low. It is worth noting at this point that if the pH of a solution containing an acid is above the $pK_A$ of that acid then the conjugate base form will predominate, whereas if the pH is below this value the acid will exist mainly in its undissociated form. This point is extremely important when reactions which are pH-dependent are under consideration.

The strength of an acid, defined thus in terms of an equilibrium constant, is a thermodynamic property related to the free energy change for the overall proton-transfer process. This tells us very little about the rate at which the process occurs, although the equilibrium constant is a function of the rates of the forward and reverse processes. In fact, most workers consider that the rates of acid–base reactions are limited mainly by the rates of diffusion of the species involved, rather than by any factors involving energies of activation.

Just as we can compare the strengths of various acids by considering their reactions with water, we may also compare the strengths of bases in this way. In this case we consider the water acting as an acid. Thus for the base ammonia we may define a constant $K_B$:

$$NH_3 + H_2O \rightleftharpoons NH_4^{\oplus} + OH^{\ominus}$$

$$K_B = \frac{[NH_4^{\oplus}][OH^{\ominus}]}{[NH_3]}.$$

It would be possible to compare different bases in terms of their different values of $K_B$, that is comparing their abilities to remove a hydrogen ion from water. In order to produce a more unified system most workers compare base strengths in terms of the $K_A$ values of their conjugate acids. It can be shown that for a base and its conjugate acid

$$pK_A = 14 - pK_B,$$

where $K_A$ refers to the conjugate acid. Thus it may be seen that a strong base will give rise to a conjugate acid which is a weak acid.

Just as water may act both as an acid and as a base, so also may other species, including amines. Thus, ethylamine $C_2H_5NH_2$ may donate a proton and become the conjugate base, $C_2H_5NH^{\ominus}$. This anion is a far stronger base than the ethylamine itself. Similarly, ethyl alcohol $C_2H_5OH$

may in some reactions act as an acid, but its protonated form $C_2H_5OH_2^{\oplus}$ is a far stronger acid.

## Acid–Base Catalysis

When an acid catalyses a chemical reaction it does so by donating a proton to one of the reactants. In general, the strongest acid present in high concentration in a system will be the conjugate acid derived from the solvent, the lyonium ion (for example, if the solvent is water the lyonium ion is $H_3O^{\oplus}$). It is possible to divide acid catalysis into two categories according to whether, as determined by kinetic studies, the catalysis is attributable to the lyonium ion alone or to all of the acids present in the system. Where the lyonium ion appears to be the only effective catalyst the reaction is said to be subject to *specific lyonium-ion catalysis*. The other case, where all of the acids present contribute to the catalytic effect, is known as *general acid catalysis*. Similarly a base catalyses a reaction by removing a proton from one of the reacting species. *Specific lyate-ion* catalysis is said to be involved when only the conjugate base of the solvent is responsible for the catalysis (in the case of water this is the hydroxide ion $OH^{\ominus}$), while for *general base* catalysis, all of the bases present make a contribution.

Although many examples of specific lyonium- and specific lyate-ion catalysis have been reported (for example, [12], [13]) the majority of acid- or base-catalysed reactions appear to be subject to the general forms of catalysis. General acid catalysis leads to a more complicated rate equation than does specific lyonium-ion catalysis. Thus if a reaction between species X and Y in water is subject to specific acid catalysis the rate equation will be:

$$\text{Rate} = k[\text{X}][\text{Y}][\text{H}_3\text{O}^{\oplus}]. \qquad (1.23)$$

However, if general acid catalysis were the case, the rate equation would consist of many terms, one for each of the acidic species present. Each of these acids makes a contribution according to its catalytic constant $k$.

$$\text{Rate} = [\text{X}][\text{Y}](k_{\text{H}_3\text{O}^{\oplus}}[\text{H}_3\text{O}^{\oplus}] + k_{\text{HA}}[\text{HA}] + k_{\text{HA}'}([\text{HA}'] + k_{\text{HA}''}[\text{HA}''] + k_{\text{HA}'''}[\text{HA}'''] \ldots) \qquad (1.24)$$

Clearly the distinction drawn between specific and general acid catalysis is a purely experimental one. The assignment of a particular case as being specific lyonium-ion catalysis may simply mean that the catalytic constants for all the acids present other than the lyonium ion were too small to be measured at the time. It may be that all the reactions now thought to be subject to lyonium-ion catalysis will at some future date be shown to be subject to general acid catalysis, if more sophisticated techniques are used to detect the very small catalytic constants.

Since acids catalyse reactions by donating protons and the strength of an acid is a measure of its ability to donate protons, it would seem likely that the stronger the acid the better it would catalyse a reaction (that is, the greater its catalytic constant). With certain limitations this is, in fact, the case. For a reaction which may be catalysed by a series of similar acids it is found that the catalytic constants ($k_a$) are related to the acid strengths in a log–log expression

$$\log k_a = a \log K_A + b. \tag{1.25}$$

This linear relationship between $\log k_a$ and $\log K_A$ is known as the Brønsted Catalysis law [14]. The limitations referred to above are simply that the acids being compared must be of the same type (for example, carboxylic acids, phenols or $\beta$-diketones) and that in the comparison of monobasic with dibasic acids (for example, fumaric acid and benzoic acid) due allowance must be made for the number of acidic groups per molecule. It is not merely coincidence which leads to this linear relationship. The log of the dissociation constant $K_A$ is a function of the free energy change involved in the transfer of a hydrogen ion. Log $k_a$ is proportional to the free energy of activation for the reaction. The linear relationship may therefore be regarded as one between free energy terms and as such may have much significance.

## CHEMICAL PROCESSES IN ORGANIC CHEMISTRY

In the preceding pages we have discussed general principles which may be applied to all chemical reactions irrespective of their mechanisms. We turn now to the chemical processes themselves and we shall examine these processes in the light of those general principles.

Chemical reactions involve either the formation, or the breakage, of bonds between atoms. They range in complexity from the simple unimolecular process where one bond is broken or the bimolecular process where one is formed to complicated rearrangements in which many bonds are formed and many broken. In organic chemistry the most common method of bond cleavage is heterolysis. A covalent bond may be considered in the simplest terms as involving two electrons which are both associated more or less equally with each of the atoms involved in the bond. In a heterolytic cleavage both of the electrons move to one of the atoms as shown:

$$R{-}Y \longrightarrow R^{\oplus} + Y^{\ominus}.$$

Since the two electrons were more or less equally 'shared' between the two species, R obtains a single positive charge and Y a single negative charge. The arrow ↷ used to represent the direction of movement of the electrons refers specifically to the movement of *two* electrons. The species $Y^{\ominus}$ thus obtains what is known as an *unshared pair* of electrons.

A covalent bond may be formed by the reverse of this process, where an unshared pair of electrons (represented by two dots **:**) are contributed

$$X\!:\; R \longrightarrow X^{\oplus}\text{—}R^{\ominus}.$$

Such a process is exemplified by reaction of fluoride ion with boron trifluoride

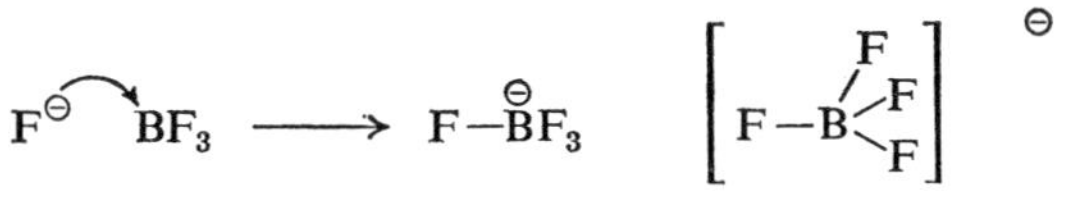
(1.26)

In this case the negative change initially produced on the boron atom is dispersed over all of the five atoms of the molecule. This dispersion of charge leads to a stabilization of the system.

The electronic arrangement which is referred to as a covalent bond is very stable, that is, it represents a situation of low energy. The stability is greatest when the distance between the two atoms involved in the bond is at a certain optimum value, this value being approximately constant for any given pair of atoms. This most stable distance between atoms is known as the bond length. Any situation which leads to an increase in the distance between the atoms, that is stretching of the bond, must therefore lead to an increase in the energy content of the system.

It is usual to refer to a species which donates its pair of electrons as a nucleophile, that is a positive-centre-seeking species, and that which accepts the pair of electrons as an electrophile (an electron-seeker). A nucleophilic substitution involves the formation of one bond and the usually-simultaneous cleavage of another, as shown by the reaction which we have already mentioned, that between chloride ion and methyl bromide

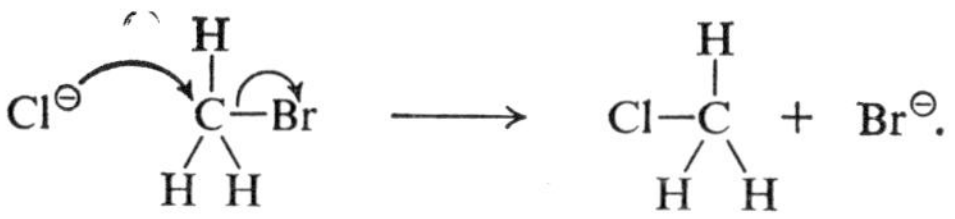

Such a process is designated $S_N2$ because in addition to being a substitution effected by a nucleophile the reaction is also bimolecular. We can envisage this process as involving a flow of electrons in such a direction that a new bond is formed between the carbon and chlorine atoms and the old carbon–bromine bond is broken. The idea that a reaction may be considered as an electron flow is shown more clearly by the reaction of diethylamine with α-methylallyl chloride where the new bond formed involves one carbon atom and the bond broken involves a different one [15]. In such a case, often designated $S_N2'$, passage through the transi-

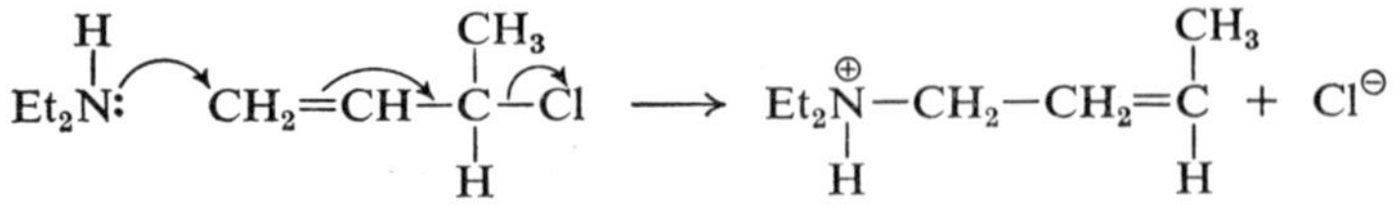

tion state involves an electron flow extending from the nitrogen atom over three carbons to the chlorine atom.

## Energy of Activation for Nucleophilic Substitution

We have already said that a covalent bond represents a stable electronic situation, that is an energy minimum. Breakage of a bond requires an increase in energy of the system. In other words energy is *needed* for a bond to be broken, whereas energy is *released* when a bond is formed. It is easy to understand therefore that the energy of activation of a nucleophilic substitution such as (1.27) depends to a great extent on the nature of the leaving group Y. In order for the reactants to proceed to the transition

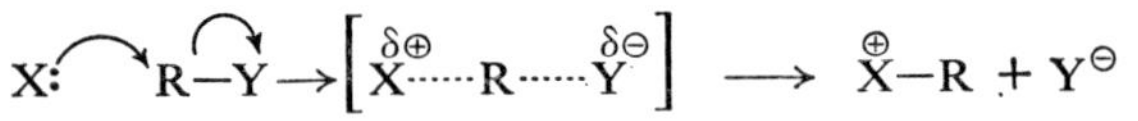

(1.27)

state as shown, the R—Y bond must be partially broken, and the height of the energy of activation 'barrier' depends on the amount of energy which is required for this partial cleavage. That is, it depends on the strength of the bond and the extent to which the bond is broken at the stage of the transition state. Similarly the X—R bond has been partially formed and this has released energy, causing a stabilization of the transition state. In other words the energy released by the partial formation of the X—R bond may be used to offset the energy required for bond cleavage. However, it must not be assumed that the energy of activation for such a reaction is attributable only to the energy required for the bond cleavage. Other factors are involved and in some cases may be extremely important. For example, when the groups X and R are bulky there would be great resistance to the close approach of X and the energy of activation would be increased by an amount corresponding to the repulsive forces operating.

The above generalization is limited, but it can help us in our understanding of the principles involved in reaction rates.

The proton transfers which take place in acid–base reactions also involve heterolytic bond cleavage and we may represent the reaction between water and acetic acid as a nucleophilic substitution:

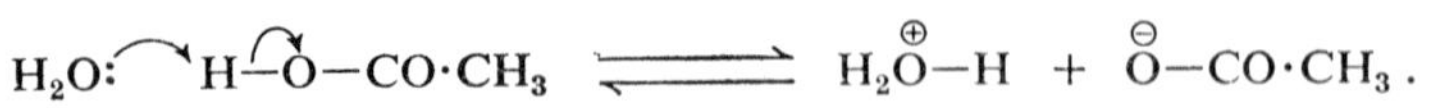

In this reaction the base is acting as a nucleophile and the leaving group is the conjugate base of the acid. The reaction is reversible and the acetate ion may itself initiate a nucleophilic reaction. This relationship between bases and nucleophiles leads to the more general definition proposed by Lewis [16] in which a base is defined as any species which is able to donate its unshared pair of electrons. This definition includes not only all bases in the Lowry–Brønsted sense but also all nucleophiles. Similarly the Lewis definition of an acid, a species which is able to accept a pair of electrons, includes all Lowry–Brønsted acids as well as all electrophiles. Species which while not being acids in the Lowry–Brønsted definition are nevertheless effective electrophiles, such as boron trifluoride as in reaction (1.26), are sometimes referred to as Lewis acids.

### The Nucleophile or Lewis Base

A good nucleophile is one which participates in a rapid nucleophilic substitution. It is one which donates its electrons readily and forms a strong bond. A clear relationship might be expected to exist between the *nucleophilicity* of a reagent as measured by the rate of its reaction and its *basicity* as measured by its $K_B$ (or inversely by the $K_A$ of its conjugate acid). Such a comparison will have certain limitations because it involves a kinetic property, associated with a rate, and a thermodynamic one, which is based on an equilibrium. In many cases a good correlation between base strength and nucleophilicity does exist, especially if when making comparisons we confine ourselves to chemically-similar molecules. For example, if we compare nucleophiles in which the attacking atom is oxygen we would find the order of nucleophilicity thus:

$$OH^{\ominus} > OPh^{\ominus} > OAc^{\ominus} > CH_3—C_6H_4SO_3^{\ominus} > Br.C_6H_4SO_3^{\ominus} > H_2O$$

which is the same as the order of basicity.

Having stated this generalization we find, as is often the case, that the exceptions to it are more interesting, because they give us more insight into the various modifying factors which must be considered. One important modifying factor is steric hindrance. This is clearly shown in the reaction of the methylpyridines with methyl iodide. Although 2-methylpyridine and 4-methylpyridine have approximately the same base

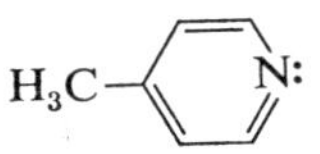

4-methylpyridine

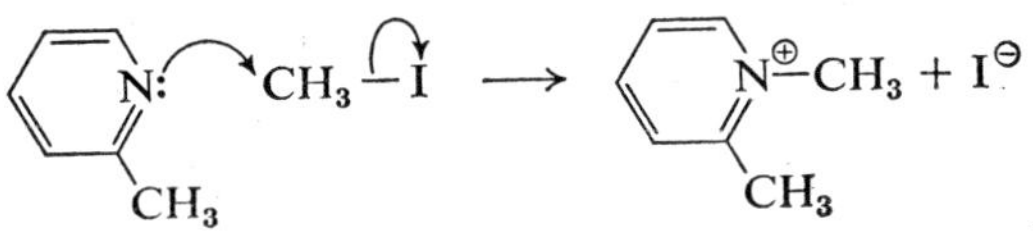

2-methylpyridine

strengths ($K_B$), the reaction of the 2-methyl derivative with methyl iodide is about fifty times slower than that of the 4-methyl compound [17]. This is because the methyl group hinders the reaction in the case of the former but not in the case of the latter. In terms of the energies of activation we may consider such steric hindrance in two ways. First, because of the interfering group the nitrogen atom cannot approach closely enough to the central carbon of the iodide. Hence the new bond is not formed to a sufficient extent to offset the energy of activation. Secondly, the transition state is more crowded and this leads to an unfavourable entropy of activation term.

Other exceptions to the above generalization are afforded by such series as

$$F^{\ominus},\quad Cl^{\ominus},\quad Br^{\ominus},\quad I^{\ominus}$$
$$\text{or}\quad OR^{\ominus},\quad SR^{\ominus},\quad SeR^{\ominus}$$

in which the attacking atoms are members of the same vertical groups of the periodic table. In these cases the orders are completely reversed, that is the heavier atoms which are the weakest bases are in fact the best nucleophiles. Two factors are often suggested to account for these anomalies. The heavier atoms are considered to be more *polarizable*; the outer electrons, which are the ones involved in forming the chemical bonds, being further from the atomic nucleus and also being shielded from the nucleus by the inner electrons, are far more subject to the influences of nearby molecular species. Because of this the electron cloud is more easily distorted and bond formation occurs more readily. The more easily the electron cloud is distorted the lower the activation energy and hence, the better the nucleophile.

Solution effects are also invoked to explain the better nucleophilicity of large atoms. We have already noted (p. 7) that in a solution interaction occurs between molecules of solvent and molecules of solute. Much of this interaction is due to coulombic attraction. For example, a positively-charged sodium ion will attract to itself the more negatively-charged portions of the dipolar water molecule. The water molecules will tend therefore to arrange themselves around the cation in such a way that the oxygen atoms are nearest to it. A considerable amount of stabilization

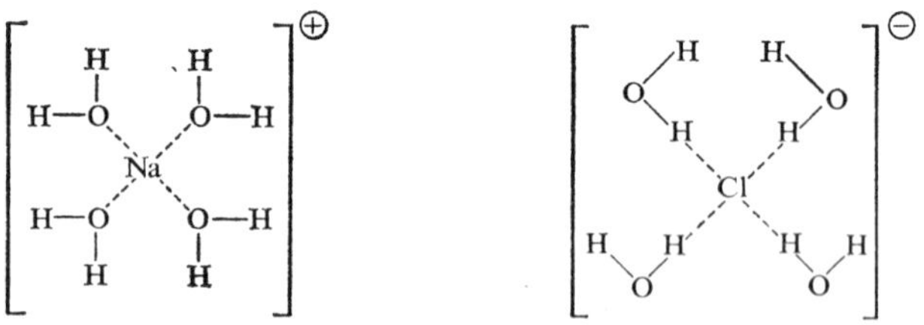

FIG. 1.5. Diagrams representing the solvation of sodium and chloride ions in water

results from this arrangement (Fig. 1.5). Similarly, around a negatively-charged solute species, for example the chloride ion, the molecules of water will tend to arrange themselves so that the slightly positive hydrogen atoms are closest to the anion. In this latter case the interaction between the hydrogens of the solvent and the anion involves 'hydrogen bonding', which might be described as a partial donation of a hydrogen ion ($H^{\oplus}$) to an electron-rich species. Hydrogen bonding makes an important contribution to the solvation of many solutes especially those such as amides, ketones, etc., which do not carry overall electrical charges.

Solvation as described by the diagrams above emphasizes an important principle in chemistry. This is that the energy associated with a charge is an inverse function of the volume over which the charge is spread. Solvation of an anion decreases the energy associated with its charge by permitting that charge to be spread or dispersed over several solvent molecules and thus over a larger volume. A small ion, which unsolvated has its charge concentrated into a very small space, gains considerable stability by solvation whereas a large ion gains less. When a solvated anion participates in a nucleophilic substitution much of the solvation is destroyed and hence some stability is lost. This loss of stability will be reflected in a higher energy of activation for the reaction. A small ion will therefore have a higher energy of activation than a large ion which has less solvation stability to lose. That such an effect really has a considerable influence is shown by the fact that the nucleophilicity of the small fluoride ion is higher in non-hydroxyl solvents where the solvation is less effective [18].

In connection with solvent effects we should consider at this point another way in which solvents may influence the rates of reaction. The force between two opposite electrostatic charges, and therefore the energy associated with them, depends on the dielectric constant of the medium which surrounds them. A high dielectric constant will mean a small force and a low energy. If a reaction involves the production of electrostatic charges in the transition state, then the energy of the transition state, and hence the energy of activation for the reaction, will be decreased if the solvent has a high dielectric constant.

Both these effects are in accordance with the *Hughes–Ingold theory* of solvent effects which states that an increase in the ion-solvating power of the medium will accelerate the creation and concentration of charges and inhibit their destruction and diffusion [19]. The ion-solvating power of a solvent is related to its dielectric constant.

### The Leaving Group

We have already stated that the cleavage of the R—Y bond in a nucleophilic substitution has a considerable influence on the energy of activation for the reaction. Clearly then the nature of the leaving group will be of great importance. To be an effective leaving group the substituent should be attached to the remainder of the molecule by only a rather weak bond. Just as one could relate the basicity of a group with its nucleophilicity, one can also find an inverse relationship between the basicity of a leaving group and its 'tendency to depart'. That is, a good leaving group will be a poor base. Alternatively we might say that the conjugate acid derived from a good leaving group will be a strong acid. Again we must point out that base strength is a thermodynamic factor, based on an equilibrium constant. This equilibrium constant is itself related to the free energy change for the proton transfer process. In other words, when discussing the base strength of a leaving group Y we are considering the relative stabilities of the two forms H—Y and $Y^{\ominus}$. If $Y^{\ominus}$ is a very stable species relative to H—Y, then H—Y will be a strong acid (easily losing its proton) and $Y^{\ominus}$ will be a weak base (since it has no stability to gain by abstracting a proton from some other acid). Clearly, if $Y^{\ominus}$ is a very stable species it will also make an excellent leaving group.

Since a strong base is a species which has a tendency to donate electrons, a very weak base is one which will have a tendency to attract electrons (and thus form the stable conjugate base). Therefore a potentially good leaving group will tend to distort the bond linking it to the rest of the molecule, that is the R—Y bond in the general case. This distortion will result in the R-portion carrying a small positive charge, even before the approach of the nucleophile. Such a situation may lead to an enhancement of the nucleophilic attack:

$$X\colon \curvearrowright R^{\delta\oplus}\cdots\cdots Y^{\delta\ominus}.$$

One way of reducing the basicity of a substituent, and thereby improving its ability as a leaving group, is by protonation, that is, conversion into its conjugate acid; this process naturally reduces its basicity. Many nucleophilic substitutions are in fact more rapid under conditions where the leaving group becomes protonated, in other words, they are acid-catalysed. Indeed, association of any Lewis acid with the leaving group can enhance the rate of reaction. This effect is dealt with in more detail later in this chapter.

### The R-group

On page 20, reference is made to the fact that a bulky group on a nucleophile may slow down the reaction of that species in two ways; first by hindering formation of the new bond, and secondly by an entropy effect on the transition state. Clearly, bulky groups on the molecule undergoing substitution can have the same effect. Thus the reaction of triethylamine (I) with isopropyl iodide (II) is considerably slower than with methyl iodide [20]. On page 7 another example has been given, the emphasis there being on the entropy of activation effect.

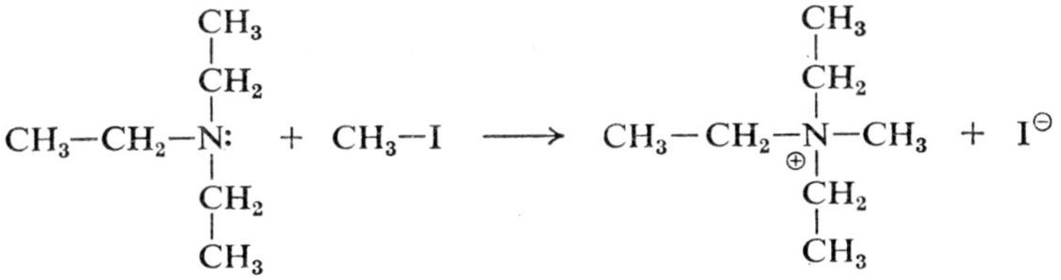

(I)

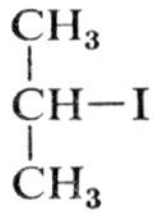

(II)

These and other examples show that the size of the groups attached to the central carbon of the R-group can have an effect which is mainly steric. The nature of these groups may have another effect, an electronic effect. It is clear that partial (or complete) cleavage of the C—Y bond will give rise to a partial (or complete) positive charge on the central carbon atom. Also, partial formation of the new X—C bond will produce a partial negative charge on this atom. The size and sign of the charge which may be considered to exist on the central atom in the transition state will therefore depend on how far each of these two processes has progressed. If the cleavage of the C—Y bond predominates, as it may do if Y is a very good leaving group, then the carbon atom will possess a positive charge. This positive charge may be stabilized by distribution over the rest of the molecule if the central carbon carries substituents which are capable of donating electrons. Stabilization of the transition state in this way leads to a decrease in the energy of activation. Similarly, if the formation of the new bond predominates in the transition state, then electron-withdrawing groups can help to stabilize the resulting negative charge on the central atom.

**Extreme Cases in Nucleophilic Substitution**

The discussion so far has been concerned with $S_N2$ reactions; reactions in which the attack of the nucleophile and the departure of the leaving group are simultaneous. Under some circumstances these two events may not occur simultaneously. If the leaving group is an especially good one and the structure of the whole molecule is favourable, the cleavage of the R—Y bond may occur spontaneously, before the nucleophile has made an approach. In such a case the overall nucleophilic substitution takes place in two clearly-defined phases, first the heterolytic cleavage of the bond

$$R{-}Y \longrightarrow R^{\oplus} + Y^{\ominus},$$

and secondly the attack of the nucleophile on to the carbonium ion produced

$$X{:}\ R^{\oplus} \longrightarrow X^{\oplus}{-}R\,.$$

The first phase is usually the rate-determining step and because this is a unimolecular process, this type of nucleophilic substitution is designated $S_N1$.

The conditions which favour an $S_N1$ process may be listed as follows: (a) a very good leaving group, for example the para-bromosulphonyl group, (b) a solvent medium which promotes the production of ionic species (high dielectric constant) and which stabilizes the ions by solvation; this factor is closely related to (a) because the weakly basic nature of the departing species will depend on the stability afforded by solvation, (c) the structure of the molecule must be such that the carbonium ion is stable, that is the charge is distributed over the whole molecule by electron-donating species; phenyl groups are able to stabilize carbonium ions in this way, as are also methyl groups.

In connection with (c) it is worth pointing out that since a carbonium ion is of necessity a planar structure any steric factor which prevents the molecule from taking up a planar configuration will also prevent its participation in an $S_N1$ process.

At the other extreme, the formation of the new bond may occur without the cleavage of the old one, this cleavage occurring at a subsequent stage. This kind of process only occurs when the nature of the molecule under attack is such that it is able to cope with the negative charge which results. This situation applies for aromatic nucleophilic substitution. In this case also, a solvent medium which favours the production of ions is needed. Meisenheimer [21] showed that the reaction of trinitroanisole (III) with ethoxide proceeds via the intermediate (IV). Here the nitro-groups and the aromatic ring stabilize the charge.

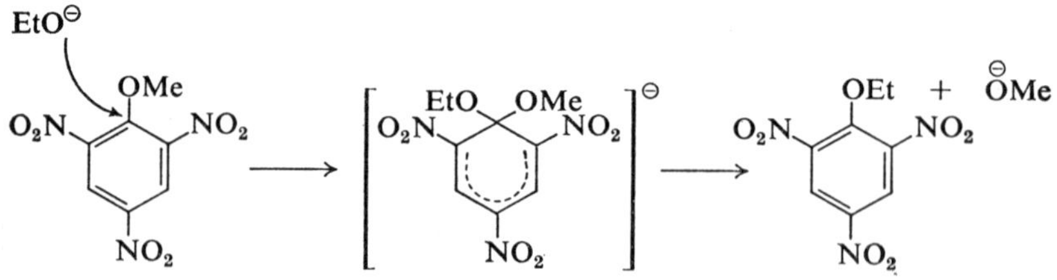


(III) (IV)

The fact that many reactions occur by the $S_N1$ and aromatic–nucleophilic mechanisms does not invalidate the idea of a chemical reaction as a flow of electrons from an electron source (usually a nucleophile) to the electron terminus or sink. They are merely extreme cases in which the electron flow occurs in two stages. Indeed, Swain considered every nucleophilic substitution as being made up of a 'pushing' and a 'pulling' action, and from this attempted to develop a quantitative treatment for the rates of these reactions [22] with some success.

**Elimination Reactions**

We have dealt with nucleophilic substitutions in some detail so that all the factors which can influence the rates of reactions could be introduced. Many other organic reactions may be considered as being essentially similar to nucleophilic substitutions, especially in that they constitute flows of electrons, a flow of electrons requiring a driving force made up of a 'push' from the electron source and a 'pull' from the electron sink. The same factors may influence these reactions also. Thus, reaction of a strong base such as ethoxide with an alkyl bromide proceeds by the following mechanism.

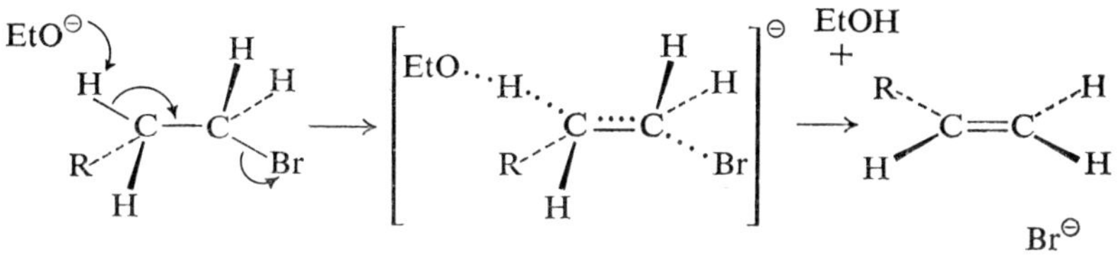

Since this is a bimolecular reaction the process may be designated E2.

In the transition state the flow of electrons extends from the oxygen atom of the ethoxide ion over the hydrogen and two carbon atoms to the bromine. As in the case of nucleophilic substitution the reaction is favoured by a strong base and a good leaving group. In addition, just as in the $S_N2$ process, the nucleophile approaches from the side opposite to that from which the leaving group departs (see page 4) so the hydrogen which is attacked by the base in an elimination reaction must be on the side opposite to that of the leaving group. It is in fact a general rule that an electron flow extends over atoms which are coplanar, and this condition is satisfied in the above reaction only if the molecule is arranged as shown.

We saw that in the case of nucleophilic substitution extreme cases might be observed in which the attack and the departure may not be simultaneous. Similar extreme cases are encountered in elimination reactions. Thus, the E1 reaction proceeds by two stages, the first of which involves production of a carbonium ion. Such a process is favoured by factors corresponding to those which favour an $S_N1$ process.

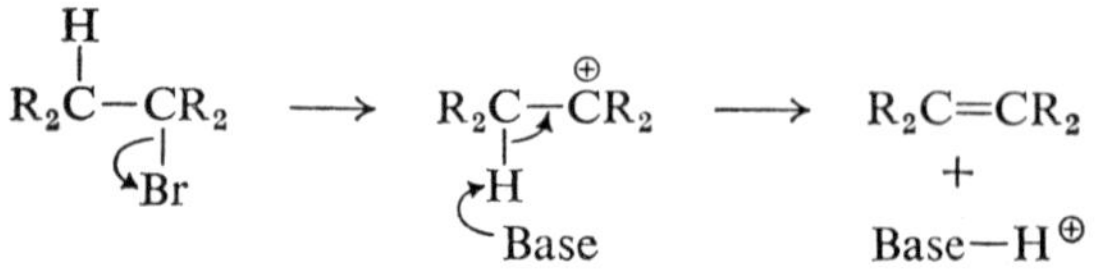

At the other extreme the base may react, abstracting a hydrogen ion and leaving a carbanion, stabilized by various factors. In a subsequent step, the leaving group departs.

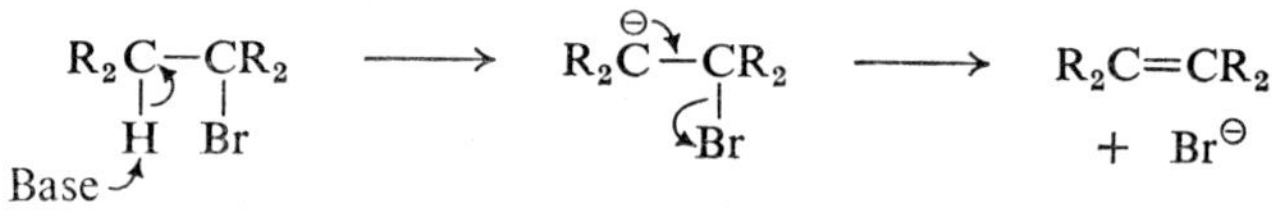

Because the conjugate base of the halide loses bromide by a unimolecular process, this reaction would be designated *E*1*c*B.

In these two extreme cases the stereochemical requirements are not very severe because the carbanion or carbonium ion produced would be free to rotate into a favourable configuration for the subsequent step.

**Reactions of Carbonyl Groups**

The oxygen of a carbonyl group attracts electrons strongly and because of this the carbon–oxygen double bond is polarized so that the carbon atom

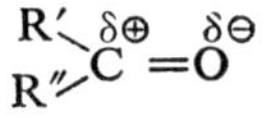

assumes a small positive charge. This renders the carbon atom susceptible to attack by nucleophilic agents. The doubly-bonded oxygen may be regarded as a leaving group so that the reaction takes the form of a nucleophilic substitution

(1.28)

Such a process is almost invariably the first stage in a reaction involving a carbonyl group. The course of the subsequent steps will be decided by the nature of the three groups present R′, R″ and X. Under some circumstances the system is relatively stable as it is, and after the oxygen anion has acquired a proton (from the solvent) the quaternary compound undergoes no further reaction. This is the situation in which hydrates and hemiacetals are produced.

If R′ or R″ is a good leaving group then an elimination similar to the second stage of an $E1cB$ reaction may take place. If this occurs the overall reaction has resulted in a nucleophilic substitution, which is similar to the

aromatic case previously mentioned, that is, a negatively-charged intermediate is formed before the departure of the leaving group. This process occurs in many of the reactions of esters and other carboxylic acid derivatives.

A third possibility exists if the original nucleophile was an amine, $RNH_2$. In this case a series of proton transfers might take place, with the solvent participating, the final step being an elimination.

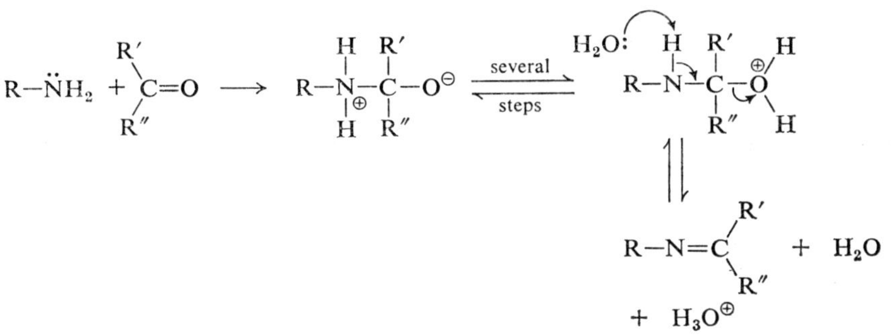

**Homolytic Cleavage of Chemical Bonds**

It would be incorrect to suggest that all reactions in organic chemistry proceed by heterolytic cleavage or its reverse. Whereas a heterolytic cleavage requires that the two electrons associated with a chemical bond both move to one of the atoms, it is possible for the bond to break homolytically, that is, one electron goes to each of the atoms. We can represent the movement of *one* electron by a broken curved arrow. The result is that two species are produced each of which carries an unpaired electron (•) and

these species are known as radicals. The location of an unpaired electron on such a species confers special reactivity upon it.

We shall come across this type of process only very infrequently in our discussion of enzyme mechanisms. It is relevant mainly to discussions of oxidation–reduction processes in which one-electron transfers are involved (see chapter 3).

## CATALYSIS

It has long been known that the rate of a chemical reaction may be influenced, favourably or otherwise, by the presence of some foreign substance, that is, some substance which does not appear in the equation which represents that reaction. In general we shall use the term *catalyst* to refer only to an agent which increases the rate of a reaction, and the term *negative catalyst* for an agent whose action decreases the rate. The specific term *inhibitor*, which is sometimes used by other writers as a synonym for negative catalyst, will be used here in a special sense, meaning a reagent which prevents a catalyst from exerting its catalytic influence. Many catalysts, especially enzymes, are subject to the action of inhibitors.

A *substrate* is the chemical species whose reaction is influenced by the catalyst. Usually a small number of catalyst molecules can effect the conversion of many times that number of molecules of substrate. This is because, no matter how a molecule of catalyst may be involved, it is regenerated at the end of the process, and so is able to participate again with another molecule of substrate. The fact that a catalyst must be regenerated (which of course leads to its not being included in the reaction equation) is of great importance and is regarded as a criterion for catalysis.

We have already seen that for a reversible reaction such as (1.5) the equilibrium constant is a function of the standard free energy change, $\Delta G^0$, which is the change in free energy when one mole of each of the reactants reacts to give one mole each of the products. If such a process is catalysed by the addition of a small quantity of an agent C, the reaction equation actually becomes

$$C + X + Y \rightleftharpoons P + Q + C. \tag{1.29}$$

Since the catalyst is unchanged by the reaction, that is it appears in the same state chemically and energetically in the products as it did in the reactants, its presence in the system can have no effect on the value of $\Delta G^0$ for the reaction and hence on the equilibrium constant. The ratio [X][Y]/[P][Q] in the system at equilibrium will therefore be unchanged. The catalyst influences the rates of both the forward and the reverse processes.

Equation (1.16) shows that the rate constant for any reaction depends entirely on the free energy of activation. The pathway of a catalysed reaction must therefore involve a free energy of activation which is lower than that of the non-catalysed reaction. We may envisage the catalyst having its effect in two ways. It may modify the reactants so that the energy of activation is decreased or it may provide an entirely different, alternative pathway for the reaction whose energy of activation is lower than that for the original pathway. Whichever of these possibilities may apply, the catalysed reaction will probably involve more stages than the uncatalysed process. For example if a one-stage process in which a molecule of species S is converted into a molecule of P

$$S \rightleftharpoons P$$

is catalysed by the participation of species C, then the simplest sequence of events is probably as shown:

$$S + C \rightleftharpoons C\text{—}S \rightleftharpoons C\text{—}P \rightleftharpoons P + C. \tag{1.30}$$

In this sequence, the substrate S associates with the catalyst in some way, giving a catalyst–substrate complex C—S. This undergoes chemical transformation to the catalyst–product complex C—P which then dissociates to give the product, regenerating the catalyst. The energy profile diagram for such a system might be as shown in Fig. 1.6. The full line represents the uncatalysed reaction and $\Delta G_u^*$ is the free energy of activation. The broken

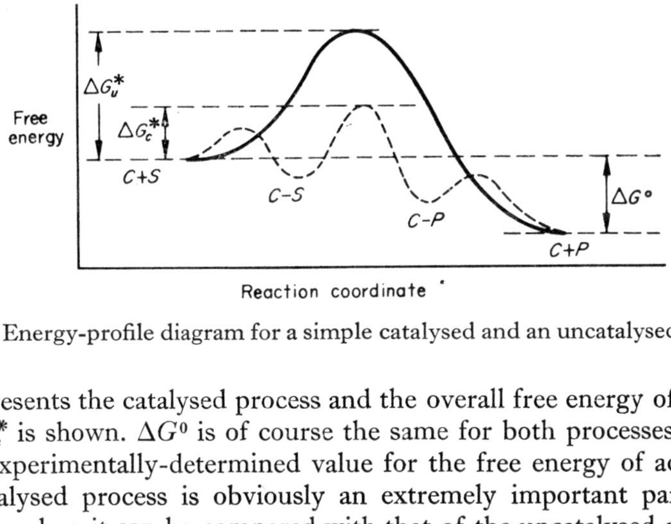


FIG. 1.6. Energy-profile diagram for a simple catalysed and an uncatalysed reaction

line represents the catalysed process and the overall free energy of activation $\Delta G_c^*$ is shown. $\Delta G^0$ is of course the same for both processes.

The experimentally-determined value for the free energy of activation of a catalysed process is obviously an extremely important parameter, especially when it can be compared with that of the uncatalysed reaction. Ideally a discussion of the mechanism by which any catalyst influences a reaction should include consideration of the heats and entropies of activation for each stage in the reaction sequence.

Table. 1.1 shows some experimentally-obtained activation parameters. The figures presented emphasize the great effectiveness of enzymes as catalysts.

Table 1.1. *Activation energies for chemical reactions*

| Reaction | Catalyst | Energy of activation (kcal/mol) |
|---|---|---|
| $H_2O_2$ decomposition[a] | None | 18·0 (ΔH*) |
| | Platinum | 11·7 (ΔH*) |
| | Catalase (1.11.1.6) | 5·5 (ΔH*) |
| p-Nitrophenyl acetate hydrolysis[b] | Water (i.e. no added catalyst) | 21·9 (ΔG*) |
| | $H^{\oplus}$ | 18·0 (ΔH* est.) |
| | $OH^{\ominus}$ | 16·2 (ΔG*) |
| | Imidazole | 15·9 (ΔG*) |
| Casein hydrolysis[a] | $H^{\oplus}$ | 20·6 (ΔH*) |
| | Trypsin (3.4.4.4) | 12·0 (ΔH*) |
| Ethyl butyrate[a] hydrolysis | $H^{\oplus}$ | 13·2 (ΔH*) |
| | Lipase (3.1.1.3) | 4·2 (ΔH*) |

[a] From Lineweaver, *J. Am. chem. Soc.* **61**, 403 (1939). [b] Ref. [43].

**Catalysis by Acids and Bases**

We have already seen that for an $S_N2$ reaction one of the factors which may determine the energy of activation is the nature of the leaving group, and that decreasing the basicity of this group will increase the reaction rate. This end may be achieved by protonation. For example, the hydroxyl or alkoxyl ions are very poor leaving groups, the anions being strong bases, and nucleophilic displacements of the type shown are very rare.

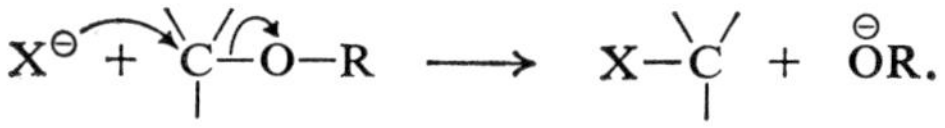

In the presence of acids, however, such compounds exist in the form of their conjugate acids in significant amounts and nucleophilic substitution is enhanced considerably.

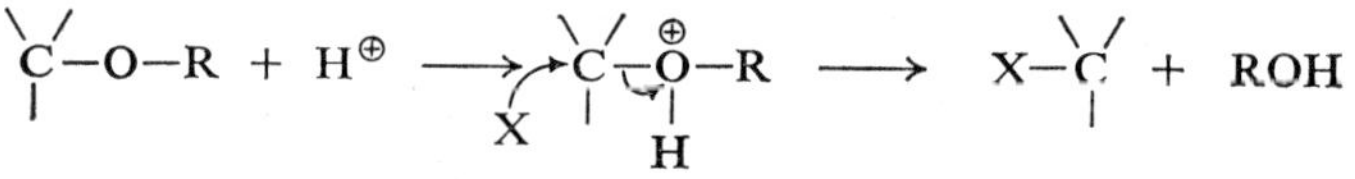

The positively-charged oxygen atom attracts electrons and provides a better electron sink. In this way the electron flow, and therefore the reaction, is facilitated.

The cleavage of ethers by iodide ion and the formation of alkyl bromides from alcohols are reactions which must be carried out under acidic conditions. It may be important to note that these reactions may not come into the category of acid catalysis in the strict sense, because the hydrogen ion may not be regenerated. The acid is simply a reagent. However, these examples do show clearly how association with a hydrogen ion may modify the reactant so that its reaction rate is increased and some reactions, especially those in which the solvent molecule itself acts as a nucleophile, may be examples of acid catalysis in the true sense. For example, in water the hydroxonium ion $H_3O^{\oplus}$ will be the major acidic species and a *solvolysis* may be catalysed as shown.

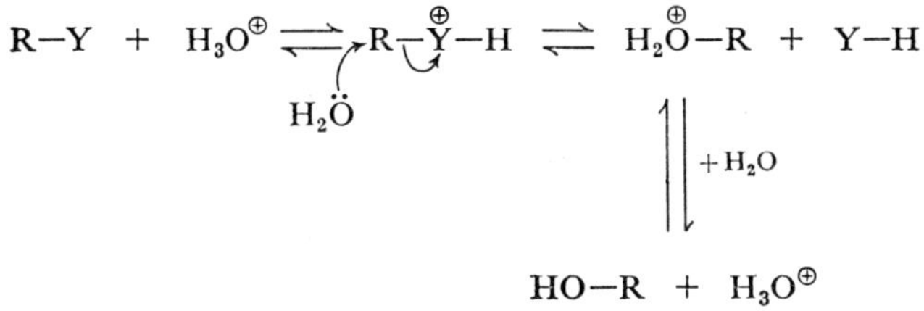


The regeneration of the acidic species therefore satisfies the criterion for catalysis. The solvolysis of some alkyl fluorides has been reported to be subject to acid catalysis in this way [23].

Just as protonation (or conversion into the conjugate acid form) improves the departing tendency of a leaving group, so also deprotonation (conversion into the conjugate base form) improves the nucleophilicity of the attacking group. Thus, the hydroxide ion reacts with alkyl halides more rapidly than does the neutral water molecule. Specific lyate-ion catalysis involves this type of effect.

In the cases cited above the catalysis is effected by means of a complete proton transfer. For acid catalysis the conjugate acid of the substrate is formed and it is this which undergoes reaction. It is possible for only a partial donation of a proton to occur, in other words a hydrogen bond may be formed. This too may bring about catalysis. The reaction of pyridine with methyl bromide in benzene solution affords an example of this effect [24]. The reaction rate is enhanced by the presence of small quantities of hydroxylic compounds such as alcohols or phenols. The catalyst appears to form a loose association with the departing bromide ion by hydrogen bonding, and this assists the departure (see ref. [2], p. 272):

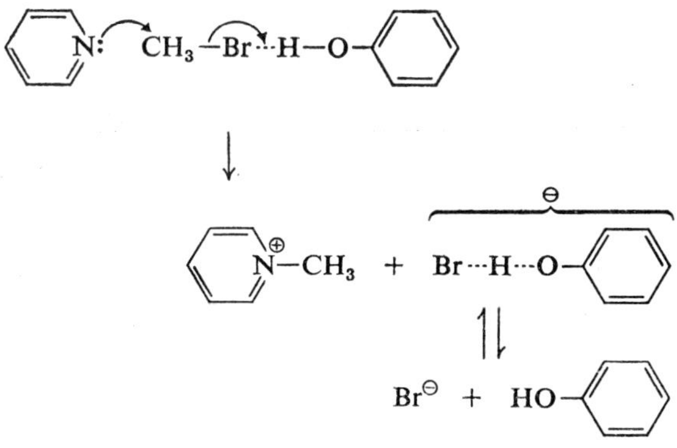

A similar process is involved in reactions which are subject to general base catalysis in water. Any base present may enhance the nucleophilicity of the water molecule by hydrogen bonding, without converting it into a hydroxide ion. An example is given later in connection with aldehyde hydration.

The reaction of pyridine with methyl bromide as stated above can be influenced by the presence of small quantities of hydroxylic compounds. Obviously this effect will be many times more important if the solvent itself is hydroxylic.

Acid (or base) catalysis as described above for nucleophilic substitution, is an important factor for other types of reactions. Reactions at carbonyl groups may be viewed as being catalysed by acids either via formation of the conjugate acid of the carbonyl compound:

$$\gt C{=}O \;+\; H{-}A \;\rightleftharpoons\; \gt C{=}\overset{\oplus}{O}{-}H \;+\; A^{\ominus}$$

or by hydrogen bonding:

$$\gt C{=}O \;+\; H{-}A \;\rightleftharpoons\; \gt C{=}O\cdots H\cdots A$$

In either case an effective electron terminus is provided for the reaction. The simplest of these reactions, the hydration of acetaldehyde, was shown by Bell [25] to be subject to general acid catalysis, and the following mechanism was suggested:

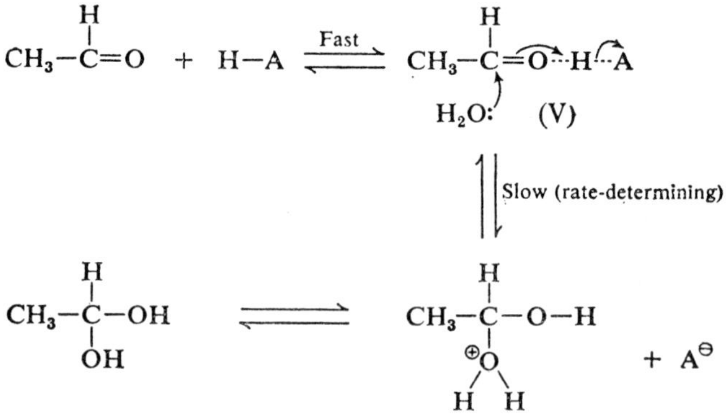


The rate-determining step for the theoretically uncatalysed process will be the attack by a water molecule on the acetaldehyde. Naturally this will be far less rapid than the attack of water on the protonated or hydrogen-bonded species (V).

Water itself can catalyse this reaction and so when discussing the theoretically uncatalysed reaction we should consider the reaction taking place in a solvent such as benzene with only a small amount of water present as reagent.

The hydration of acetaldehyde is also subject to general base catalysis. General base catalysis means that every basic species present contributes to the process. The mechanism therefore is assumed to involve attack by water molecules rendered more nucleophilic by hydrogen bonding with the bases present.

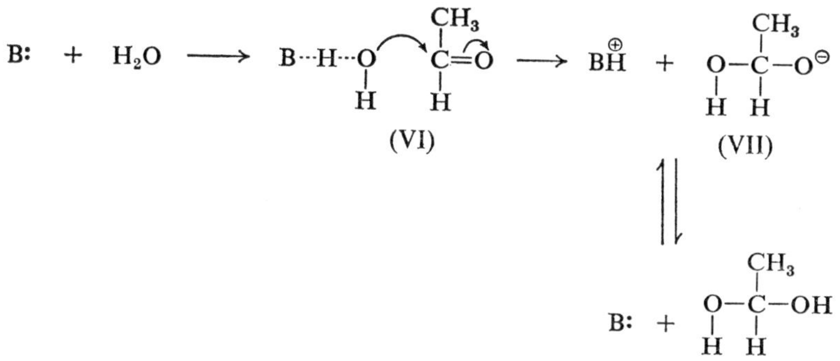

Interestingly, in the very similar case of the hydration of acetone the base catalysis is found to be specific [26], that is, only the hydroxide ion appears to be effective as a nucleophile. Presumably any water-base complex is too weak a nucleophile to have a measurable effect.

Although we have represented the general base catalysed hydration of acetaldehyde above as involving water as a reagent only, in all probability neutral water molecules act also as acidic catalysts by forming hydrogen-bonded complexes with the acetaldehyde (VIa) and by solvating the anion (VIIa). In fact it is almost impossible to overemphasize the importance of the solvent in such reactions and indeed in the whole of organic chemistry. In many of the reaction mechanisms discussed below we shall represent solvent molecules assisting in reactions, without further comment.

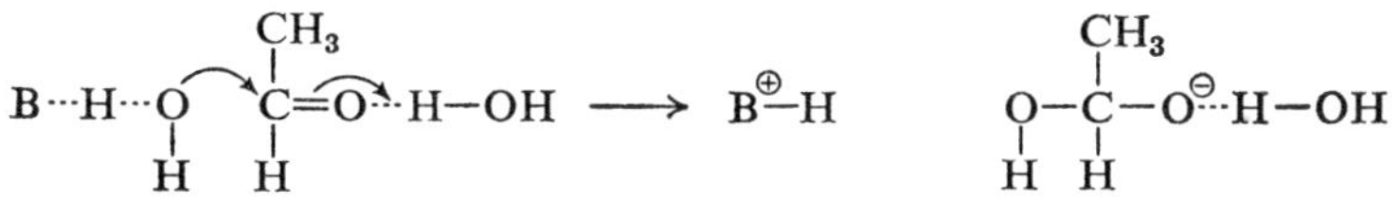

(VIa) (VIIa)

The aldol condensation is another example of a base-catalysed reaction involving attack on a carbonyl. In this case the conjugate base of the potential nucleophile is formed by a complete proton transfer.

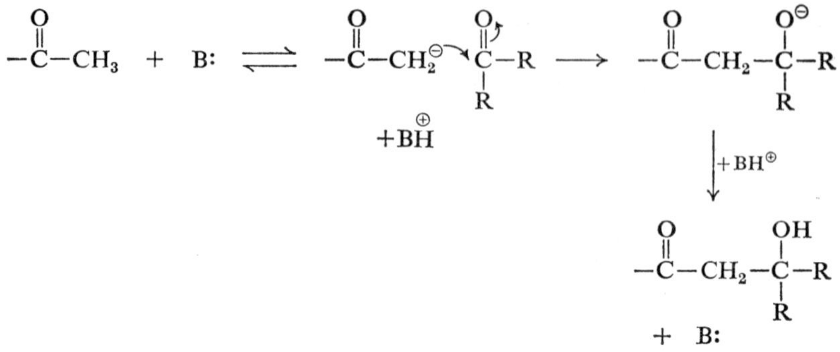

The fact that cyanohydrin formation with HCN is base-catalysed suggests that the rate-determining step is the attack of cyanide ion ($CN^{\ominus}$) on the carbonyl carbon [27].

A catalyst influences both directions of a reversible reaction. The reverse of hydration of carbonyl compounds is in fact an elimination and we may represent the base-catalysed dehydration of acetaldehyde hydrate as follows:

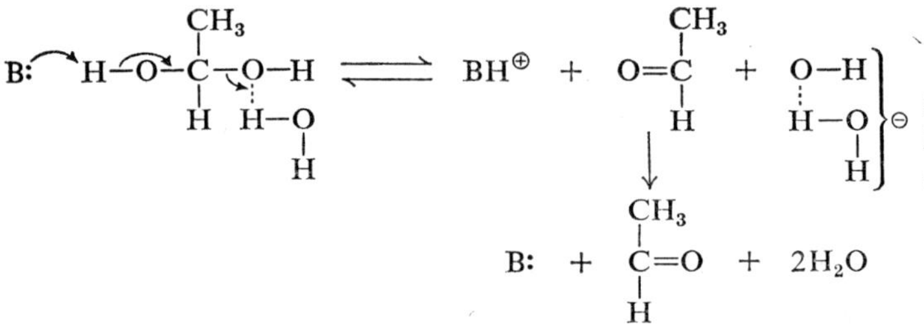


Close examination of this mechanism shows that it is exactly the reverse of the solvent-assisted base-catalysed hydration (except that B: and $OH^{\ominus}$ are interchanged). Therefore the reactions comply with the requirements of the principle of microscopic reversibility.

Many reactions in organic chemistry involve addition to a carbonyl group, followed by an elimination reaction which is formally comparable with the dehydration process. The formation of carbonyl derivatives such as oximes, semi-carbazones, etc., and also the reactions of various carboxylic acid derivatives are examples. Naturally we might expect these reactions too to be subject to acid or base catalysis or both. The formation of oximes is indeed acid-catalysed [28] and the various stages of the reaction may be represented as follows.

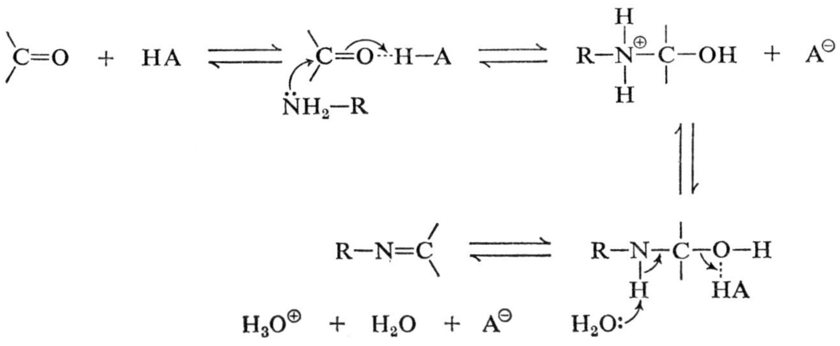

For this and similar acid-catalysed reactions a decrease in the pH will usually lead to an increase in the reaction rate. However, an excessive change may lead to a reversal of this effect, in other words in a too-strongly acidic medium the reaction will be slow. This is mainly because the nucleophile R—$NH_2$ is converted into its inert conjugate acid. For this reason it is common to find that such reactions show a pH optimum which is similar to that observed for most enzymic reactions.

The hydrolysis of carboxylic acid derivatives such as esters is similar, although the elimination process is slightly different. The hydrolysis of an ester in pure, neutral water (that is catalysed only by water itself) may be represented in the following sequence:

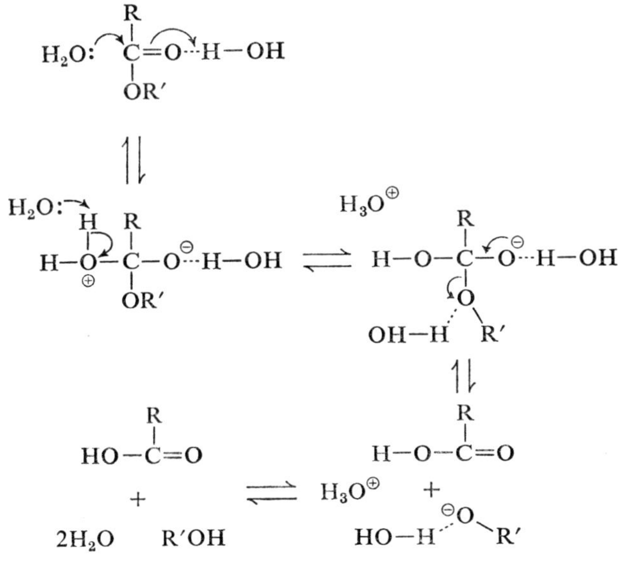

The elimination leads to the loss of solvated alkoxide ion which obtains a hydrogen ion from solvent and becomes a neutral alcohol molecule.

Catalysis by acid or base is more effective. In the following mechanism for the former, some of the essential proton transfers are not represented, for the sake of brevity and clarity.

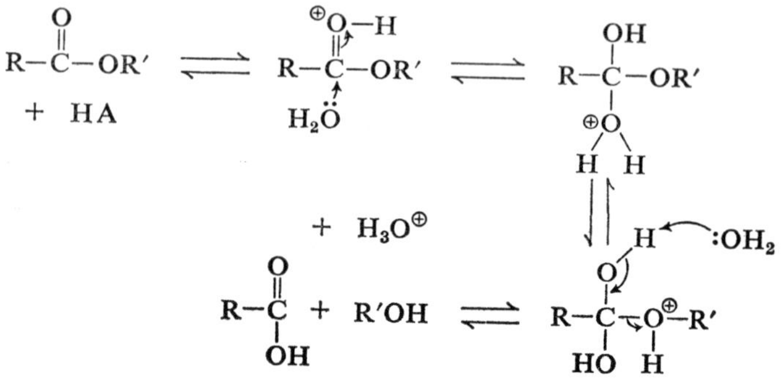


## Mutarotation of Glucose and Glucose Derivatives

The cleavage of hemi-hetals and hemi-acetals, like the dehydration of acetaldehyde hydrate, is an elimination process and may similarly be subject to both acidic and basic catalysis. Conversion of pure α- or pure β-glucose into a mixture of the two forms is known as mutarotation, and it proceeds via the open-chain aldehyde form. Essentially the process involves an elimination reaction and its reverse.

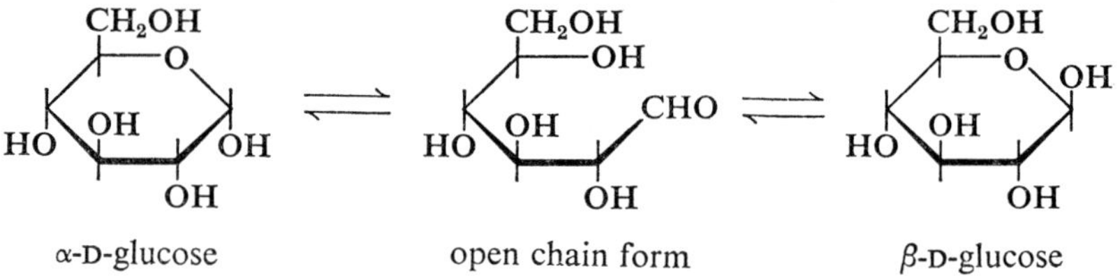


Mutarotation of glucose in aqueous systems may be general acid- or general base-catalysed and the mechanisms appear to be as shown.

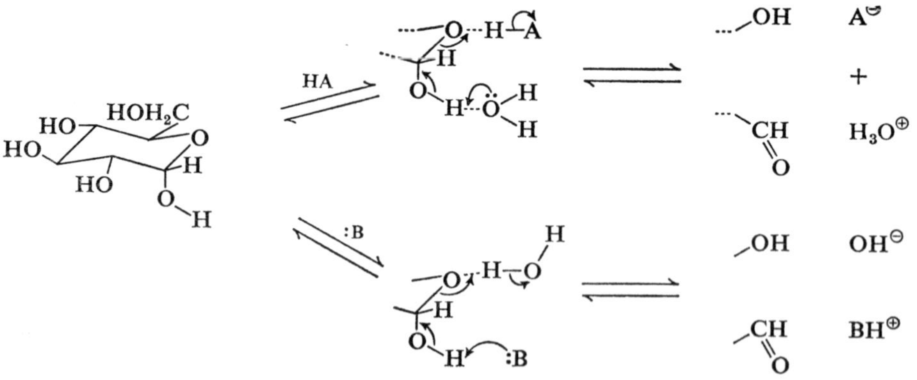

The reaction appears to require the presence of both base and acid for catalysis, and the dual nature of the water allows it to fulfil the necessary function. In the case of tetramethyl glucose in a solvent such as benzene, however, the solvent cannot play these parts. It is found that both an acid and a base must be added for catalysis. In the presence of either the base pyridine, or the acid cresol, the reaction proceeds only very slowly. If a mixture of these two is added the process occurs readily.

In 1952, Swain and Brown [29] reported an observation in connection with this reaction which has had a profound effect on our views about the catalytic process. They found that the bifunctional compound 2-hydroxypyridine (VIII) is a powerful catalyst for the mutarotation of tetramethyl glucose. Its effect lies in the fact that it possesses both an acidic and a basic group and that these are arranged in space in exactly the correct positions to perform what is now known as a concerted acid–base catalysis.

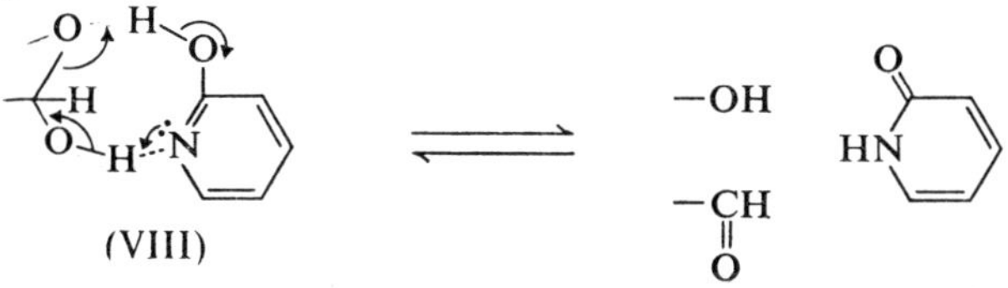

(VIII)

The reaction is first order in both tetramethyl glucose and 2-hydroxypyridine. The catalyst was found to be 7000 times more effective than a mixture of pyridine and cresol at the same concentrations, even though it is a weaker base than pyridine and a weaker acid than cresol. Swain and Brown suggested that catalysis by enzymes might involve similar concerted processes, and at the present time it seems likely that in at least some examples this is indeed the case.

**Decarboxylation of β-ketoacids**

Westheimer and Jones [30] suggested that the decarboxylation of 2,2-dimethylacetoacetic acid (IX), which is an elimination reaction of a rather

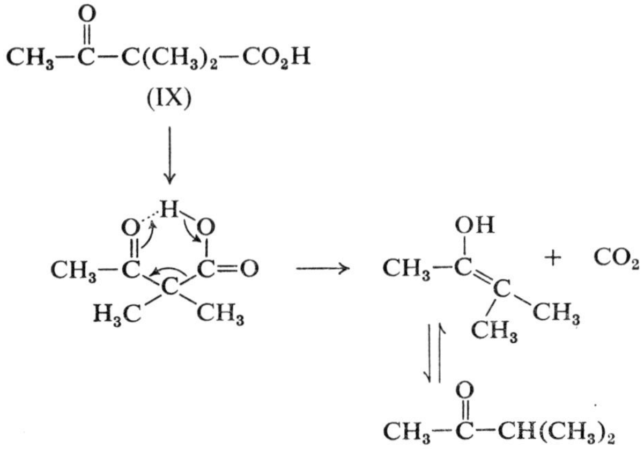

more complex type, proceeds via an intramolecular proton transfer. The electron flow is as shown, the immediate product being the enol which undergoes tautomerism to the ketone. We may view this reaction as a concerted intramolecular acid–base catalysis in which the carbonyl oxygen, acting as a base, removes a proton from the carboxyl group which must be the electron source. At the same time the carboxyl, as an acid, protonates the carbonyl and improves its ability to function as the electron sink.

**Metal-ions**

So far, in considering acid–base catalysis, we have limited ourselves to the Brønsted–Lowry definition, that is, we have considered only proton transfers. However, all acids and bases, including those which fall into the Lewis definition, will be able to influence the rates of chemical reactions by mechanisms similar to those mentioned above. Coordination of any potential leaving group or electron terminus with a Lewis acid will enhance the electron flow and thus increase the reaction rate. Among the most common Lewis acids available to the chemists are metal derivatives including metal cations, and many cases are known in which reactions are catalysed by these agents.

Many metal cations form coordination compounds with other molecules which are able to donate electrons to form a chemical bond. The species which donates the electrons, the Lewis base, is known in this situation as a ligand. The formation of a metal-ion–ligand complex is subject to the equilibrium constant for the process.

$$\mathrm{M} + \mathrm{L} \rightleftharpoons \mathrm{ML}$$

$$\mathrm{K} = \frac{[\mathrm{ML}]}{[\mathrm{M}][\mathrm{L}]}$$

The equilibrium constant for the reaction expressed in this way is often referred to as the *stability constant* and it does reflect the strength of the metal-ion–ligand bond. One of the complicating features of ligand–metal-ion interaction is the fact that each metal-ion is able to accommodate more than one ligand and so a series of stability constants may be obtained for the association of a particular ion with the ligand molecules.

$$\mathrm{M} + \mathrm{L} \rightleftharpoons \mathrm{ML}; \quad \mathrm{ML} + \mathrm{L} \rightleftharpoons \mathrm{ML_2};$$

$$\mathrm{ML_2} + \mathrm{L} \rightleftharpoons \mathrm{ML_3} \ldots \mathrm{ML}_{n-1} + \mathrm{L} \rightleftharpoons \mathrm{ML}_n$$

$$K_1 = \frac{[\mathrm{ML}]}{[\mathrm{M}][\mathrm{L}]}; \quad K_2 = \frac{[\mathrm{ML_2}]}{[\mathrm{ML}][\mathrm{L}]};$$

$$K_3 = \frac{[\mathrm{ML_3}]}{[\mathrm{ML_2}][\mathrm{L}]} \ldots K_n = \frac{[\mathrm{ML}_n]}{[\mathrm{ML}_{n-1}][\mathrm{L}]}$$

In general the stability constants for complexes decrease as the number of ligands in the complex increases. In other words $K_1 > K_2 > K_3 > K_n$.

For example, the complexes formed from imidazole and the cupric ion are such that $\log K_1 = 4{\cdot}36$, $\log K_2 = 3{\cdot}57$, $\log K_3 = 2{\cdot}85$ and $\log K_4 = 2{\cdot}66$ [31]. Thus addition of an extra ligand to a complex becomes increasingly more difficult and this is especially so when the ligands carry a negative charge. Potential incoming ligands will then be repelled by electrostatic forces.

Metal-ions are usually able to accommodate two, four or six ligands in their complexes. Two ligands take up a linear arrangement, four may form either planar or tetrahedral complexes with the metal-ion at the centre. Six ligands dispose themselves around the central metal-ion at the corners of an octahedron. The natures of the metal-ion and the ligands will determine which of these arrangements will be produced in any particular case.

Just as the strength of a Brønsted–Lowry acid, that is, its ability to donate a proton, is reflected in the equilibrium constant, so the stability constant for a metal-ion–ligand complex may be regarded as a measure of the ability of the metal-ion to accept electrons. It might therefore be expected that those metal-ions which form the strongest complexes will also be the most effective in catalysing chemical reactions. This is borne out in practice. The metal-ions from groups IA, IIA and IIIA of the periodic table form relatively weak complexes with the following cited as their order of stability [32].

$$Cs^{I} < Rb^{I} < K^{I} < Na^{I} < Li^{I} < Ba^{II} < Sr^{II} < Cu^{II} < Mg^{II} < La^{III} < Y^{III} < Al^{III}$$

Of these the first five do not appear to be effective at all as catalysts.

The transition elements form far more stable complexes and this is generally attributed to the combined effects of the small diameters of the cations, their +2 and +3 charges and the unfilled orbitals which are able to accommodate the donated electrons. The transition element cations may therefore be expected to catalyse chemical reactions efficiently, and it is interesting to observe that the most common metal-ions found to be associated with enzymes are transition elements, especially Mn, Fe, Co, Ni, Cu and Zn.

The stability of a ligand–metal-ion complex depends to a certain extent on the nature of the ligand, especially in one respect. The complex will usually be far more stable if two ligand groups are joined together in the same molecule than if the two ligands were entirely separate. For the association of ammonia with the cupric-ion it was found that $\log K_1 = 4{\cdot}2$ and $\log K_2 = 3{\cdot}5$. However, for the ethylenediamine–cupric-ion complex (X) $\log K = 10{\cdot}6$ [33]. The *bidentate* ligand ethylenediamine binds better to the metal ion than do two separate molecules of ammonia. Also, glycine binds better to a metal-ion than does a mixture of acetic acid and ammonia (XI). In such complexes, known as *chelates*, ring structures are formed

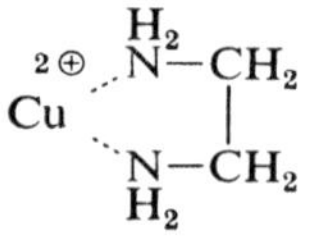


(X)

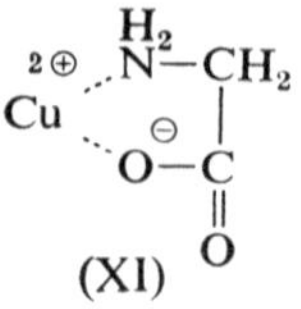


(XI)

and it is a feature of organic chemistry as a whole that the formation of a five- or six-membered ring, in a compound or in a transition state, is highly favoured.

It is interesting to examine some metal-ion-catalysed reactions in the light of these comments. Although the ligand–metal-ion bonds are usually quite strong, and are essentially similar to covalent bonds, we shall represent them as broken lines, as in (X) and (XI), for the sake of clarity.

*Decarboxylation.* Many reactions involving carbonyl groups are catalysed by hydrogen ions. It is difficult to demonstrate catalysis of decarboxylation of $\beta$-ketoacids by added acid, because under conditions where the carbonyl group may be protonated the carboxyl group cannot exist in the form of the anion and this is an important prerequisite for the reaction. It is usually assumed that an internal proton transfer occurs in the 'uncatalysed' process. The manganese ion has been shown [34] to be a very effective catalyst for the decarboxylation of dimethyloxaloacetic acid.

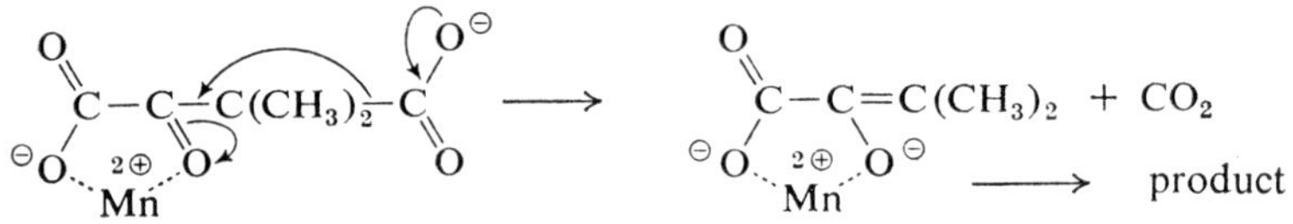


Association of the cation with the carbonyl group provides a suitable electron terminus for the process. In this situation an acid (a Lewis acid) is made available at the carbonyl group under conditions (neutral pH) which permit the presence of the carboxylate anion in the other part of the molecule.

The great stability of the chelate complex is in evidence here, and indeed this appears to be essential to the reaction. Metal-ions are unable to catalyse the decarboxylation of either acetoacetic acid or the ester $C_2H_5OCO.CO.C(CH_3)_2.CO_2H$. In these cases formation of catalytically useful chelates is impossible. As a result the manganese ion is able to catalyse the decarboxylation of some $\beta$-ketoacids and not others; it may therefore be said to be a selective catalyst.

*Ester hydrolysis.* Hydrogen-ion catalysis of ester hydrolysis has already been mentioned (page 36). It is not surprising that this reaction may also be catalysed by metal-ions [35]. Hydrolysis of amino-acid esters, in particular those of glycine and phenylalanine, has been shown to be influenced strongly by cupric ions in glycine buffers [36]. The following mechanism was proposed:

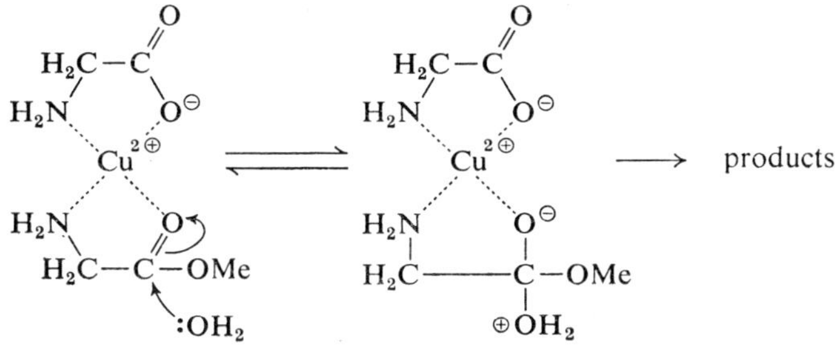

In this case also the substrate is a bidentate ligand, the complex therefore being very stable. A molecule of glycine from the buffer fills the other available sites on the metal-ion. Meriweather and Westheimer have shown that the hydrolysis of amino-acid amides may also be catalysed in this way [37].

An extremely interesting hydrolytic reaction is brought about by the cobalt complex, cis-hydroxyaquotriethylenetetramine cobalt(III) [Co-trien($H_2O$)($OH^{\ominus}$)$^{2\oplus}$] which is represented as XII. This reagent reacts with peptides in such a way that the N-terminal peptide bond is hydrolysed, releasing the free N-terminal amino-acid, which forms a complex with the cobalt ion, as indicated by the examples below [38, 39].

Co-trien(OH)($H_2O$) + Gly.Gly.Gly → CO-trien Gly + Gly.Gly

Co-trien(OH)($H_2O$) + Phe.Gly → Co-trien Phe + Gly

Co-trien(OH)($H_2O$) + Gly.Phe.$NH_2$ → Co-trien Gly + Phe.$NH_2$

It was suggested that the first stage of the reaction involves replacement of the water ligand by the N-terminal amino-group, and that this is the rate-determining step. The next step may be the replacement of the hydroxyl ligand by the carbonyl group of the peptide bond to be cleaved (XIII). The attack by hydroxide ion on the carbonyl group is thus enhanced. Evidence that metal-ion-catalysed hydrolyses of carboxylic acid derivatives involve nucleophilic attack by hydroxide ion has been obtained by Angelici and Leach [40].

Although the reagent (XII) is not in the strict sense a catalyst because the active form is not regenerated and one mole is required for the conversion of one mole of peptide, the process is obviously of significance to the study of catalysis. Of great interest is the specificity which this reagent demonstrates in bringing about the cleavage of only the N-terminal peptide bond. This is similar to the action of the amino-peptidases and this type of compound may well become the forerunner of many synthetic enzyme-like catalyses.

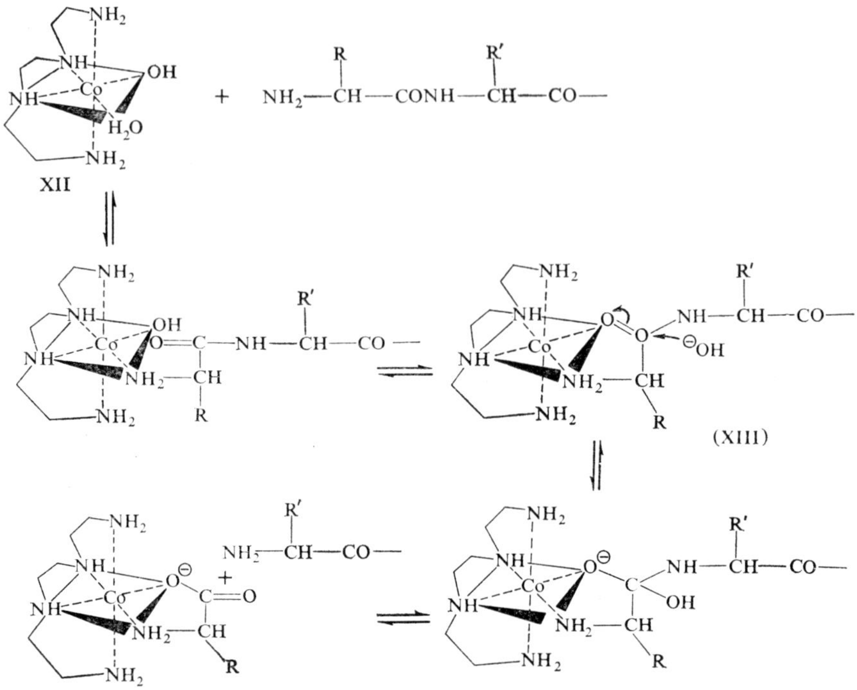


It is probable that the metal-ions which are essential for the action of the amino-peptidases and the carboxypeptidases perform similar functions to those described above (see chapter 4).

Another biologically important reaction which may be catalysed by metal-ions is the hydration of carbonyl compounds. Pocker and co-worker [41] have shown that the hydration of 2-pyridinealdehyde (XIV) is catalysed by both zinc(II) and cobalt(II) ions. A possible mechanism is shown. It is noteworthy that these ions are far less effective for the hydration of 4-pyridinealdehyde (XV). In this context it should be pointed out that the enzyme carbonate hydro-lyase (carbonic anhydrase, E.C. 4.2.1.1),

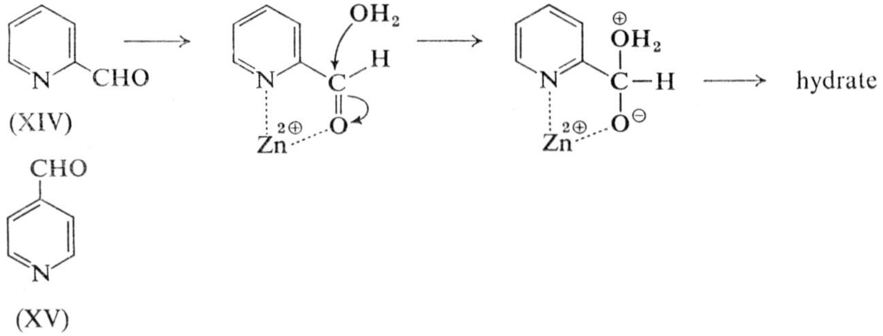

which catalyses the dehydration of carbonic acid and also the hydration of many aldehydes, requires zinc ions for its activity.

The many reactions of the coenzyme pyridoxal and its derivatives are catalysed by metal-ions, and these are discussed later in the appropriate chapters.

*Nucleophilic substitutions.* Before leaving the subject of metal-ion catalysis it is worth mentioning that nucleophilic substitutions themselves may be influenced in this way. Thus, the Friedel–Crafts reaction may be regarded as a substitution catalysed by the powerful Lewis-acid aluminium chloride which assists the departure of a chloride ion from a reagent such as methyl chloride.

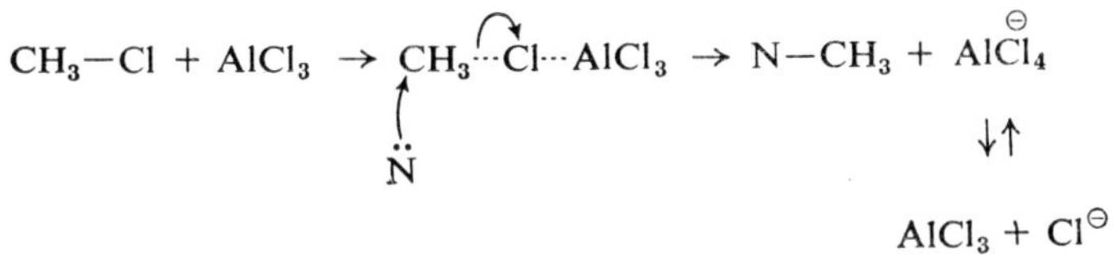

Other nucleophilic substitutions on alkyl halides may be influenced by silver(I) or mercury(II) ions which demonstrate great affinity for halide leaving groups and so assist their departure. In some of these cases the assisted departure of the leaving group may result in the reaction having the properties of an $S_N1$ process.

## Alternative Pathway Catalysis

The catalytic processes described in the previous section involve modification of the reactants so that the speed of the rate-determining step is increased. The pathways of the catalysed and uncatalysed reactions are almost identical; the same atoms are involved, the same bonds formed or broken. No new or different species are involved and the catalyst exerts its influence by the relatively simple process of association. We may call this process *catalysis by modification.* There is, of course, another way in which a catalyst may influence a reaction rate, and this is by providing an alternative pathway for the conversion of the reactants into products. Such a pathway would involve the formation of bonds and the production of chemical species which are entirely different from those observed in the uncatalysed process. This type may be called *alternative pathway catalysis* and many reactions in organic chemistry are subject to its effect.

The following example provides a convenient demonstration of the processes which may be involved in an alternative pathway reaction. The reaction between water and methyl bromide is ordinarily rather slow, producing methanol and bromide ions by a direct nucleophilic substitution.

 (1.31)

Introduction of iodide ions into the reaction medium has a profound effect. In the first place iodide is a better nucleophile than is water and so the reaction (1.32) will take place more rapidly than (1.31). Now iodide is an unusual species because it is not only a good nucleophile but also a

$$I^{\ominus} + CH_3{-}Br \rightarrow I{-}CH_3 + Br^{\ominus} \qquad (1.32)$$

good leaving group. Therefore the reaction (1.33) will also be more rapid than (1.31). The overall sequence (1.32) and (1.33) involving production

$$H_2O\colon\; CH_3{-}I \rightarrow H_2\overset{\oplus}{O}{-}CH_3 + I^{\ominus} \xrightarrow{H_2O} HO{\cdot}CH_3 + I^{\ominus} + H_3\overset{\oplus}{O} \qquad (1.33)$$

and destruction of methyl iodide turns out to be faster than the reaction of water with methyl bromide and thus reactants are converted into products by a faster process by means of this alternative pathway. The iodide is regenerated during the process and in practical terms a small amount of iodide can assist in the conversion of a large amount of methyl bromide. This, therefore, is catalysis in the strictest sense.

*Esters*. The hydrolysis of esters affords several examples of this type of catalysis. Imidazole catalyses the hydrolysis of the much-studied p-nitrophenyl acetate (XVI) [42, 43], the acetylimidazole (XVII) being formed

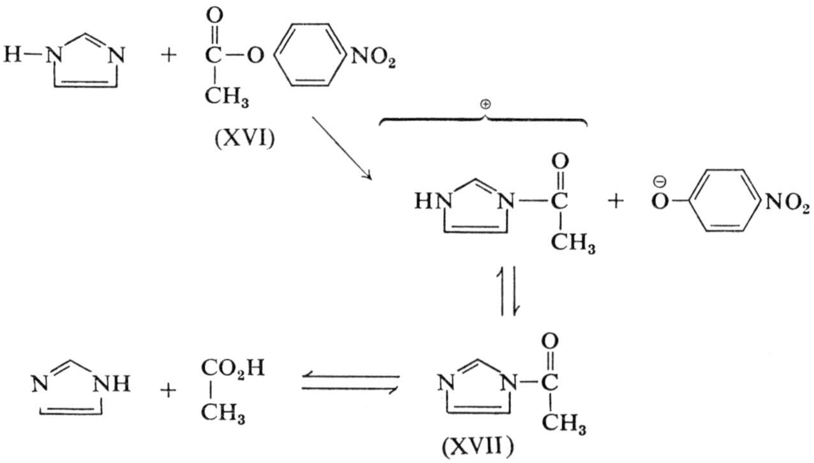


(XVI)

(XVII)

as an intermediate. This intermediate has been isolated and studied separately [44]. It is interesting that imidazole may act not only as a nucleophilic catalyst in this way but also as a basic catalyst. The hydrolysis of acetylimidazole itself may be catalysed by imidazole [44]. This clearly cannot involve nucleophilic attack because starting material would merely be regenerated. The probable mechanism is as shown, the imidazole increasing the nucleophilicity of water by hydrogen bonding (XVIII). In this

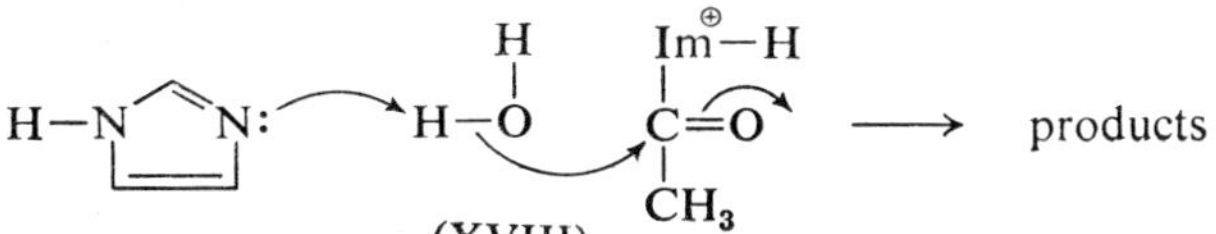

(XVIII)

diagram the acetylimidazole is represented as being protonated as suggested by Bruice and Benkovic [4, p. 66].

The ability to distinguish experimentally between the two types of catalysis in which imidazole may be involved—general base and alternative pathway—is clearly very important in any study of mechanisms. In general two important criteria are available for this distinction. The first is the detection, or better the isolation, of the intermediate produced by nucleophilic catalysis. Thus in the reaction of p-nitrophenyl acetate catalysed by imidazole the intermediate acetylimidazole was detected spectroscopically and also isolated and identified independently. The imidazole-catalysed hydrolysis of ethyl acetate on the other hand has afforded no detectable intermediate.

The second criterion which may be considered is the possibility of an isotope-effect. Thus, if the hydrolysis of ethyl acetate is catalysed by imidazole by a general base process as shown

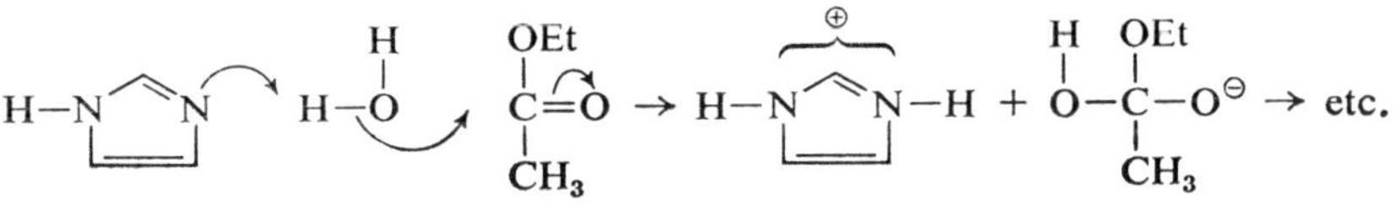

then the reaction involves transfer of a hydrogen ion in the rate-determining step. However, if the reaction is carried out in deuterium oxide instead of water a deuterium ion transfer is involved.

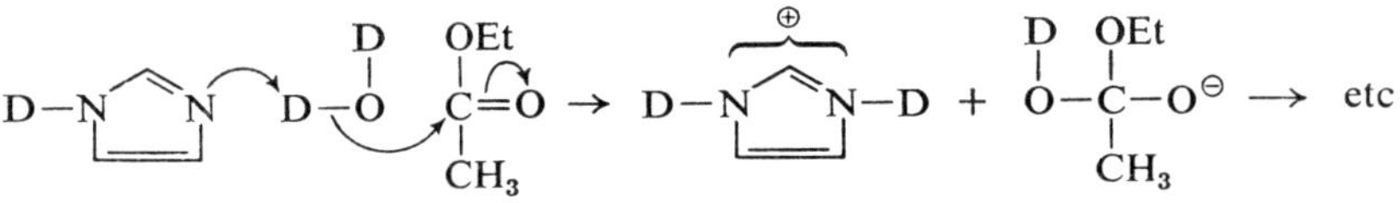

Any reaction involving the transfer of a deuterium ion is normally two to three times slower than the corresponding reaction in which a hydrogen ion is transferred. Because the acidic hydrogen of almost all acids is rapidly exchanged with the deuterium, all reactions involving acid or base catalysis by a proton-transfer reaction in the rate-determining step will be subject to this effect.

It has been found that the catalysis of ethyl acetate hydrolysis by imidazole is subject to such an isotope effect. This, coupled with the fact that an intermediate has not been isolated or detected, indicates the presence of general base catalysis. The imidazole-catalysed hydrolysis of m-nitrophenyl acetate, however, is not subject to an isotope-effect [45] and clearly this reaction proceeds via a nucleophilic attack by the imidazole.

The ability of the imidazole group to participate in such catalytic reactions has been studied intensively because it has been found that histidine residues (which possess this group) are present at the active sites of many enzymes, and are closely involved in the function of these enzymes.

Many other tertiary amines are able to catalyse the hydrolysis of esters by pathways involving acyl-intermediates, the general case being as shown:

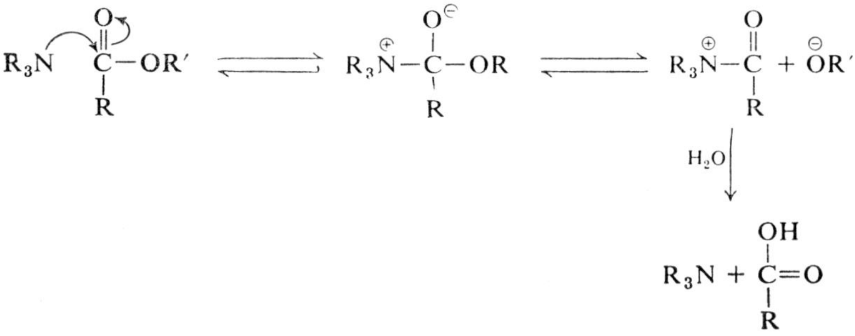

Such amines as triethylamine, pyridine and picoline may be effective in this way [42].

In the examples mentioned above the nucleophiles have been nitrogen atoms. Oxygen anions may also participate in this way. Thus, the hydrolysis of 2,4-dinitrophenyl benzoate is catalysed by acetate ions, the intermediate being the mixed anhydride (XIX) [46].

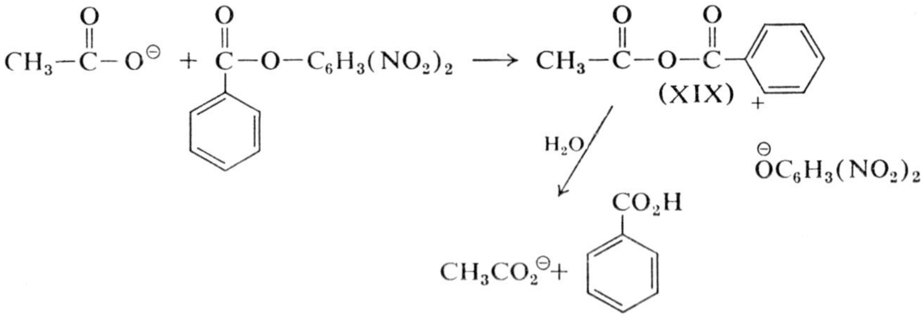

However, as in the case of imidazole, not all cases of acetate-ion catalysis involve such an intermediate. For example, the hydrolysis of p-nitrophenyl acetate is almost certainly catalysed by acetate ion acting as a basic catalyst. Hydrolysis of acetic anhydride itself is catalysed by acetate ions.

*Reactions involving carbonyl groups.* We have already seen that reaction of a carbonyl compound with an amine leads to the production of an imine. If hydroxylamine is used, the product is an oxime. The rate of formation of oximes may be increased by the addition of a primary amine such as aniline and it has been shown that the catalysed reaction proceeds through an aniline–Schiff's base [47].

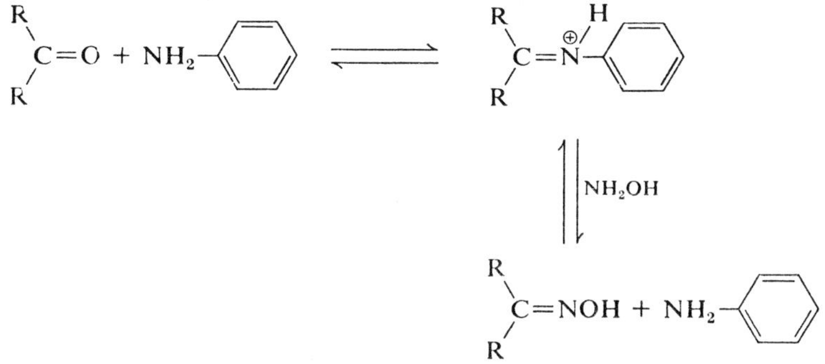

This reaction is itself subject to acid catalysis. Clearly aniline reacts with the carbonyl group more rapidly than does the hydroxylamine, and the protonated aniline–Schiff's base is more susceptible to attack by the hydroxylamine than was the original ketone. The production of similar derivatives such as semi-carbazones is also influenced in this way.

Schiff's bases are formed as intermediates in many reactions catalysed by amines. For example, $\beta$-ketoacids undergo rapid decarboxylation in the presence of primary amines [48].

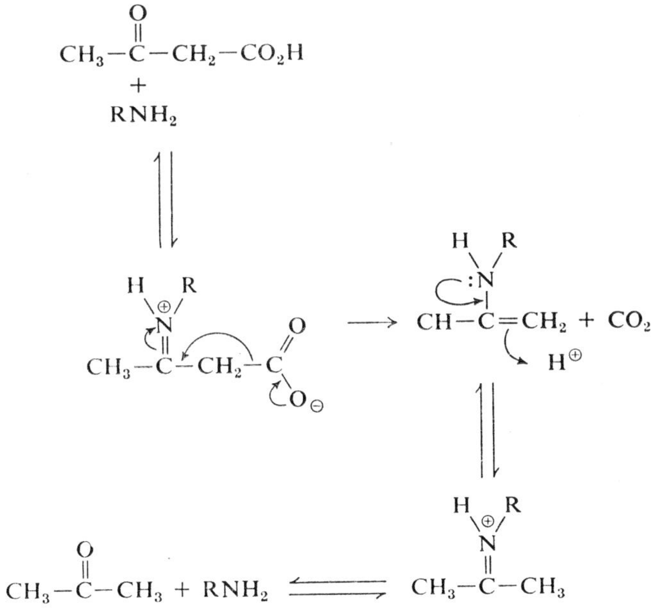

This type of process appears to be important for the action of the decarboxylating enzyme acetoacetate carboxylase (acetoacetate decarboxylase, E.C. 4.1.1.4) where the formation of a Schiff's base between enzyme and substrate has been demonstrated [49].

By a similar, but more complicated, process amines may bring about the decarboxylation of α-ketoacids. Langenbeck [50] carried out this type of reaction in phenolic solvents, the phenol providing acidic and basic groups where necessary. The first stage involves the acid-catalysed formation of a Schiff's base. Thus, in the decarboxylation of phenylglyoxylic acid in phenol in the presence of the amine R $CH_2NH_2$, the Schiff's base (XX) is formed.

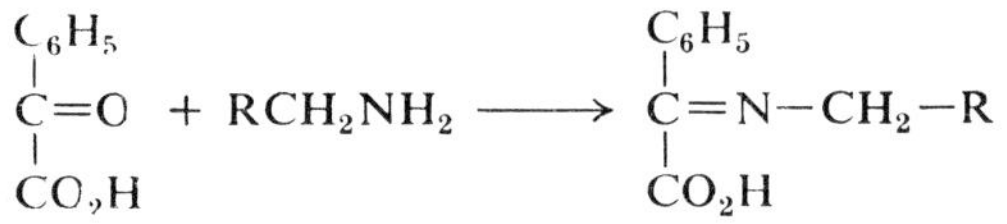

(XX)

The next step is migration of the double bond, effected by a means of two proton-transfer processes involving both a base and an acid (a prototropic shift). This produces a β,γ-unsaturated acid (XXI) which undergoes decarboxylation readily. This decarboxylation is catalysed by both acid and base as shown.

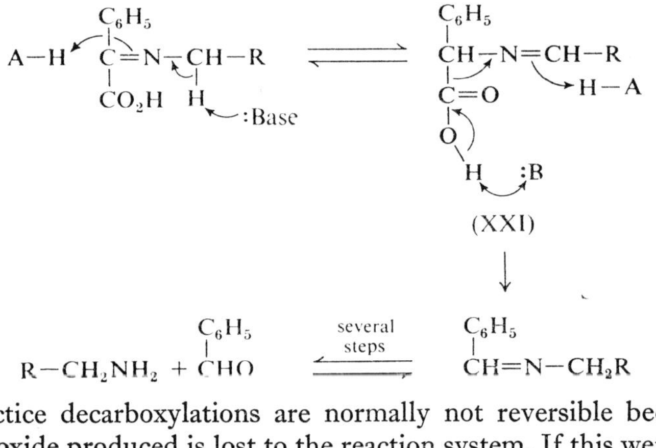


In practice decarboxylations are normally not reversible because the carbon dioxide produced is lost to the reaction system. If this were not the case we might expect that the reverse process would also be catalysed by amines. Decarboxylation in biological terms may be regarded as a degradative process. Synthetic processes, that is processes in which carbon–carbon bonds are formed, may also be catalysed via the intermediacy of imines. These include acyloin and aldol condensations as described below.

Under the conditions which favour the decarboxylation of α-ketoacids such as pyruvic acid, acyloins may also be formed [50]. The reaction almost certainly involves condensation of the Schiff's base of the α-ketoacid with a molecule of aldehyde already formed by decarboxylation. A further decarboxylation follows. As in the previous case the reaction requires acidic and basic catalysts [46, p. 195].

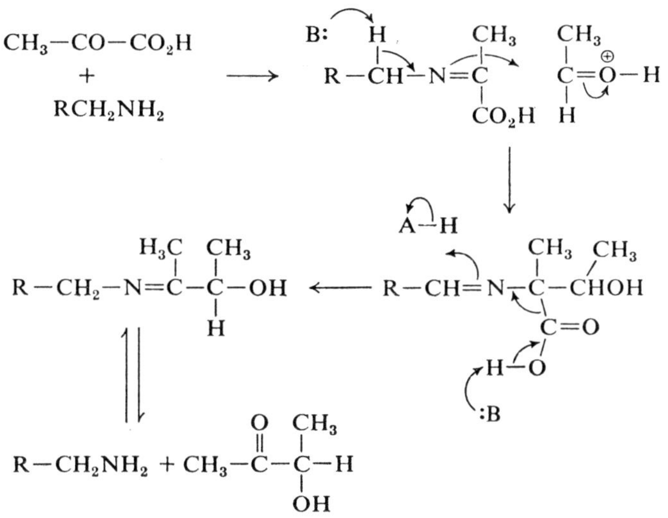

Aldol condensations, when catalysed by primary amines, proceed via an interesting prototropic shift producing an enamine (XXII) [51]. The 'aldolization' of acetone may be represented as follows:

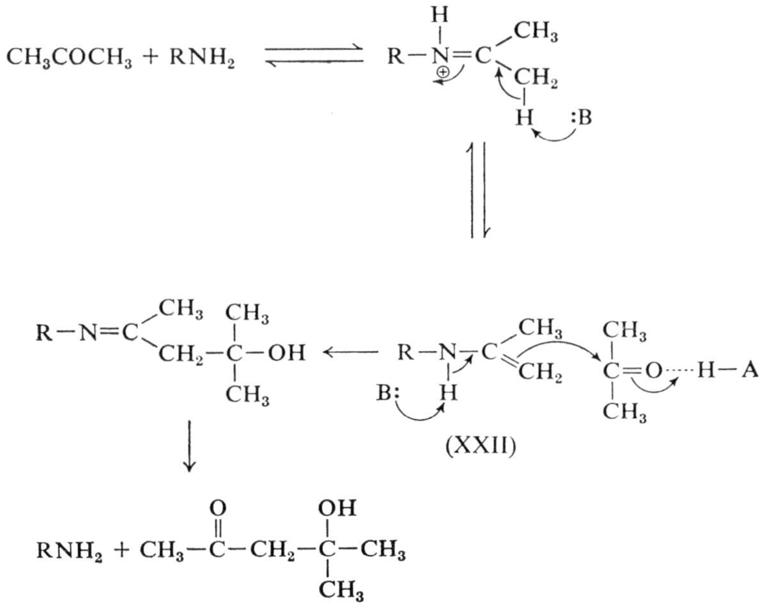

For acetone, the equilibrium constants are such that the synthetic reaction is not favoured and experimentally the 'dealdolization' is easier to study. The principle of microscopic reversibility has been invoked to suggest the above mechanism. The kinetics of butyraldehyde condensation were studied by Yasnikov [51] with essentially the same results. Since the enzyme fructose-1,6-diphosphate-D-glyceraldehyde-3-phosphate lyase (fructose-diphosphate aldolase; often called aldolase E.C. 4.1.2.13) which catalyses a similar reaction has been shown to form imines with the substrate [52], it is assumed that a similar mechanism is involved.

Primary amines are not the only agents in organic chemistry which are able to catalyse reactions involving carbonyl groups by providing alternative pathways. The cyanide ion is very versatile in this respect. Thus, it may assist in the conversion of aldehydes to acyloins [53], the mechanism for this process having been suggested by Lapworth [27]. It involves the formation of a carbanion (XXIII) which is stabilized by the presence of

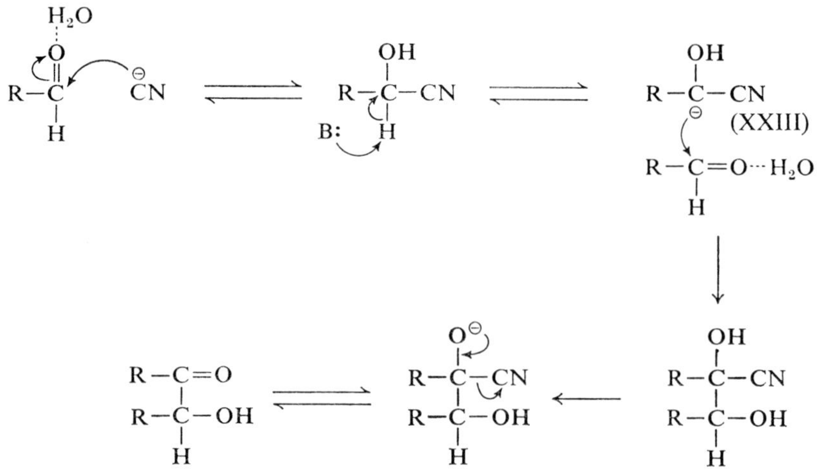


the nitrile group adjacent to it. When R is $C_6H_5$ this reaction is called the benzoin condensation. It may also be catalysed by other species. These include the thiazolium ion (XXIV) which acts in the same way, that is by providing a stabilizing factor for the carbanion required (XXV and XXVI) [54].

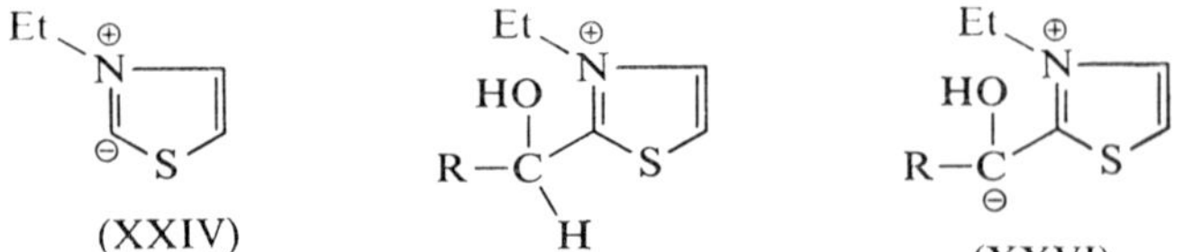

(XXIV) (XXV) (XXVI)

Decarboxylation of α-ketoacids is another reaction in which cyanide may participate [55]. Here again a carbanion is stabilized by the presence of the nitrile group (XXVII).

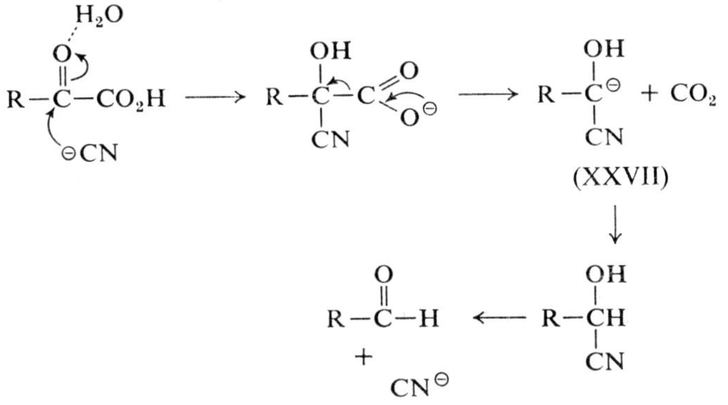

In such reactions the ability of the nitrile group to stabilize carbanions is the critical factor. It is this, and not merely its nucleophylicity, which renders the cyanide ion effective. This point is clearly illustrated in the case of the reactions of α-diketones. Here the two nucleophiles cyanide ion and hydroxide ion catalyse entirely different reactions. In alcoholic solution, (XXVIII) is converted by cyanide into an aldehyde and an ester [56]. The reaction involves attack on a carbonyl group by a neighbouring oxyanion, leading eventually to a nitrile-stabilized anion (XXIX).

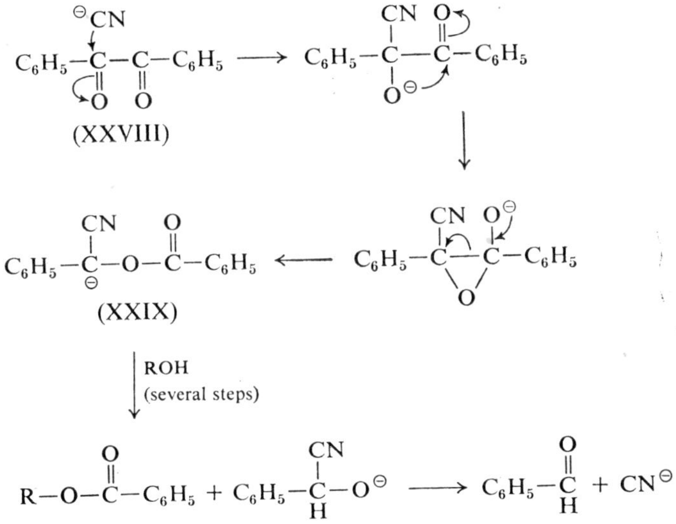

By contrast, hydroxide ion brings about the conversion of benzil into benzilic acid (XXX). In this case an unusual rearrangement takes place involving migration of a carbon–carbon bond.

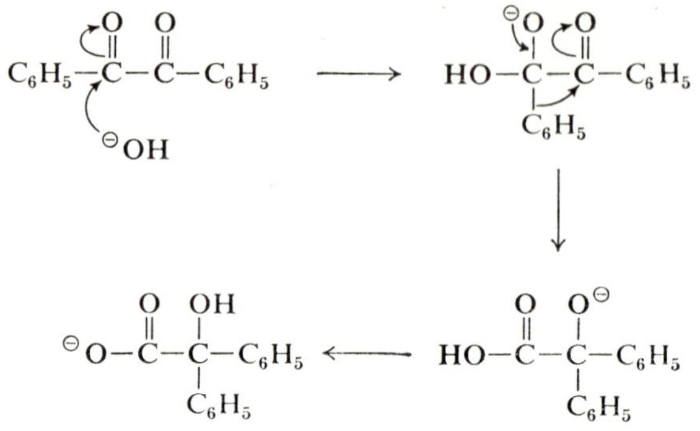

(XXX)

Cyanide and hydroxide ions produced different products, too, in their reactions with tropolone (XXXI). Hydroxide, being a strong base, removes a hydrogen ion producing the tropolone anion (XXXII) which is very stable, and no further reaction takes place. Cyanide, however, is not a powerful base but nevertheless is a good nucleophile and under its action tropolone is converted into benzoic acid [57].

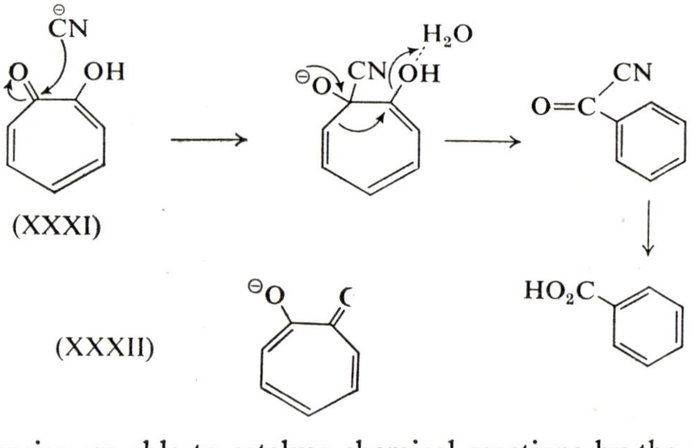

(XXXI)

(XXXII)

Other species are able to catalyse chemical reactions by the alternative pathway process, and among these perhaps the most important are thiamine and pyridoxal and their derivatives. Studies on these compounds have led to great advances in our knowledge of the mechanisms of enzymic catalysis. Discussion of these and other coenzymes is left to those chapters which deal with the enzymes individually.

**Proximity Effects**

At this stage we consider a phenomenon which, although not itself a form of catalysis, does have considerable influence on the rates of chemical reactions. This phenomenon, the proximity effect (sometimes referred to as propinquity catalysis), is best discussed in terms of some examples.

The hydrolysis of phenyl acetate (XXXIII) may be catalysed by trimethylamine (XXXIV) as described on page 46, the overall process being first order in both amine and ester and having a bimolecular rate constant $k_B$. The hydrolysis of the ester (XXXV) is much faster; it is first order with respect to the amino-ester and its unimolecular rate constant $k_u$ is about $10^3$ times greater than $k_B$ [58]. Now the nucleophilicity of the amino-group in compound (XXXV) is unlikely to be very different from that in (XXXIV); any electrical effects due to the ester carbonyl group will be negligible when the groups are separated by a chain of four carbons. Similarly the susceptibility to nucleophilic attack of the two ester groups will

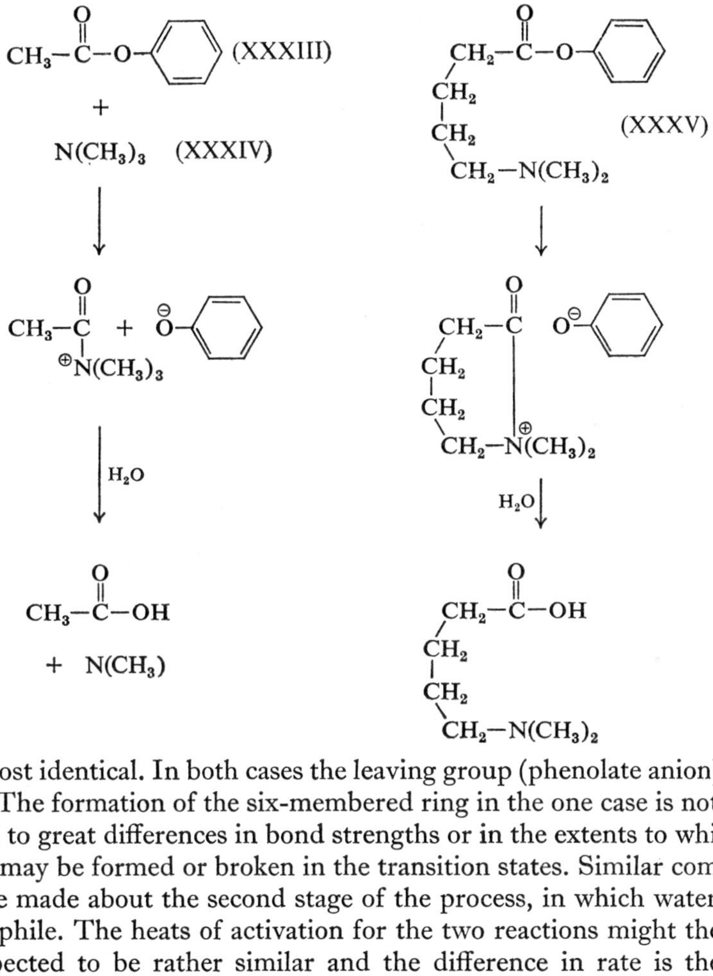


be almost identical. In both cases the leaving group (phenolate anion) is the same. The formation of the six-membered ring in the one case is not likely to lead to great differences in bond strengths or in the extents to which the bonds may be formed or broken in the transition states. Similar comments may be made about the second stage of the process, in which water is the nucleophile. The heats of activation for the two reactions might therefore be expected to be rather similar and the difference in rate is therefore likely to be due, in the main, to differences in the entropies of activation. The unimolecular process (XXXV) has a higher entropy of activation (a lower negative value) than that of the bimolecular process.

At first sight this appears to be a surprising result. The transition state for the reaction of (XXXV) must be a cyclic, relatively rigid system and therefore it must have, associated with it, a low entropy value, whereas the transition state for the reaction of the ester (XXXIII) with the amine (XXXIV) will not be so rigid. However, the absolute value for the entropy associated with a transition state is not the important parameter. What does matter is the *difference* in entropy between the transition state and the starting materials. The system (XXXV) already has a much lower entropy built into it, being a more restricted molecule compared with the system (XXXIII and XXXIV) where the two molecules are independent of each other and free to move unimpeded throughout the solvent medium. Therefore the change in entropy for the unimolecular process may be more favourable.

Bruice and Benkovic [58] suggest that the entropy of activation for a reaction is related closely to the order of the reaction (see page 6). Thus for a series of displacements on phenyl esters they obtained the empirical relationships [4, Vol. 1, p. 123]:

$$\frac{(\text{experimentally determined } T\,\Delta S^*)}{4\text{–}5\ (\text{kcal/mole})} = \text{kinetic order}$$

Therefore if two very similar reactions are compared, the main difference being that one reaction has an order which is one less than that of the other, then the values of $T\,\Delta S^*$ for the two reactions will differ by 4–5 kcal/mole. The reaction which has the lower order will have the lower negative entropy of activation and hence will have the higher rate. A decrease of 4–5 kcal/mole in the overall free energy of activation will on average give rise to a rate enhancement of a thousand-fold. Even in cases where the steric factors involved lead to changes in the heat of activation the proximity effect may still cause significant rate enhancements.

In the case which we have used as an example, it is clear that the hydrolysis of the phenyl ester species, whose rate is already enhanced by the presence of the trimethylamine via an alternative pathway, is made even more rapid because of the close proximity of the amino-group in compound (XXXV).

Many examples of the proximity effect are known. One which is of special interest is the formation of acyl-imidazoles from p-nitrophenyl esters. The rate of release of p-nitrophenolate anion from ester (XXXVI) is far greater than that of the reaction between imidazole and p-nitrophenyl acetate [59, 60]. In fact the rate of reaction of XXXVI is actually slightly greater than that of p-nitrophenyl acetate with the enzyme chymotrypsin (E.C. 3.4.4.5) which leads to the formation of acetyl-chymotrypsin and liberation of the p-nitrophenolate ion [61] (see Chapter 4). Since acyl-imidazoles such as those represented above react with water to give the

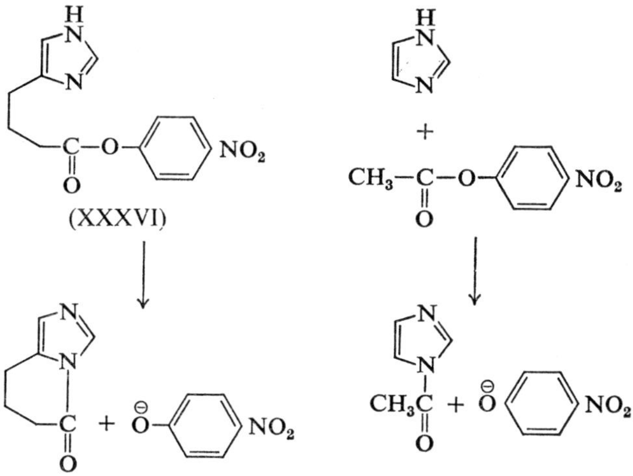


(XXXVI)

free acids, they afford a further example of a catalysis of hydrolysis further enhanced by the proximity effect.

We have already seen that oxygen anions may participate in alternative pathway catalysis, and there are examples of such catalysis being influenced by the proximity effect. The hydrolysis of acetylsalicylic acid is brought about by the attack of the nearby carboxylate ion in this way [62].

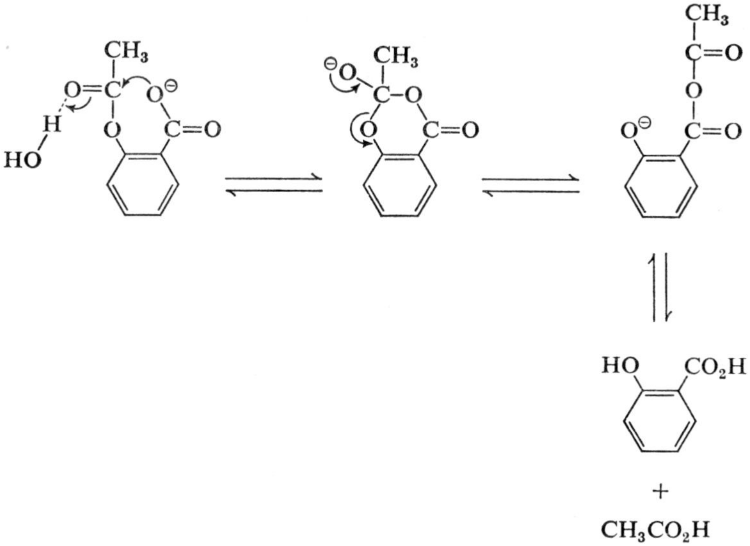


The intermediate anhydride is rapidly broken down.

### Strain

The proximity effect is of course not in itself a mechanism for catalysis. It is a feature built into the structure of the molecule, and by virtue of this feature the reactivity of a group is considerably higher than it would be if the group were present in a molecule which did not possess this feature. A similar molecular feature is *strain.* If a molecule possesses a degree of strain then any reaction which will release that strain will tend to proceed more rapidly than that in an unstrained molecule. This is because in releasing the strain, energy is liberated which offsets the energy of activation for the reaction. In this way the reactivity of a group in the molecule may be considered to be enhanced. For example, a $\gamma$-lactone which is highly strained because of its four-membered ring structure is hydrolysed at a more rapid rate than is a $\delta$-lactone which, having a five-membered ring, is less strained. Thus the susceptibility of the ester group to hydrolysis is enhanced by this particular molecular feature.

## CONCLUSION

We have seen in this chapter that the rate of a chemical reaction is entirely dependent on the free energy of activation for that reaction. This free energy of activation may be considered as being made up of two terms, the heat and the entropy of activation. The strengths of the bonds which are broken and formed during the reaction, and the extent to which such bond cleavage or formation has come about by the time the situation of maximum energy has been reached, determines the heat of activation, at least to a first approximation. The entropy of activation will be determined by changes in the overall randomness and rigidity of the system including three-dimensional arrangements in the molecules. It should be noticed that steric effects can influence in two ways, firstly by affecting the entropy of activation, and secondly by favouring or hindering bond formation at the transition-state phase, thus influencing the heat of activation.

A catalyst influences the rate of a reaction either by modifying the reactants or by providing an alternative pathway, the energy of activation in either case being lower than that for the uncatalysed reaction. In addition, we might envisage a catalyst having its effect by so arranging the system that the rate-determining step has a lower kinetic order than that of the uncatalysed reaction, by changing the dielectric constant of the medium or by somehow introducing a degree of strain into the molecule.

It is the purpose of the remaining chapters of this book to give a broad review of the chemical information which is available concerning the enzymes and hence to discuss the mechanisms by which the enzymes may function.

## REFERENCES


1. Everett, D. H., *An Introduction to the Study of Chemical Thermodynamics* (Longmans, 1960)
2. Gould, E. S., *Mechanism and Structure in Organic Chemistry* (Holt, Rinehart and Winston, 1959)
3. Hine, J., *Physical Organic Chemistry* (McGraw-Hill, 1962)
4. Bruice, T. C. and Benkovic, S. J., *Bioorganic Mechanisms* Vols. 1 and 2 (Benjamin, 1966)
5. Harcourt. A. V., *J. chem. Soc.* **20,** 460 (1867)
6. Arrhenius, S., *Z. physik. Chem.* **4,** 226 (1889)
7. Glasstone, S., Laidler, K. J. and Eyring, H., *The Theory of Rate Processes* (McGraw-Hill, 1941)
8. Crowell, T. I. and Hammett, L. P., *J. Am. chem. Soc.* **70,** 3444 (1948)
9. Dunbar, P. M. and Hammett, L. P., *J. Am. chem. Soc.* **72,** 109 (1950)
10. Hine, J. and Brader, W. H., jr., *J. Am. chem. Soc.* **75,** 3964 (1953)
11. Spryksov, A. A., *J. gen. Chem. U.S.S.R.* **17,** 1309 (1947) (*Chem Abst.* **42,** 1921 (1948)); *J. gen. Chem. U.S.S.R.* **16,** 1059 (1946) (*Chem. Abst.* **41,** 2720 (1947))
12. Brønsted, J. N. and Wynne-Jones, W. F. K., *Trans. Faraday Soc.* **25,** 59 (1929)
13. Coombs, E. and Evans, D. P., *J. chem. Soc.* 1295 (1940)
14. Brønsted, J. N. and Pederson, K., *Z. physik. Chem.* **108,** 185 (1924)
15. DeWolfe, R. H. and Young, W. G., *Chem. Rev.* **56,** 769 (1956); Kepner, R. E., Winstein, S. and Young, W. G., *J. Am. chem. Soc.* **71,** 115 (1949)
16. Lewis, G. N., *Valence and the Structure of Atoms and Molecules* (Chemical Catalogue Co., New York, 1923)
17. Brown, H. C. and Cann, A., *J. Am. chem. Soc.* **77,** 1715 (1955)
18. Hine, J., *Physical Organic Chemistry*, p. 160 (McGraw-Hill, 1962)
19. Hughes, E. D. and Ingold, C. K., *J. chem. Soc.* 244 (1935)
20. Brown, H. C. and Eldred, N. R., *J. Am. chem. Soc.* **71,** 445 (1949)
21. Meisenheimer, J., *Ann.* **323,** 205 (1902)
22. Swain, C. G., *J. Am. chem. Soc.* **70,** 1119 (1948); Swain, C. G. and Scott, C. B., *J. Am. chem. Soc.* **75,** 141 (1953); Swain, C. G. and Mosely, R. B., *J. Am. chem. Soc.* **77,** 3727 (1955); Swain, C. G., Mosely, R. B. and Brown, D. E., *J. Am. chem. Soc.* **77,** 3731 (1955)
23. Chapman, N. B. and Levy, J. L., *J. chem. Soc.* 1677 (1952)
24. Swain, C. G. and Eddy, R. W., *J. Am. chem. Soc.* **70,** 2989 (1948); Pocker, Y., *J. chem. Soc.* 1279 (1957)
25. Bell, R. P. and Higginson, W. C. E., *Proc. R. Soc.* **197A**, 141 (1949); Bell, R. P. and Darwent, B. de B., *Trans. Faraday Soc.* **46,** 36 (1950)
26. Cohn, M. and Urey, H. C., *J. Am. chem. Soc.* **60,** 679 (1938)
27. Lapworth, A., *J. chem. Soc.* **83,** 995 (1903)
28. Barrett, E. and Lapworth, A., *J. chem. Soc.* **93,** 85 (1908); Lapworth, A., *J. chem. Soc.* **91,** 1133 (1907)
29. Swain, C. G. and Brown, J. F., jr., *J. Am. chem. Soc.* **74,** 2534, 2538 (1952)
30. Westheimer, F. H. and Jones, W. A., *J. Am. chem. Soc.* **63,** 3283 (1941)

31. MAHLER, H. R. and CORDES, E. H., *Biological Chemistry* (Harper and Row, New York, 1967)
32. WILLIAMS, R. J. P., *The Enzymes*, Eds. P. D. Boyer, H. Hardy and K. Myrback, p. 391, Vol. 1 (Academic Press, New York, 1959)
33. MARTIN, R. B., *Introduction to Biophysical Chemistry*, p. 337 (McGraw-Hill, 1964)
34. STEINBERGER, R. and WESTHEIMER, F. H., *J. Am. chem. Soc.* **73,** 429 (1951); PRUE, J. E., *J. chem. Soc.* 2331 (1952); see also PEDERSON, K., *Acta chem. Scand.* **6,** 285 (1952)
35. KROLL, H., *J. Am. chem. Soc,* **74,** 2036 (1952)
36. BENDER, M. L. and TURNQUEST, B. W., *J. Am. chem. Soc.* **79,** 1889 (1957)
37. MERIWEATHER, L. and WESTHEIMER, F. H., *J. Am. chem. Soc.* **78,** 5119 (1956)
38. BUCKINGHAM, D. A., CALLMAN, J. P., HAPPER, D. A. and MARZILLI, L. G., *J. Am. chem. Soc.* **89,** 1082 (1967)
39. COLLMAN, J. P. and BUCKINGHAM, D. A., *Chem. Engng News* **41,** 38 (1963)
40. ANGELICI, R. J. and LEACH, B. E., *J. Am. chem. Soc.* **90,** 2499 (1968)
41. POCKER, Y. and MEANY, J. E., *J. Am. chem. Soc.* **89,** 631 (1967), and references cited therein.
42. BENDER, M. L. and TURNQUEST, B. W., *J. Am. chem. Soc.* **79,** 1656 (1957)
43. BRUICE, T. C. and SCHMIR, G. L., *J. Am. chem. Soc.* **79,** 1663 (1957)
44. JENCKS, W. P. and CARRIUOLO, J., *J. biol. Chem.* **234,** 1280 (1959)
45. BRUICE, T. C. and BENKOVIC, S. J., *J. Am. chem. Soc.,* **86,** 418 (1964)
46. BENDER, M. L. and BEVEU, M. C., *J. Am. chem. Soc.* **80,** 5388 (1958)
47. CORDES, E. H. and JENCKS, W. P., *J. Am. chem. Soc.* **84,** 826 (1962)
48. PEDERSON, K. J., *J. Am. chem. Soc.* **60,** 595 (1938), and references quoted therein.
49. WESTHEIMER, F. H., *Proc. chem. Soc.* 253 (1963)
50. LANGENBECK, W., *Die Organischen Katalysatoren*, 2nd edn (Springer, Berlin, 1949)
51. YASNIKOV, A. A., BOIKO, T. S., VOLKOVA, N. V. and MEL'NICHENKO, I. V., *Biokhimiya* **31,** 969 (1966) (*Chem. Abst.* **66,** 8418q (1967))
52. GRAZI, E., CHENG, T. and HORECKER, B. L., *Biochem. biophys. Res. Commun.* **7,** 250 (1962); GRAZI, E., ROWLEY, P. T., CHENG, T., TCHOLA, O. and HORECKER, B. L., *Biochem. biophys. Res. Commun.* **9,** 38 (1962)
53. IDE, W. S. and BUCK, J. S., *Org. React.* **4,** 269 (1948)
54. BRESLOW, R., *Chemy Ind.* 893 (1957)
55. FRANZEN, V. and FIKENTSCHER, L., *Ann.* **613,** 1 (1958)
56. KWART, H. and BAEVSKY, M. H., *J. Am. chem. Soc.* **80,** 580 (1958)
57. FORBES, E. J. and WARRELL, D. C., private communication.
58. BRUICE, T. C. and BENKOVIC, S. J., *J. Am. chem. Soc.* **85,** 1 (1963)
59. BRUICE, T. C. and STURTEVANT, J. M., *Biochim. biophys. Acta* **30,** 208 (1958)
60. BRUICE, T. C. and STURTEVANT, J. M., *J. Am. chem. Soc.* **81,** 2860 (1959)
61. SPENCER, T. and STURTEVANT, J. M., *J. Am. chem. Soc.* **81,** 1874 (1959)
62. GARRETT, E. R., *J. Am. chem. Soc.* **79,** 3401 (1957), and references quoted therein.

# CHAPTER 2

# ENZYMES AS CATALYSTS

## INTRODUCTION

In the first chapter we considered how various organic compounds might catalyse chemical reactions. In this chapter we proceed to a consideration of a group of organic molecules whose ability in the direction is truly remarkable, and indeed vital. The enzymes are proteins which catalyse the chemical reactions which take place in living organisms. Their catalytic abilities are extraordinary not only in that they are far more efficient than any other catalysts studied but also in that many exhibit exceptional properties of specificity. For example, the enzyme mannitol-1-phosphate dehydrogenase (D-mannitol-1-phosphate:NAD oxidoreductase 1.1.1.17) catalyses the oxidation of only one of the five hydroxyl groups in the substrate. The 5-hydroxyl group is oxidized and the product is D-fructose-6-phosphate.

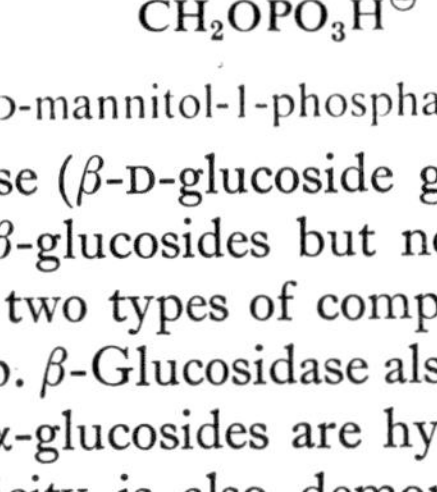

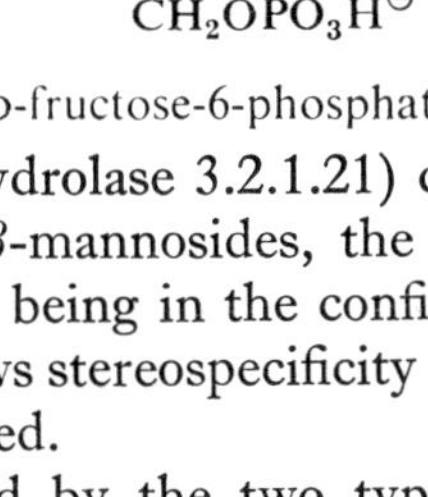

D-mannitol-1-phosphate    D-fructose-6-phosphate

$\beta$-Glucosidase ($\beta$-D-glucoside glucohydrolase 3.2.1.21) can catalyse the hydrolysis of $\beta$-glucosides but not of $\beta$-mannosides, the only difference between these two types of compounds being in the configuration of one hydroxyl group. $\beta$-Glucosidase also shows stereospecificity in that $\beta$-glucosides but not $\alpha$-glucosides are hydrolysed.

Stereospecificity is also demonstrated by the two types of enzymes, steroid-11-$\alpha$-hydroxylase (1.99.1.6) and steroid-11-$\beta$-hydroxylase (1.99.1.7). Whereas the former oxidizes steroids at the 11-position to give 11-$\alpha$-hydroxy compounds, the latter catalyses the oxidation of the same substrates to give 11-$\beta$-hydroxy products (see Fig. 2.1).

It must be pointed out that not all enzymes show such very narrow specificity, some indeed catalyse reactions for a broad range of substrates.

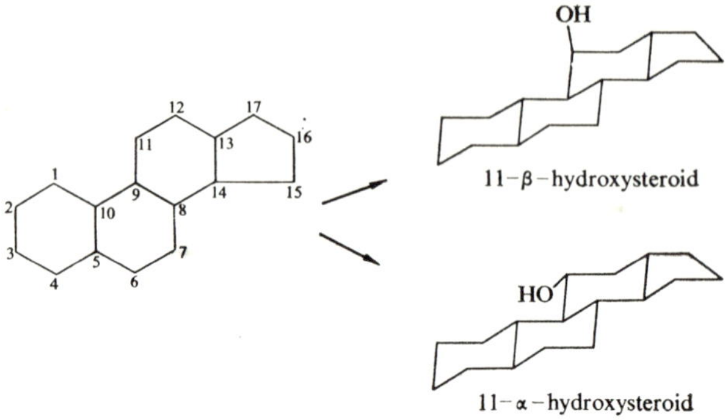


FIG. 2.1. Outline of steroid structure showing hydroxylation at 11-α- and 11-β-positions

Since all enzymes are proteins it is obvious that the nature of proteins and the chemical reactions which they can undergo must be central to any discussion of the way in which enzymes function. Proteins are extremely complex molecules and the complexities arise ultimately from the large number of different amino-acid residues which may be linked together to form the peptide chains which comprise the protein molecule. Table 2.1 gives a list of the twenty-one amino-acids commonly found in proteins, showing the great variety of their side-chain groups. The amino-acids may be classified according to the chemical nature of their side-chains. Thus, alanine, valine, leucine and isoleucine all possess aliphatic side-chains, while phenylalanine, tyrosine and tryptophan have side-chains which are aromatic. In addition to being aromatic, tyrosine also possesses a hydroxyl group which can function as an acidic group, can form hydrogen bonds and can be acylated. Serine and threonine are hydroxyamino-acids and possess the properties expected of hydroxy compounds. Another group consists of the dicarboxylic amino-acids and their amides (aspartic and glutamic acids, asparagine and glutamine) while yet another group possesses basic side-chains (lysine, arginine and histidine). The imidazole group in the histidine side-chain has interesting properties; it can function as an acidic or a basic catalyst and in addition can be acylated with ease. Cysteine and methionine possess sulphur-containing side-chains. Cysteine is often found in proteins as the dimer cystine which is obtained from cysteine by oxidation. The sulphur atoms in both cysteine and methionine have nucleophilic properties and in addition cysteine is able to form hydrogen bonds, while the side-chain of methionine is hydrophobic.

The amino-acids, therefore, manifest a variety of functional side-chains which are able to form various types of bonds. Interaction of acidic and basic groups can produce ionic (electrostatic) bonding while hydrogen bonds can be produced with the hydrophilic groups, the amides and the

Table 2.1. *Amino-acids with the general structure* $NH_2-CH(R)-CO_2H$

| Amino-acid | R |
|---|---|
| Glycine (Gly) | $-H$ |
| Alanine (Ala) | $-CH_3$ |
| Serine (Ser) | $-CH_2OH$ |
| Threonine (Thr) | $-CH(CH_3)OH$ |
| Cysteine (Cys) | $-CH_2SH$ |
| Methionine (Met) | $-CH_2CH_2SCH_3$ |
| Valine (Val) | $-CH(CH_3)_2$ |
| Leucine (Leu) | $-CH_2CH(CH_3)_2$ |
| Isoleucine (Ile) | $-CH(CH_3)CH_2CH_3$ |
| Phenylalanine (Phe) | $-CH_2C_6H_5$ |
| Tyrosine (Tyr) | $-CH_2C_6H_4OH$ |
| Glutamic acid (Glu) | $-CH_2CH_2CO_2H$ |
| Glutamine (Gln) | $-CH_2CH_2CONH_2$ |
| Aspartic acid (Asp) | $-CH_2CO_2H$ |
| Asparagine (Asn) | $-CH_2CONH_2$ |
| Lysine (Lys) | $-CH_2CH_2CH_2CH_2NH_2$ |
| Arginine (Arg) | $-CH_2CH_2CH_2NH-C(=NH)-NH_2$ |
| Histidine (His) | 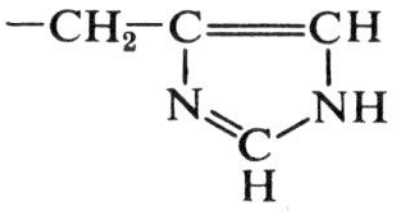 |
| Tryptophan (Try) | 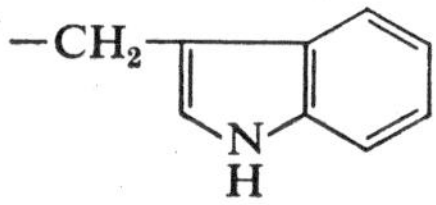 |

Also

| Amino-acid | Structure |
|---|---|
| Proline | 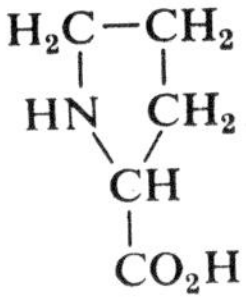 |
| Hydroxyproline | 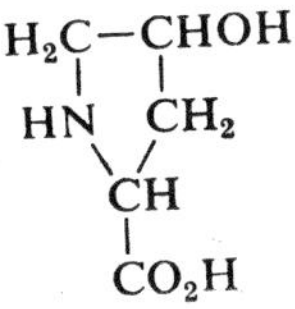 |

hydroxy compounds. Aliphatic and aromatic side-chains may interact to form the so-called *non-polar* bond.

The stability of a non-polar bond arises in two ways. First, the $\sigma$- and especially the $\pi$-electrons in the non-polar groups overlap forming what may be regarded as partial covalent bonds. The second factor is due to the presence of the solvent. A polar solvent such as water possesses a considerable degree of structure due to intermolecular forces. The presence of a non-polar group in the water disrupts many potential intermolecular bonds and therefore leads to a certain amount of instability. The instability of two separate non-polar groups in water can be reduced if these two groups approach each other closely enough to 'squeeze' out any water molecules from between them. The result is a smaller non-polar region and stability is gained when the water molecules re-form their structures.

The overall structure of a protein is considered in terms of four degrees. The *primary* structure refers to the sequence of the amino-acid residues in the protein chain. Examples of primary structures are given in later chapters. The *secondary* structure relates to any regularities in the three-dimensional molecule, and is mainly concerned with hydrogen bonds formed between the carbonyl and N—H groups of the peptide backbone, as shown below. From a consideration of these hydrogen bonds, Pauling *et al.*

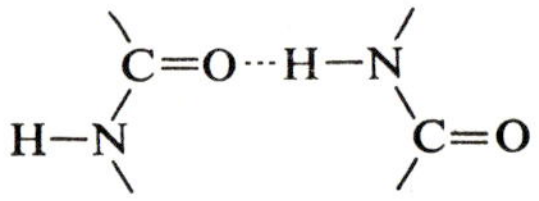

[1] showed that helical structures could be formed from *intra*chain bonds and that sheet structures could be obtained if *inter*chain bonds were formed. The most stable helical structure is the *α-helix* in which a carbonyl oxygen is hydrogen bonded to the amide nitrogen three residues away in the peptide chain. Some enzymes have been shown to contain regions of helical and sheet structures.

The next degree of complexity of a protein is its *tertiary* structure. This refers to the way in which the peptide chain is coiled and re-coiled about itself. This complex three-dimensional structure arises from interactions between the side-chains of the amino-acid residues, or between side-chains and the peptide backbone itself. The tertiary structure is therefore determined entirely by the primary sequence of the amino-acid residues in the chain. Figures 4.2, 5.7 and 4.12 show how the peptide chains of three enzymes are coiled and re-coiled in complex three-dimensional structures.

In some cases, individual protein sub-units associate to form dimers, trimers, tetramers, etc., giving what is known as the *quaternary* structures. Here again, binding between sub-units is mainly due to the residue side-chains which are exposed on the surface of each sub-unit.

The result of all these factors for any protein is a very large complex molecule in which the three-dimensional structure contains regions where certain amino-acid side-chains are brought together. Thus on the surface, or in the body, of the enzyme there will be regions containing several charged groups giving rise to highly polar *micro-environments*. In other regions the micro-environments may contain non-polar residues and will therefore be hydrophobic, while other sites in the enzyme molecule will possess functional groups capable of nucleophilic reactions, of forming hydrogen bonds or of acidic and basic catalysis. It is in terms of this type of arrangement, unique for each individual enzyme, that the origin of enzymic catalysis is to be found.

In addition to the various reactive groups available as the side-chain groups of amino-acids, certain enzymes make use of other chemical species in the form of coenzymes and metal-ions. These coenzymes, which are discussed in more detail in the relevant chapters, add to the versatility of protein molecules by providing even more varied reactive groups. For example, the coenzyme NAD possesses a molecular system capable of hydride transfer which is otherwise unavailable to a pure protein. Similarly FAD provides a different type of oxidation–reduction system (see chapter 3). Pyridoxal provides an aldehyde group in conjunction with an aromatic heterocyclic species, which can take part in various deaminations, transaminations and other reactions involving amino-acids. The vitamin $B_{12}$ coenzyme is capable of methyl-group transfer, while thiamine pyrophosphate is a system highly suited to certain transformations in carbohydrate metabolism.

Some enzymes require metal-ions for full activity and the possible role of a metal-ion in the enzymes function is of great interest. In general one may envisage three possible roles for a metal-ion. First, it may interact with various ligands from the protein molecule itself and in doing so help to ensure that the three-dimensional structure of the enzyme which is essential for activity is maintained. Secondly, it may contribute to the binding of the substrate (or a coenzyme) to the enzyme by coordinating with groups from both the protein and the substrate (or coenzyme) and thus forming a bridge between them. A third possibility is that by virtue of its properties as a Lewis acid it may assist in the catalytic process, in a manner similar to that described in chapter 1 for metal-catalysed reactions. These possible roles for a metal-ion are not, of course, mutually exclusive; indeed they are unlikely to be so. A metal-ion which contributes to the binding of a substrate, for example, by coordinating with the carbonyl oxygen of a peptide, will by virtue of this coordination polarize the carbonyl and thus enhance its reaction. This is exactly the situation envisaged for the zinc atom in carboxypeptidase (3.4.2.1).

Among the side-chain groups which may be expected to coordinate with metal-ions are the thiol group of cysteine, the carboxylate group of aspartic and glutamic acids, the imidazole species of histidine, the ε-amino-group of lysine and the α-amino and α-carboxylate groups of the terminal residues. Other possibilities are the hydroxyl groups of serine and threonine and the sulphur atom of methionine. In carboxypeptidase, for example, the side-chains of a glutamic acid and two histidine residues are coordinated to the zinc atom which is essential to the activity of the enzyme [2, 2a].

## THE ACTIVE SITE

The enzymes are large molecules which catalyse the reactions of molecules much smaller than themselves. Even in those cases where the substrate of an enzyme is itself a very large molecule, such as the case of ribonuclease (2.7.7.16) which catalyses the depolymerization of ribonucleic acid, it is clear that at any one time the enzyme is involved with only a very small portion of the substrate. In any case, ribonuclease has been shown to catalyse the hydrolysis of very small molecules related to ribonucleic acid. Similarly proteolytic enzymes interact with only a small portion of their protein substrates, and are also able to hydrolyse small peptides and similar molecules. The great difference in size between enzyme and substrate therefore means that during the catalytic process only a very small portion of the enzyme is in contact with the substrate. For our purposes we shall refer to this portion of the enzyme which is in contact with, or in very close proximity to, the substrate during the reaction as the *active site*. Much of the work which has been carried out with enzymes has been directed to the elucidation of the nature of the active site and hence to the interpretation of the catalytic process in terms of this nature.

An essential feature of the enzymic reaction is that the first step in the process is the association of the enzyme and the substrate into an enzyme–substrate complex. The formation of this complex is reversible, subject to an equilibrium constant $K_s$, and the binding between enzyme and substrate is generally non-covalent involving ionic and non-polar forces in addition to hydrogen bonds. Again carboxypeptidase provides a useful example. The substrate is bound to the enzyme by a variety of forces. The carboxylate ion of the substrate forms an ion pair with an arginine side-chain, the phenyl-group is bound in a hydrophobic ‘pocket’ and the amide nitrogen is probably hydrogen-bonded to a tyrosine residue. These forces, in addition to the coordination of the zinc ion with the carbonyl oxygen, provide a good framework of binding between enzyme and substrate. Similarly, in the case of chymotrypsin a non-polar region is available at the active site to accommodate the aromatic side-chain of the substrate. These binding forces also provide at least one key to the specificities of the enzymes. The presence of non-polar regions in the active site of both carboxypeptidase and chymotrypsin controls the specificities of these enzymes for aromatic amino-acids. In the case of trypsin, whose specificity is for amino-acid derivatives with basic side-chains, the region which in chymotrypsin is non-polar, possesses the carboxylate side-chain of aspartic-177 which can form an ion pair with the side-chain of the substrate.

Substrate specificity, however, is due in some cases to more than exact complementarity between enzyme and substrate. Koshland [3] proposed, in his *induced-fit* theory, that the binding of substrate to enzyme might cause a conformational change in the enzyme and bring about a delicately-controlled geometrical arrangement of the catalytic site. Such a conformation change has been detected in the X-ray crystallographic studies on carboxypeptidase [2].

In some cases the binding between substrate and enzyme may be covalent, for example the active lysine lyases, and in others a covalent link between substrate and coenzyme may be formed; for example, the pyridoxal and thiamine pyrophosphate enzymes. In these cases, the formation of the covalent enzyme–substrate complex is preceded by a non-covalent complex.

## Chemical modification

A considerable number of chemical studies have been carried out in order to determine which amino-acid residues may be present at the active sites of the enzymes. For the most part these have consisted of treating the enzymes with chemical reagents which react with certain amino-acid side-chains and testing the modified enzyme for activity. If the modification has destroyed or decreased the activity of the enzyme then it has usually been deduced that the amino-acid residue reacted was present at the active site. Such a deduction is not always justified. Destruction or modification of an amino-acid residue of enzyme may cause loss of activity for several reasons. The amino-acid residue may for example contribute to the maintenance of the enzyme in its correct three-dimensional structure; destruction of such a residue would destroy activity. In order to eliminate this possibility the chemical modification should also be carried out in the presence of the substrate or of some other compound which binds to the active site (e.g. a competitive inhibitor). If the substrate or inhibitor protects the enzyme from the inactivation, relative to the situation in the absence of such a compound, then it may be assumed with more justification that the reaction occurred at the active site. Even where this has been established it must be borne in mind that an amino-acid residue shown to be present at the active site, and also shown to be essential for the enzymic activity, may play one of two roles in the process. First, it may be required for the binding of the substrate to the enzyme. If this is the case then modification may not totally destroy activity. The dissociation constant $K_s$ may be affected but in the presence of a large concentration of substrate the maximum velocity $V$ (see later) may be unchanged. Secondly, the amino-acid residue may be involved in the actual catalytic process. After modification the enzyme would be expected to be totally inactive, but may still bind the substrate.

The wide variety of chemically reactive side groups in amino-acids allows the chemist to use a number of chemical reagents to bring about modification of an enzyme. For the most meaningful results a highly-selective reagent is to be preferred, so that each type of amino-acid can be studied individually. However, the ideal of a completely selective reagent is very rarely achieved.

The side-chain imidazole group of histidine is of great importance in the chemistry of enzymes. This is mainly due to its ability to take part in acid–base reactions, with a $pK_a$ of about 6. This means that at the pH values where most enzymes function, this residue can act as a donor or an acceptor of protons. Its presence at the active sites of many enzymes has been established by a variety of reactions, which are based on the properties of the imidazole as an aromatic heterocyclic species. Thus, the nitrogens of imidazole are nucleophilic and histidine residues may be alkylated by appropriate halogeno-compounds, and coupled with diazo-compounds.

An interesting example of the inactivation of an enzyme by alkylation at histidine is afforded by ribonuclease [4]. In this enzyme two histidine residues appear to be present at the active site, and either but not both of them can react with iodoacetate. Iodoacetamide has no effect. 1- or 3-Carboxymethylhistidine could be isolated from the product. The inactivation

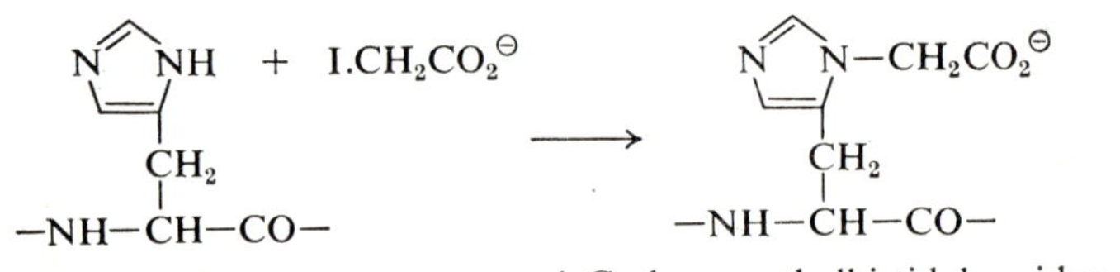

1-Carboxymethylhistidyl residue

of ribonuclease by iodoacetate was prevented by the presence of phosphate which binds to the active site.

The most important halogen alkylating agents used for modification of enzymes at histidine are those which are closely related to the substrates of the enzymes. Schoellmann and Shaw [5] used chloroketones related to ester substrates of chymotrypsin and trypsin (see Chapter 4). TPCK (*q.v.*) totally inactivated chymotrypsin by alkylation of the histidine at position 57 in the primary sequence. $\beta$-Phenylpropionate and benzamide, both competitive inhibitors of chymotrypsin, protect the enzyme to a certain extent from inactivation by TPCK. The chloroketone inactivates efficiently because, due to its similarity to a substrate, it is first bound to the active site of the enzyme by the usual non-covalent forces. This brings the chloroketone group into close proximity with the histidine residue which is present at the active site. TPCK does not inactivate trypsin, whose specificity is not for derivatives of phenylalanine, but for lysine derivatives. The corresponding chloroketone TLCK (*q.v.*) inactivates trypsin, also by alkylation at an essential histidine.

The importance of specific inhibitors of this type, based on substrate analogues, cannot at present be overemphasized not merely for the study of enzymes, but mainly for the pharmaceutical sciences since these inhibitors may have great therapeutic value [6].

A histidine residue at the active site of carbonic anhydrase (4.2.1.1) may be alkylated by a reagent derived not from a substrate but from a specific inhibitor. Acetazolamide binds strongly to the enzyme at the zinc ion which is at the active site. Chloro-acetazolamide, which is an active halogen compound derived from it, irreversibly inhibits the enzyme, by first binding at the zinc and then reacting with a histidine residue [7].

The histidine side-chain also reacts with diazo-compounds such as diazosulphanilic acid (I). This couples with the imidazole ring giving a

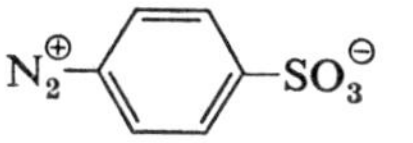


(I)

chromophore which absorbs at 450 nm, allowing the nature of the modified protein to be identified. Diazosulphanilic acid inactivates fructosediphosphate aldolase (4.1.2.13) the inactivation being prevented by the presence of substrate fructosediphosphate [8]. It is probable that this inactivation is due to reaction at a histidine although diazosulphanilic acid can also react with tyrosine and lysine side-chains [8].

Photo-oxidation, which is oxidation by activated oxygen, light-induced in the presence of a photosensitizer such as methylene blue or rose bengal, is another method of modifying proteins at histidine residues. Unfortunately this method is not entirely specific for histidine; tyrosine, tryptophan, methionine and cysteine are also oxidized in some cases [8, 9, 10]. However, occasionally photo-oxidation may be specific. Westhead [9] showed that in the presence of rose bengal only one histidine residue in enolase (4.2.1.11) was destroyed, and that this caused loss of activity. The enzyme was protected by magnesium ions which are essential to the enzymic reaction, and Westhead concluded that the histidine residue was involved in binding metal-ion to the enzyme. In other cases, even where the photo-oxidation is not specific it may be possible to attribute losses in activity to destruction of histidine. Koshland *et al.* [10] showed that, while histidine and methionine in chymotrypsin were both oxidized, destruction of the histidine caused total inactivation while oxidation of the methionine only caused partial loss of activity.

Photo-oxidation of fructosediphosphate aldolase [8, 11] caused destruction of 50–70% of the histidine residues and this was accompanied by a change in the properties of the enzyme. Hoffee *et al.* [11] interpreted their results as indicating that a histidine residue was involved in an essential proton-transfer process (see chapter 7).

The thiol group of a cysteine residue is another which can be modified chemically. This group is easily alkylated with halogeno-compounds like

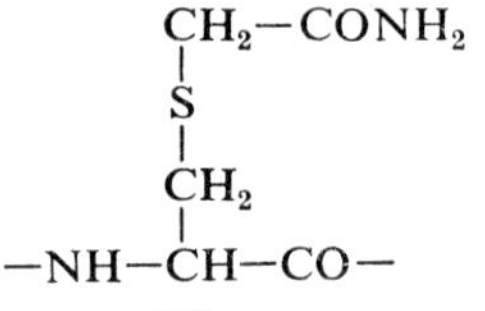

(II)

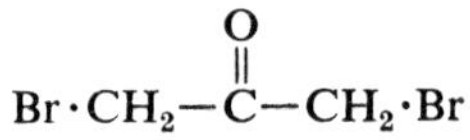

(III)

those which alkylate histidine. The essential cysteine residue of papain reacts with iodoacetamide giving a carboxamidomethylcysteine (II) [12]. The chloroketone TPCK, which inactivates chymotrypsin by reaction with histidine, also inactivates papain, but by reaction with the cysteine [13]. Husain and Lowe [14] treated papain with the bifunctional 1,3-dibromoacetone (III) and showed that this reagent reacted with both a cysteine and a histidine, forming a bridge between the two.

The —SH group of cysteine also adds to certain unsaturated compounds, for example N-ethylmaleimide (IV). N-Ethylmaleimide inhibits

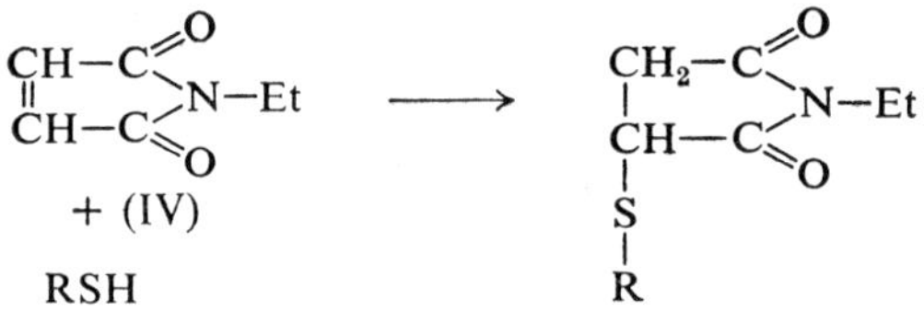


myosin ATPase (see Chapter5) and benzyl alcohol dehydrogenase (1.1.1.–) [15], and a derivative, N-(N-acetyl-4-sulphamoylphenyl) malcimide, inactivates L-glutamate dehydrogenase (1.4.1.2) [16] and also pyrophosphatase (3.6.1.1) [17].

The great affinity of metal-ions, especially mercury, for thiol groups leads to their use as inhibitors of thiol-containing enzymes. Mercaptide formation is often valuable in allowing the chemist to determine quantitatively the number of —SH groups in the protein. Various titrimetric and amperometric methods have been developed [18, 19]. In addition, spectrophotometric methods are available [20]. Mercaptide formation by p-mercuribenzoate is accompanied by an increase in ultraviolet absorption at 250 nm which can be used as a measure of the —SH groups present. Various mercury compounds, especially p-chloromercuribenzoate (PCMB), have been used as inhibitors of sulphhydryl enzymes. For example, the 'active cysteine' enzymes papain and ficin are inactivated [12]. Other enzymes in which an —SH group may play an important role, and which are inactivated by mercurials, are certain acyltransferases [21], myosin ATPase (3.6.1.3) [22], phospholipase D (3.1.4.4.) [23] and many others. Although many enzymes are inactivated by thiol-group reagents it is not possible to conclude from such a result that the thiol group is present at the active site and involved in the catalytic process [20]. The results of studies in which enzymes are inactivated by such reagents must be treated with care. Indeed, the special case of carboxypeptidase provides an example of a molecule which, although it reacted with heavy metal-ions, did not in fact possess a thiol group at all (see Chapter 4).

Methionine is another amino-acid residue important in the chemistry of enzymes, and methionine residues have been shown to be present at the active sites of several enzymes. The photo-oxidation of chymotrypsin has already been mentioned. This leads to oxidation of a methionine residue at the active site, and causes a decrease in the affinity of the enzyme for the substrate. This methionine may also be alkylated, and the resulting chymotrypsin derivatives also show a decrease in affinity for the substrate [23, 24]. A methionine at the active site of isocitrate dehydrogenase (1.1.1.4.1) may be alkylated with iodoacetate; isocitrate protects this enzyme from inactivation [25]. Oxidation of the methionine residue at position 13 of ribonuclease quite clearly influences the way in which the S-peptide fragment binds to the S-protein [26]. This suggests that the role of methionine in some enzymes may be to help in maintaining the three-dimensional structure of the active molecule.

The involvement of a tyrosine residue at the active site of carboxypeptidase A has been demonstrated by means of various chemical modifications which influence the enzymic activity. These modifications may either involve the ring (iodination, nitration) or the hydroxyl group (acetylation, acylation) [27]. Pugh and Horecker [28] treated frustosediphosphate aldolase with acetyl imidazole, said to be specific for tyrosyl residues, and showed that the product possessed decreased ability to cleave fructosediphosphate. Removal of terminal tyrosyl residues by carboxypeptidase caused the same result [29]. Acetylation, iodination and coupling with diazo-compounds also inactivated pepsin [30] probably by reaction at tyrosine. Pontremoli *et al.* [31] reported that acetylation of tyrosine residues with acetyl imidazole, and also coupling with diazosulphanilic acid (again at tyrosine) inactivated fructosediphosphatase (3.1.3.11).

The identification of an unusually reactive serine residue at the active sites of certain hydrolytic enzymes was carried out by chemical reactions involving the hydroxymethyl group. In chapter 4 some of these reactions are discussed in detail.

In some enzymes the carboxylate groups of aspartic and glutamic acids have been identified at the active sites. Usually the chemical modification involves production of esters of the carboxylate group. p-Bromophenacyl bromide inactivates pepsin by forming an ester with the side-chain carboxyl of an aspartyl residue [32]. Certain aliphatic diazo-compounds have also been used to identify the functional group at the active site of pepsin. Thus, 1-diazo-6-phenyl-3-toluenesulphonamidobutan-2-one (the diazo-ketone analogue of the chloroketone TPCK) esterifies the carboxyl and inactivates the enzyme [32a]. In the case of β-glucosidase (3.2.1.21) the presence of a carboxyl group at the active site has been indicated by another reagent which is capable of forming an ester. This is conduritol B-epoxide, which is similar in structure to the substrate and which first binds to the active site and then esterifies the carboxyl group [33].

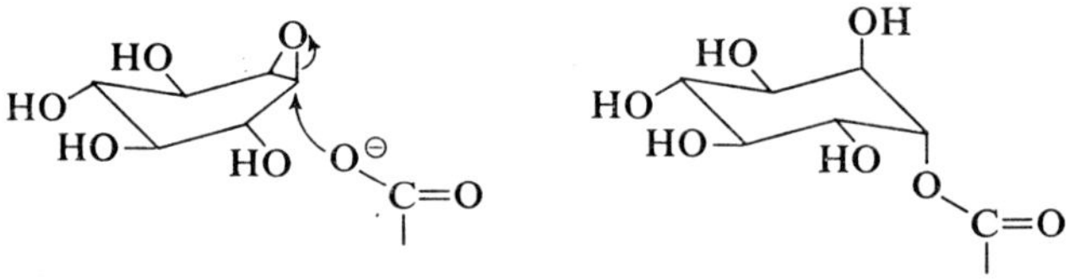


Conduritol-B-epoxide

Ester of (+)-Inositol

In many of these cases, where esters are formed causing inactivation the ester link may be cleaved with hydroxylamine.

A lysine residue has been identified as a probable binding site for substrate in ribonuclease. Dinitrofluorobenzene (V) reacts with the side-chain

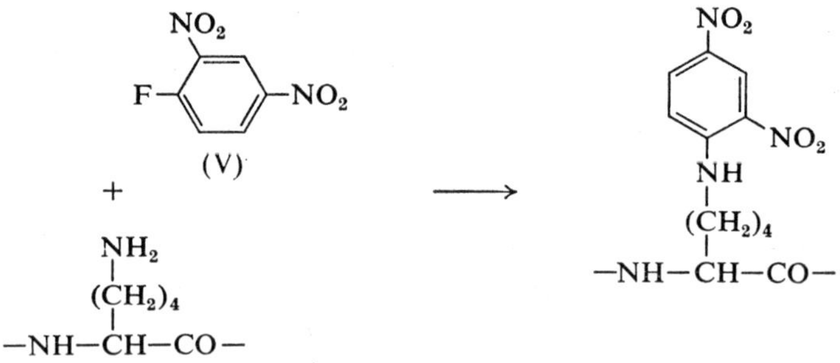


amino-group of lysine-41 and inactivates the enzyme [34]. A lysine residue has also been implicated at the active site of myosin ATPase by reaction with trinitrobenzene sulphonate [35].

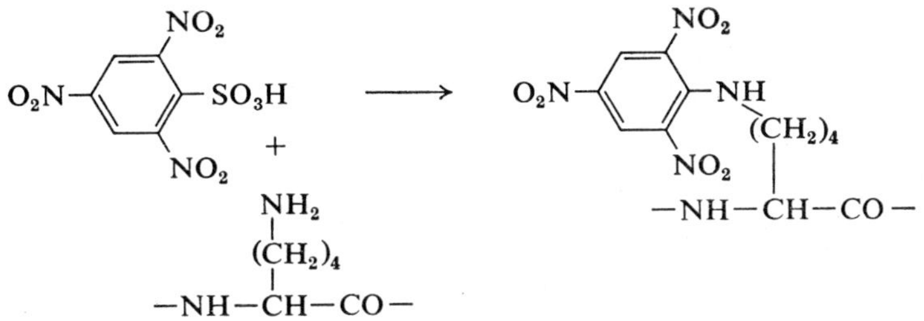

Lysine residues have been detected at the active sites of a number of enzymes by a rather different type of chemical modification. In most cases the chemical reaction which leads to inactivation is carried out in the absence of substrate, which if present would protect the enzyme. However, the opposite is the case for the 'active lysine' lyases. Sodium borohydride causes irreversible inactivation only when substrate is present. For example, when acetoacetate decarboxylase (4.1.1.4) was treated with borohydride in the presence of substrate the enzyme was inactivated, but in the absence of substrate no activity loss was observed [36]. Similar results were obtained for fructosediphosphate aldolase, transaldolase (2.2.1.2) and 2-keto-3-deoxy-L-arabinonate dehydratase (4.2.1.–) [37, 38, 39]. A similar inactivation was observed when borohydride acted upon pyridoxal phosphate enzymes but only in the presence of the pyridoxal phosphate [40]. In all of these cases a Schiff-base is formed between the ε-amino-group of a lysine and the carbonyl group of the substrate (or the aldehyde group of the pyridoxal coenzyme). Normally this process is reversible but on reduction with borohydride the Schiff base is reduced to a secondary amine which cannot dissociate. In these cases, as in many others, the identity of the

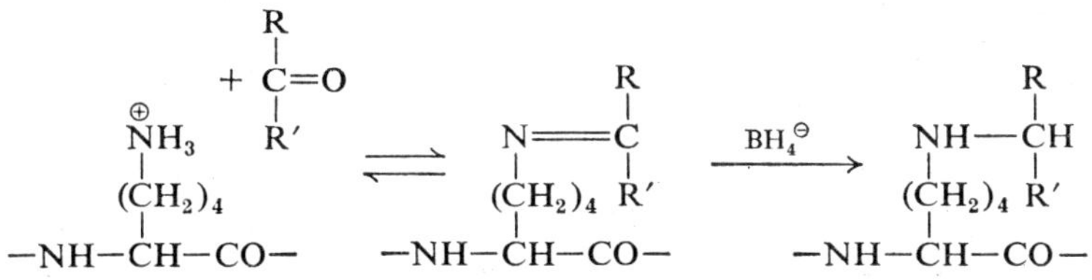

residue involved in the reaction is conveniently established if a radioactively-labelled group is attached to it. After degradation, by hydrolysis for example, the labelled aminoacid derivative is easily identified.

Proteolytic enzymes may sometimes be used to identify amino-acid residues involved in the activity of enzymes. The removal of terminal tyrosyl residues from fructosediphosphate aldolase by carboxypeptidase is an example [29]. Perhaps the best example of the use of an enzyme in this way is that of ribonuclease. Treatment of this enzyme with subtilopeptidase A (3.4.4.16) caused cleavage of the protein chain at position 20 and the product could be separated into two portions, one eicosapeptide (S-peptide), the other the remainder of the enzyme (S-protein). Separately the portions were inactive but on recombination full activity was restored. Therefore both portions possessed areas which were required for the full active site [41]. Chemical modifications [42] and the synthetic experiments of Hofmann *et al.* [26] helped to establish that the histidine at position 12 was essential to the catalytic activity of the molecule.

Many conclusions have been drawn about the identities of amino-acid residues at the active sites of various enzymes, based on studies of the way in which the properties of the enzymes vary according to the pH (see later). However, interpretations of the results of such studies must be made with great care.

## KINETICS OF ENZYME-CATALYSED REACTIONS

Kinetic studies have probably contributed more to our knowledge of the mechanisms of enzyme-catalysed reactions than any other method of approach. Rate equations have been derived relating to all the possible reaction mechanisms and by performing appropriate measurements of reaction rates, the chemist is usually able to eliminate many of these possible mechanisms. The sometimes-complex derivations of these rate equations are outside the scope of this book, although an attempt will be made to show in an introductory manner how kinetic studies may help in the elucidation of the most probable reaction mechanism for any given enzyme. For detailed discussions of the derivations of rate equations the reader is referred to the publications of Dixon [43], Haldane [44], Dalziel [45], Cleland [46], Alberty [47], King and Altman [48], Briggs and Haldane [49], Gutfreund [50], Dixon and Webb [51], Wong and Hanes [52] and many others.

However, it is necessary for us to consider some aspects of the kinetic theory of enzymic catalysis if only to establish certain definitions and useful concepts. For a single-substrate, enzyme-catalysed reaction, the Michaelis–Menten [53] approach may be stated briefly as follows. The first step is envisaged as the reversible formation of an enzyme–substrate complex (equation 2.1, where E, S and ES represent enzyme, substrate and enzyme–substrate complex). The rate of formation of products is given by the equation (2.2).

$$E + S \underset{k_{-1}}{\overset{k_{+1}}{\rightleftharpoons}} ES \xrightarrow{k_{+2}} \text{products} \tag{2.1}$$

$$\text{Rate } (v) = k_{+2}\,[ES]. \tag{2.2}$$

The basic assumption of the Michaelis–Menten theory is that $k_{+2}$ is sufficiently small that the second step in the process does not influence the equilibrium formation of the ES complex. Under these circumstances, for a constant enzyme concentration, when the substrate concentration is low only a fraction of the total number of enzyme molecules present will be in the form of the enzyme–substrate complex. As the substrate concentration is increased, the concentration of ES also increases and therefore the rate of the overall reaction rises. In the presence of very high substrate concentrations the equilibrium will be such that virtually all the enzyme molecules would be present as the ES complex, i.e. the enzyme is saturated with substrate. Once this situation is reached an increase in substrate concentration will not produce any further increase in reaction rate. Therefore, for the given enzyme concentration there is a maximal velocity attainable. This is shown by Fig. 2.2.

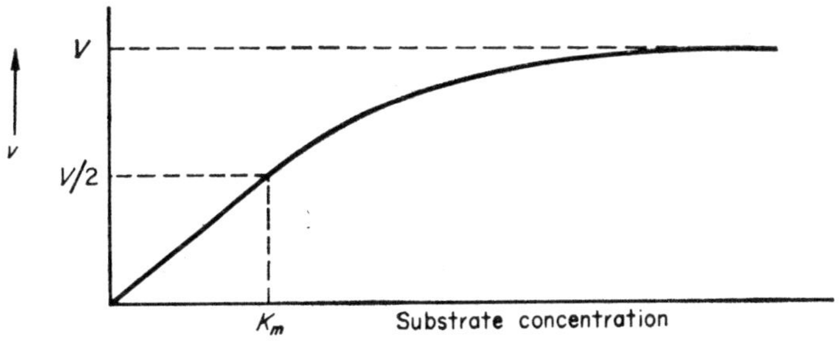


FIG. 2.2. Plot of rate of reaction $v$ against substrate concentration at constant enzyme concentration. At high substrate concentration the rate approaches a maximum value $V$. $K_m$ is equal to the substrate concentration at half maximal rate

If we define $K_s$ as the equilibrium constant for the dissociation of the enzyme–substrate complex into enzyme and substrate, then if [E′] is the concentration of enzyme initially added, i.e. the total amount of enzyme present, the concentration of free enzyme E will be ([E′] − [ES]).

Therefore, $$K_s = \frac{[\mathrm{E}][\mathrm{S}]}{[\mathrm{ES}]} = \frac{([\mathrm{E}'] - [\mathrm{ES}])[\mathrm{S}]}{[\mathrm{ES}]}. \tag{2.3}$$

Hence, $$K_s = \frac{[\mathrm{E}'][\mathrm{S}] - [\mathrm{ES}][\mathrm{S}]}{[\mathrm{ES}]} \tag{2.4}$$

and $$[\mathrm{ES}] = \frac{[\mathrm{E}'][\mathrm{S}]}{K_s + [\mathrm{S}]}.$$

From equation (2.2)

$$v = \frac{k_{+2}[\mathrm{E}']}{(1 + K_s/[\mathrm{S}])}. \tag{2.5}$$

When the concentration of substrate is very large, i.e. when the enzyme is saturated with substrate, then equation (2.5) gives the maximal velocity V and approximates to:

$$V = k_{+2}[E'].$$

Inserting this expression for $V$ into (2.5) gives:

$$v = \frac{V}{(1 + K_s/[S])}. \tag{2.6}$$

This is the Michaelis–Menten equation.

If [S] happens to be equal to $K_s$ then we see that the rate of the reaction will be equal to $V/2$. This gives rise to the most common definition of the Michaelis constant $K_m$ as the substrate concentration which produces half maximal velocity. Under the conditions of the Michaelis–Menten assumption $K_m = K_s$. Values for $K_m$ may be obtained experimentally by the use of equation (2.6). Although it is possible to measure, by means of a graph such as Fig. 2.2, the maximal velocity and hence the substrate concentration which would give half maximal velocity, a rather inaccurate value would be obtained, mainly due to the difficulty of determining a value which is approached asymptotically. In practice many experimentalists prefer to use equation (2.7) which may be derived from (2.6):

$$1/v = 1/[S].(K_m/V) + 1/V. \tag{2.7}$$

Thus, a plot of $1/v$ against $1/[S]$, known as a double reciprocal plot or Lineweaver–Burk [54] plot, should be a straight line with intercept on the ordinate $1/V$ and slope $K_m/V$. The intercept on the abscissa gives $-1/K_m$ [51]. Figure 2.3 shows the type of result obtainable.

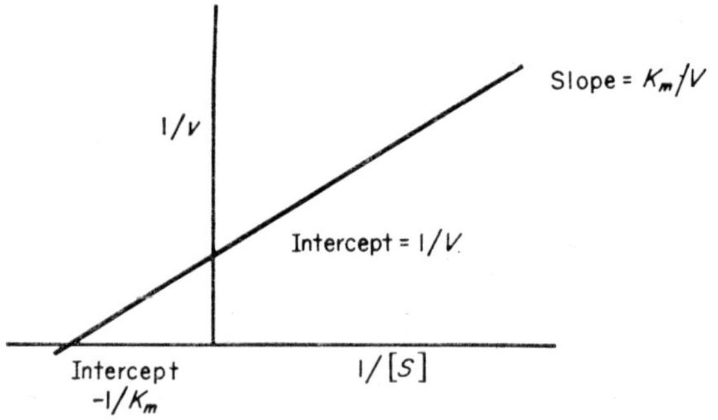


FIG. 2.3. Plot of $1/v$ against $1/[S]$ (double reciprocal plot) for an enzyme-catalysed reaction obeying Michaelis–Menten kinetics

The $K_m$ values obtained from such experiments are important properties of the enzymes. Since under Michaelis–Menten kinetics $K_m = K_s$, then the $K_m$ for any enzyme with a given substrate is a direct measure of the strength of the binding between the enzyme and the substrate.

Double reciprocal plots are useful for examining another phenomenon in enzyme chemistry, that of inhibition. In this context an inhibitor is a chemical species whose presence causes a decrease in the ability of an enzyme to catalyse its particular reaction. An inhibitor may be *reversible* or *irreversible*, *competitive* or *non-competitive*. An irreversible inhibitor usually functions by means of a chemical reaction at the active site of the enzyme, and either prevents the substrate from binding or removes from an essential amino-acid side-chain the ability to function normally in the catalytic process. The inhibitor species in this type of inhibition is usually bound to the enzyme via a covalent bond, does not dissociate away from the enzyme and cannot be removed by a process such as dialysis. The chloroketone TPCK and the organophosphorus compound diisopropyl fluorophosphate (chapter 4) are examples of irreversible inhibitors. It is important to note that a chemical reagent which brings about inactivation of an enzyme by a non-specific degradation of the protein is not regarded as an irreversible inhibitor. In addition, the *effect* of certain irreversible inhibitors may be reversed by chemical means. For example, acetylation of carboxypeptidase causes loss of activity and treatment of the acetyl-enzyme with hydroxylamine removes the acetyl group and reactivates the enzyme. The original acetylating agent must nevertheless be regarded as an irreversible inhibitor.

A reversible inhibitor is usually bound to the enzyme by non-covalent forces and an equilibrium situation prevails, the dissociation of the complex EI being subject to the equilibrium constant $K_I$;

$$\mathrm{E} + \mathrm{I} \rightleftharpoons \mathrm{EI}$$

$$K_I = [\mathrm{E}][\mathrm{I}]/[\mathrm{EI}],$$

where I represents the inhibitor.

When the inhibitor is closely related to the substrate of the enzyme, the nature of the inhibition is often found to be competitive. This means that both substrate and inhibitor are bound at the same site on the enzyme and they therefore compete for that site. The presence of such an inhibitor has the effect of reducing the number of molecules of enzyme available for binding with the substrate and therefore a greater concentration of substrate would be needed to ensure half-saturation of the enzyme with substrate. Hence a competitive inhibitor changes the apparent $K_m$ of the enzyme and substrate. However, if sufficiently large concentrations of substrate are used, it would still be possible to saturate the enzyme with substrate and therefore the maximal velocity $V$ would not be affected. Double reciprocal plots in the absence and the presence of a competitive inhibitor would therefore have the form shown in Fig. 2.4.

A non-competitive inhibitor is envisaged as binding reversibly to a site on the enzyme which is not the binding site for the substrate. Even

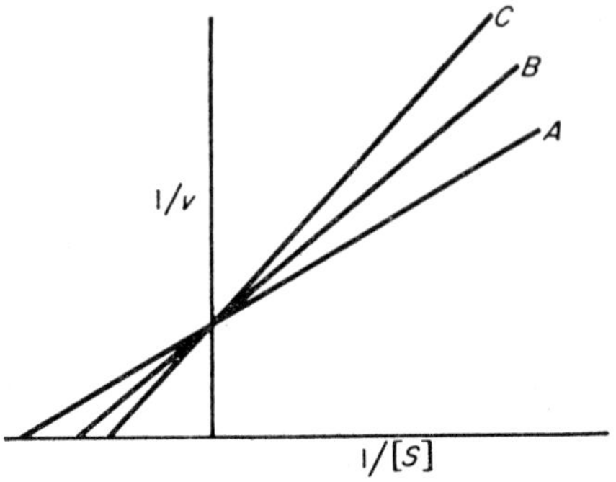


FIG. 2.4. Double reciprocal plots for a reaction in the absence ($A$) and in the presence ($B$ and $C$) of a competitive inhibitor. $B$ and $C$ represent the lines obtained with different concentrations of inhibitor

though such a species does not bind at the active site it nevertheless causes decreased activity, possibly by bringing about a change in the conformation of the protein. Non-competitive inhibitors need not be related structurally to the substrate. This type of inhibitor does not compete for the binding site of the substrate and therefore in addition to forming an EI complex, it is able to form a further, ternary complex.

$$\text{ES} + \text{I} \rightleftharpoons \text{EIS}.$$
$$\text{EI} + \text{S} \rightleftharpoons \text{EIS}.$$

Since the non-competitive inhibitor does not influence the binding of the substrate then it also has no influence on $K_m$. However, even when the enzyme is saturated with substrate the inhibitor may still form the ternary complex shown above and therefore may decrease the value of $V$. Double reciprocal plots therefore have the appearance of Fig. 2.5.

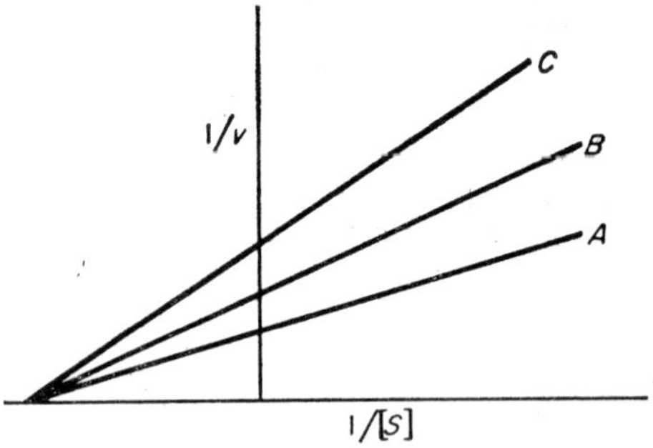


FIG. 2.5. Double reciprocal plots for a reaction in the absence ($A$) and in the presence ($B$ and $C$) of a non-competitive inhibitor. $B$ and $C$ represent the lines obtained with different concentrations of inhibitor

Therefore, by means of double reciprocal plots the chemist is able to distinguish between these two types of reversible inhibition. In some cases the inhibition is found to be mixed. This would occur, for example, if the inhibitor binding at a site other than the active site caused a conformational change in the protein which decreased the ability of the enzyme to bind the substrate. Such an inhibitor would thus bring about a change not only in $V$ but also in $K_m$.

The above discussion has been concerned only with one-substrate reactions. In fact only a very few enzyme-catalysed reactions involve one substrate. In some cases, where the second substrate is water, for example in the hydrolases, then the kinetic patterns are similar to those expected for one-substrate reactions. In such cases the water concentration is very high and therefore, in practical terms, constant. Most enzymes catalyse reactions which involve two or three substrates and a variety of possible reaction mechanisms must be considered. Rate equations for these possible mechanisms have been derived and tested for many enzymes. Again, a detailed consideration of this aspect of the study of enzymes is not within the scope of this book and the reader is referred to the previously quoted references [43–52]. However, we must consider in general terms the types of mechanism which are possible and how they may be distinguished.

A possible mechanism for the enzyme-catalysed reaction between two substrates A and B is the so-called *ping-pong* mechanism. The essence of this mechanism is that the first substrate A reacts with the enzyme and the product Q is lost before the second substrate B reacts (equations 2.8a and 2.8b). Another way of representing this type of mechanism is as shown

$$\mathrm{A + E \rightleftharpoons EA \rightleftharpoons E'Q \rightleftharpoons E' + Q} \tag{2.8a}$$

$$\mathrm{E' + B \rightleftharpoons E'B \rightleftharpoons EP \rightleftharpoons E + P} \tag{2.8b}$$

in (2.9). This is based on Cleland's method [46], in which the horizontal line represents the course taken by the enzyme through the sequence of

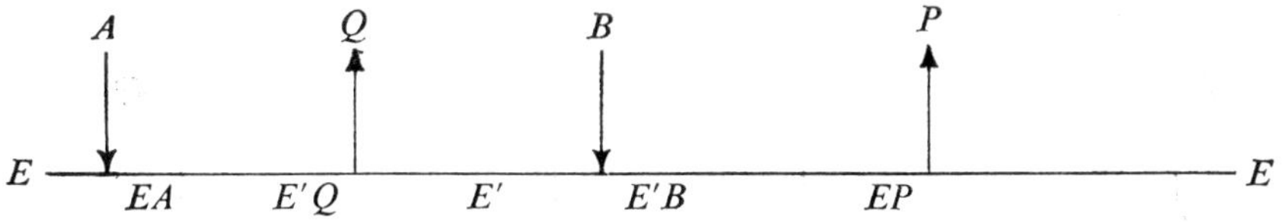


(2.9)

reactions. Addition of substrate and loss of products are shown by appropriate arrows. The reaction of the enzyme with the first substrate and the loss of the first product (equation 2.8a) converts the enzyme into a modified species E′. B reacts with this modified species and not with E itself. This type of mechanism, in which each substrate reacts with a different form of the enzyme and both substrates are never present on the enzyme at the same time, manifests itself in a characteristic kinetic pattern which is easily distinguished in a series of double reciprocal plots. It is possible to measure rate of reaction at various concentrations of A, keeping the concentration of B fixed at a certain value ($B_1$), and to make a double reciprocal plot. A series of such plots may be made using different fixed concentrations of B ($B_1$, $B_2$, $B_3$, etc.). A result as in Fig. 2.6, in which a series of parallel lines

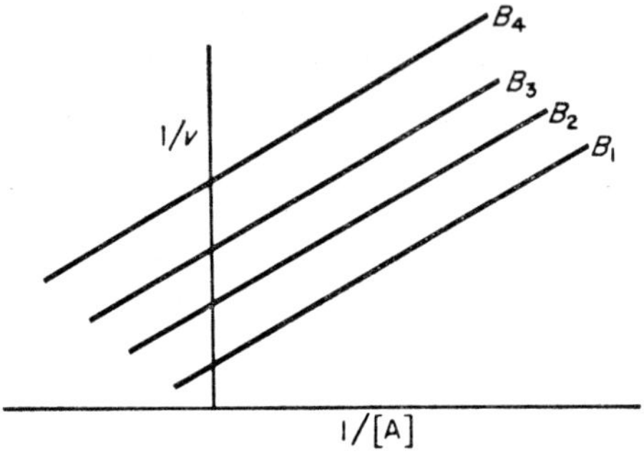


FIG. 2.6. Double reciprocal plots for a two-substrate reaction with a ping-pong mechanism. One substrate $A$ is varied while different, fixed concentrations of the other substrate $B$ are used. Plots of $1/v$ against $1/[A]$ are made at different concentrations ($B_1$, $B_2$, $B_3$, etc.) of $B$

are obtained, is diagnostic of a ping-pong mechanism. Fromm and Zewe [55] obtained this type of result when they studied hexokinase (2.7.1.1.) from brain. Hexokinase catalyses the reaction (2.10), in which a phosphate

$$\text{ATP} + \text{D-hexose} \longrightarrow \text{ADP} + \text{D-hexose-6-phosphate} \qquad (2.10)$$

group is transferred from ATP to a hexose. A ping-pong mechanism for this transfer reaction may be interpreted as indicating that the phosphate group is first transferred to the enzyme and later from the enzyme to the acceptor. The modified enzyme E′, therefore, must be a phosphorylated enzyme (EP), and the reaction represented as shown.

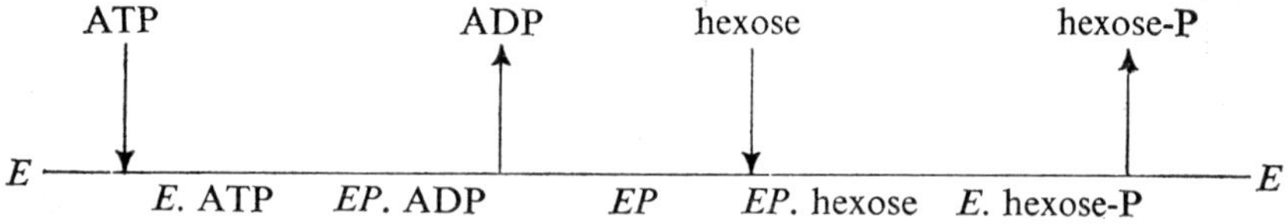


In general, where a transfer reaction is found to follow a ping-pong kinetic pattern, this must be regarded as strong evidence for a reaction scheme in which the transferred group is first transferred to the enzyme giving (usually) a covalent complex.

Another example is afforded by the case of sucrose glucosyltransferase (2.4.1.7) which catalyses the phosphorolysis of sucrose (equation 2.11), a reaction in which an α-D-glucosyl group is transferred from sucrose to phosphate. The observation of a ping-pong kinetic pattern for this reaction

$$\underset{\text{sucrose}}{\alpha\text{-D-glucosyl-1-fructose}} + P_i \longrightarrow \alpha\text{-D-glucosyl-1-}P + \text{fructose} \tag{2.11}$$

[56] suggested that a glucosyl-enzyme was involved as intermediate. Evidence of a different nature was also obtained for a glucosyl-enzyme with this enzyme [56], thus substantiating the kinetic results.

Similarly, Hersh and Jencks [57] showed that succinyl-CoA:3-ketoacid CoA-transferase (2.8.3.5) followed a ping-pong pattern and deduced from this and other evidence that the overall reaction (equation 2.12)

$$\text{succinyl-CoA} + \text{3-ketoacid} \rightleftharpoons \text{succinate} + \text{3-ketoacyl-CoA} \tag{2.12}$$

proceeded via an enzyme-CoA intermediate according to the mechanism (2.13). Hersh and Jencks [57] were able to confirm the mechanism by isolating the enzyme-CoA intermediate.

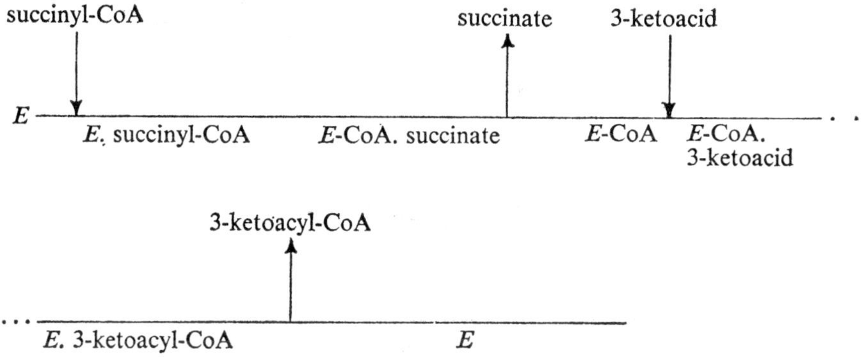


(2.13)

In the ping-pong mechanism only binary complexes are formed (EA, E′Q, E′B and EP). Other mechanisms are possible in which ternary complexes are involved. One possibility is that in which a ternary complex is formed from the enzyme and both substrates and in which the order of addition of the substrates is unimportant. Thus, the ternary complex EAB may be formed in two ways, as shown in equation (2.14). The reaction

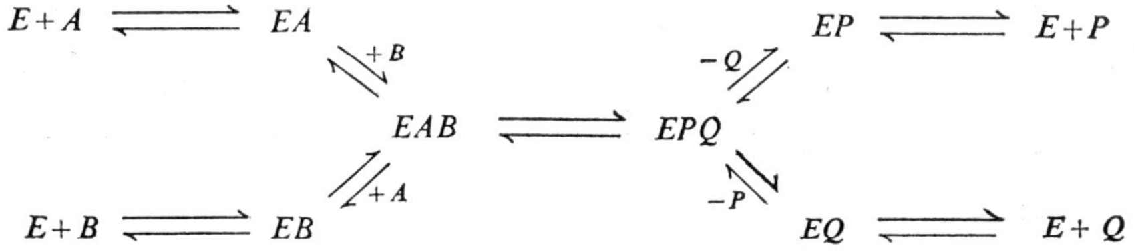


(2.14)

proceeds by the interconversion of two ternary complexes. The ternary complex (EPQ) may break down by loss either of P or of Q. Since the order in which the two substrates add to enzyme, and the order in which the two products leave, are both unimportant, the mechanism is known as a *random order* mechanism.

Alternatively it may be that the order in which the two substrates add to the enzyme is compulsory. In such a case only one route to the production of EAB is available, and similarly only one route for the breakdown of EPQ. Thus, in equation (2.15), B cannot combine with the enzyme until

$$E + A \rightleftharpoons EA \overset{+B}{\rightleftharpoons} EAB \rightleftharpoons EPQ \overset{-P}{\rightleftharpoons} EQ \rightleftharpoons E + Q \tag{2.15}$$

A is in position. Similarly Q cannot leave until P is lost first.

Double reciprocal plots in which the lines converge at a point to the left of the ordinate (see Fig. 2.7) are consistent with a compulsory order

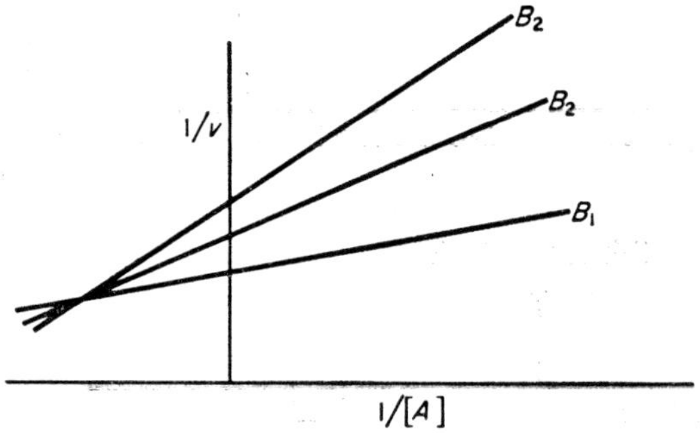


FIG. 2.7. Double reciprocal plots for a reaction with a compulsory order or random order mechanism in which both substrates add to the enzyme giving a ternary complex. Plots of $1/v$ against $1/[A]$ are made at different, fixed concentrations ($B_1$, $B_2$, $B_3$) of the second substrate $B$

or random order mechanism in which both substrates add to the enzyme giving a ternary complex. Fromm and Zewe [58] obtained converging plots as in Fig. 2.7 for hexokinase from yeast and this suggested that, in contrast to the brain enzyme, the yeast enzyme functioned via interconversion of ternary complexes. Possibly this implies the direct transfer of the phosphate group from the ATP to the hexose without the intermediacy of phospho-enzyme. The same workers [59] obtained similar results with lactate dehydrogenase (1.1.1.27) indicating that in this case the oxidation–reduction step took place in a ternary complex.

The rate of the reaction, $v$, used in the double reciprocal plot is almost invariably the initial rate of the reaction, measured before substrate concentrations can change appreciably. It is not possible, using initial rates, to distinguish between random and compulsory order reactions. Measurements of equilibrium reaction rates are able to provide information on this point [60]. All enzyme reactions are, hypothetically at least, reversible and therefore a system containing enzyme and substrates will eventually reach a situation of equilibrium where the rate of the forward reaction is equal to the rate of the reverse reaction.

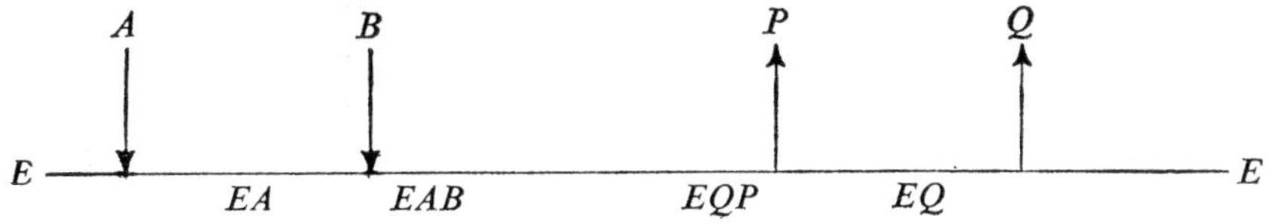


(2.16)

Equilibrium rate reactions are best measured using isotopically labelled substrates. Let us consider the enzymecatalysed reaction represented by equation (2.15), a compulsory order process. It may also be represented by (2.16). At equilibrium the concentrations of substrates and products are

$$K\,\mathrm{eq} = \frac{[\mathrm{Q}][\mathrm{P}]}{[\mathrm{A}][\mathrm{B}]} \qquad (2.17)$$

related as in equation (2.17). The equilibrium rate reaction of A, that is, the rate at which A is converted into Q, can be determined by adding to the equilibrium mixture a small quantity (too small to affect the equilibrium) of radioactively-labelled A. After intervals of time the quantity of labelled Q may be measured. Similarly the equilibrium rate of conversion of B into P can be measured. The way in which the equilibrium rate of conversion of A into Q varies according to the concentration of B and P (of course [B]/[P] must be kept constant) will indicate whether the mechanism is compulsory order or not. For example, an increase in the concentrations of B and P will cause a decrease in the rate of conversion of A into Q if a compulsory order applies [60].

Applying this method to the yeast hexokinase case Fromm *et al.* [61] showed that a compulsory order was not involved. Similarly, Joyce and Himes [62] showed that formyl tetrahydrofolate synthetase (6.3.4.3), whicd catalyses a reaction involving three substrates, does so by a mechanism involving a random order of addition of substrates giving ultimately a quaternary complex.

Other methods of ascertaining whether or not a reaction involves a random or a compulsory order of addition may require binding studies [63], or detailed comparison on a quantitative basis of the kinetic results with the rate equations derived for various reaction schemes. For example, kinetic studies [64, 65] with liver alcohol dehydrogenase (1.1.1.1.) which catalyses the reaction shown (2.18) have shown that the overall maximum rate in the

$$NAD^{\oplus} + CH_3CH_2OH \rightleftharpoons NADH + CH_3CHO + H^{\oplus} \quad (2.18)$$

forward direction may be identified with the rate of dissociation of the enzyme-reduced coenzyme complex. On the basis of this, and other kinetic results, the mechanism of action of the liver alcohol dehydrogenase is considered to be best described as a compulsory order process in which the ternary complexes are not kinetically important, i.e. they are interconverted and break down very rapidly compared with that rate at which the enzyme–coenzyme complex breaks down (equations 2.19, 2.20 and 2.21). This

$$E + NAD^{\oplus} \rightleftharpoons E.NAD^{\oplus} \quad (2.19)$$

$$E.NAD^{\oplus} + C_2H_5OH \rightleftharpoons E.NADH + CH_3CHO \quad (2.20)$$

$$E.NADH \rightleftharpoons E + NADH \quad (2.21)$$

special case of the compulsory order, ternary complex mechanism may be represented by Cleland's method as follows:

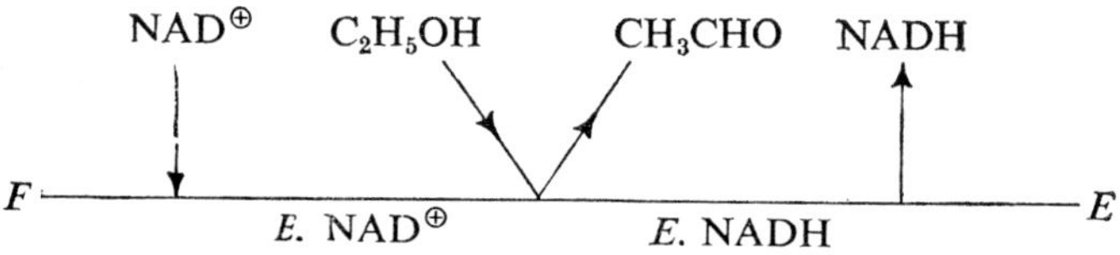


By contrast, yeast alcohol dehydrogenase appears to function by a random order mechanism (see chapter 3).

The above discussion has been brief and qualitative. In most cases, strong support for any given mechanism is obtained by quantitative comparisons of the kinetic results with the appropriate rate equations. In this way numerical values for the important rate constants and equilibrium constants may be obtained.

In addition it must be stated that the discussion has been confined to conditions where the concentrations of intermediates have reached steady values (steady-state kinetics). Considerably more information can often be obtained by analysis of the conditions which apply before the steady values have been attained (pre-steady-state kinetics).

## EFFECT OF pH ON ENZYMIC ACTIVITY†

The activity of an enzyme is almost invariably dependent on pH, each enzyme possessing a pH optimum, i.e. at a certain pH the activity of the enzyme is at a maximum and changes in pH whether to higher or lower

† Notice that under certain conditions of pH enzymes may be irreversibly inactivated. Any measurement of the variation of the activity of an enzyme with pH should be corrected for such inactivation. The present discussion of pH–activity relationships and their possible relevance to the mechanism of action of enzymes ignores this factor.

values cause significant lowerings of the activity. The pH dependence of an enzyme is usually interpreted in terms of prototropic groups at the active site although this type of interpretation can lead to oversimplifications.

If an acid HA ionizes as shown in equation (2.22), then the quantity of conjugate base $A^{\ominus}$ present in a solution will vary according to the pH of

$$HA \rightleftharpoons H^{\oplus} + A^{\ominus} \qquad (2.22)$$

the medium. Figure 2.8 represents this variation. The curve is sigmoid and the $pK_a$ represents the pH at which half of the acid is present as $A^{\ominus}$. If a certain chemical reaction depends on the compound $HA$ being in its conjugate base form, whether as a reactant or a catalyst, then the rate of the

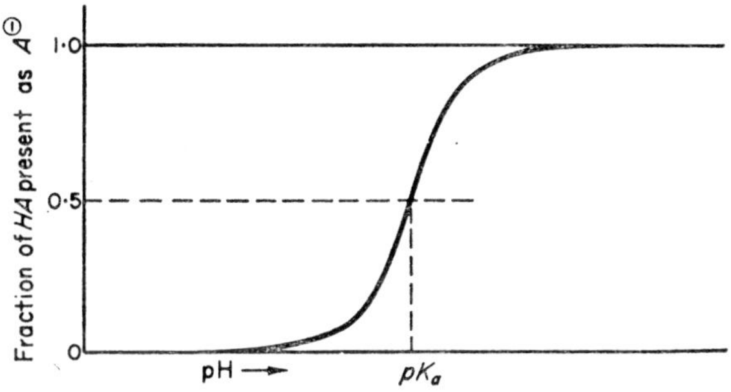


FIG. 2.8. Plot of the fraction of an acid ($HA$) present in the form of the conjugate base ($A^{\ominus}$) against pH. The pH at which the acid is half fully ionized represents the $pK_a$ of the acid. If a chemical reaction depends on $A^{\ominus}$ then the rate of the reaction will vary with pH, and this curve will also represent the rate/pH curve

reaction will vary with pH and this variation will have a similar sigmoid form. For a reaction which shows this type of variation it would be possible to determine the $pK_a$ of the compound involved by measuring the pH at which the rate of the reaction is half the maximal rate (or better by fitting a theoretical curve to that which is obtained experimentally).

Similarly, the fraction of an acid $HB$ present in the *acidic* form will vary with pH according to curve (b) in Fig. 2.9 and any reaction depending on the acidic form of this acid would show a similar rate–pH dependence.

If a reaction depends on two ionizing species $HA$ and $HB$ such that the former must be in the conjugate base form ($A^{\ominus}$) and $HB$ must be in its un-ionized form then the rate of that reaction will be proportional to the product of $A^{\ominus}$ and $HB$. Such a reaction would show a pH-rate profile ((c) Fig. 2.9) which represents the product of curves (a) and (b). This produces the bell-shaped curve typical of many enzyme-catalysed reactions.

When we consider enzyme-catalysed reactions from this point of view we must consider several aspects. First, that there are ionizable groups on the side-chains of various amino-acids. These are shown in Table 2.2 along

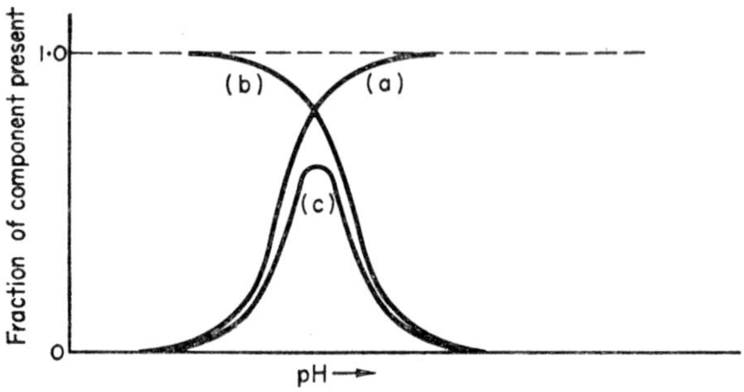


FIG. 2.9. In a system containing two acid groups, *HA* and *HB*, curve (a) represents the fraction of *HA* present as $A^{\ominus}$, (b) represents the fraction of *HB* in the unionized form, and, (c) represents the number of molecules containing both $A^{\ominus}$ and *HB*. If a chemical reaction requires *HA* in the ionized form $A^{\ominus}$, and *HB* in the unionized form, then curve (c) also represents the variation of the rate of that reaction with pH

with the $pK_a$ values measured for the free amino-acid. Note that these $pK_a$ values may be different when the amino-acid is incorporated into a protein for the reasons discussed later. Secondly, the substrate of the enzyme may itself ionize, and thirdly, if several ionizing groups are involved in the action of the enzyme then they may be involved at different stages in the reaction. For example, a particular ionizing group may be

Table 2.2. *Approximate $pK_a$ values for ionizable groups in proteins*

| Group | $pK_a$ |
|---|---|
| α-Carboxyl | 2 |
| Side-chain carboxyl (aspartic or glutamic acid) | 4 |
| α-Amino- | 9 |
| ε-Amino (lysine) | 10 |
| Imidazole (histidine) | 6 |
| Guanido (arginine) | 12 |
| Thiol (cysteine)* | 8·5 |
| Phenol (tyrosine) | 10 |

* This is the value obtained by Benesch and Benesch (*J. Am. chem. Soc.* **77**, 5877 (1955)) for the $pK_a$ of the thiol group of cysteine in which the α-amino-group remains protonated.

involved in the binding of the substrate to the enzyme, while another may of course be required for the catalytic process itself. By performing kinetic studies at various pH values and using Lineweaver–Burk plots (for example) it is possible to obtain information about the way in which both $K_m$ and $V$ for a particular enzyme–substrate system vary with pH. Dixon [66] has shown that plots of $pK_m$/pH and $pV$/pH can be interpreted in such a way that the $pK_a$ values of ionizing groups in the enzyme, substrate and enzyme–substrate complex may be determined. For example, the results of Dodgson *et al.* [67] were interpreted [51] to show that free *arylsulphatase* (3.1.6.1) has two important ionizing groups with $pK_a$ values 8·1 and 9·4, whereas the enzyme–substrate complex (enzyme-p-nitrophenylsulphate) has an ionizable group of $pK_a$ approx. 7·8.

Dixon's method allows us, therefore, to determine the $pK_a$ values of various ionizable groups at the active site of an enzyme and also to draw conclusions concerning those stages in the overall process at which these ionizable groups exert their influence. In some cases it may be possible to determine the pH-dependence of individual rate constants for a reaction. For example, detailed studies [68] showed that the chymotrypsin-catalysed hydrolysis of substrates, which proceeds *via* the formation of an acyl-enzyme (equation 2.23), had a pH-dependence as follows:

$$\underset{\text{enzyme}}{E} + \underset{\text{substrate}}{S} \overset{K_s}{\rightleftharpoons} \underset{\substack{\text{enzyme–substrate}\\\text{complex}}}{ES} \xrightarrow{k_{+2}} \underset{\text{acyl-enzyme}}{ES'} \xrightarrow{k_{+3}} E + \underset{\text{product}}{P} \qquad (2.23)$$

$K_s$ depends on a group with $pK_a$ approx. 9. This group must be in its acidic form. Both $k_{+2}$ and $k_{+3}$ depend on the basic form of a group with $pK_a$ about 6·8.

Similarly, Bender *et al.* [69] showed that the deacylation step in the reaction of papain depended on a basic group with $pK_a$ between 4 and 5.

Once the $pK_a$ of an ionizable group at the active site of an enzyme has been determined it is very tempting to identify positively the side-chain involved. Thus, when a group of $pK_a$ 6–7 is found in an enzyme, this has invariably been identified as the imidazole group of a histidine. Similarly, a group with $pK_a$ 4 is assumed to be the side-chain carboxyl of aspartic or glutamic acid. Although in many cases these attempted identifications have been confirmed by other studies, it is important that their limitations be appreciated. Any protein is extremely complex and a particular ionizable group may find itself in an environment which has a profound effect on its $pK_a$. Such an environment may be polar or non-polar; the ionizable group may be in the vicinity of species which can form hydrogen-bonds, or which by virtue of their charges, stabilize or destabilize a particular state of ionization. These factors may have the effect of changing the $pK_a$ value of a group by 2 pH units or more. Indeed, it has been suggested [70] that the active site of an enzyme should be considered not as a site containing groups which are more or less independent but as a unique entity in itself, the properties of this unique system being the resultant of many factors. Until all the factors which contribute to the active site are known it is an over-simplification to identify an apparent ionizable group in the active site with a certain amino-acid side-chain. This viewpoint is supported by the results of X-ray crystallography with chymotrypsin [71]. The active nucleophilic group in chymotrypsin is the hydroxyl group of a serine residue which is influenced by an imidazole group. The X-ray studies showed that this imidazole is itself influenced by the carboxylate group of an aspartyl residue which is in a hydrophobic region of the protein. Thus, the serine, histidine and aspartyl side-chains, as well as those groups which contribute to the non-polar nature of the environment, all contribute to the pH-dependence of the chymotrypsin reaction.

## ALLOSTERIC EFFECTS

Enzymic activity is often influenced by the presence of small molecules. Inhibition, which may be competitive or non-competitive, has already been mentioned in this chapter. In general, competitive inhibitors have their effect by becoming bound to the active site of the enzyme. When the active site is thus occupied it cannot accommodate the substrate. Other molecules, which may be either inhibitors or activators, appear to interact with the

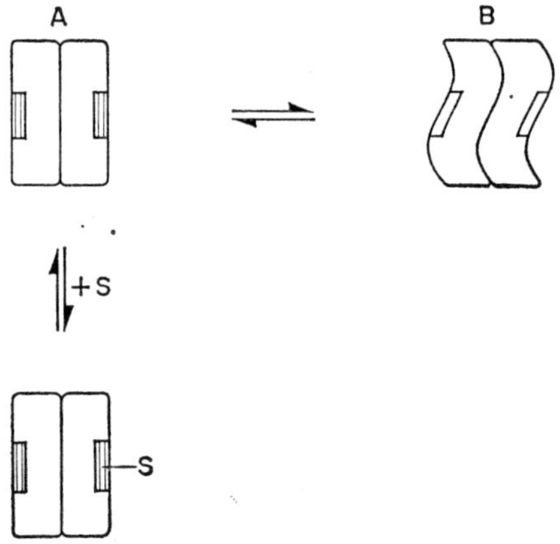


FIG. 2.10. Model to illustrate homotropic allosteric activation. A and B are the active and distorted forms, respectively, of the enzyme which consists of two protomers. The vertically hatched area represents the substrate binding site. In B this is distorted and ineffective as a binding site for substrate, and is shown unhatched

enzyme at a site quite far removed from the active site and their effects are not eliminated at very high substrate concentrations. This means they are not competitive with the substrate. Molecules which function in this way are known as *allosteric* effectors and they play an important role in the control of enzymic activity in biological metabolism.

In some cases, substrates themselves can function as allosteric effectors in which case the process is described as *homotropic*. In such a situation the variation of initial rate with substrate concentration would not be as shown in Fig. 2.2, but would probably be sigmoidal. Where the allosteric effector is not related to the substrate the effect is *heterotropic*. Whether an effect is homotropic or heterotropic, it is usually associated with, and attributed to, changes in the ternary or quaternary structure of the enzyme.

In most cases, the most satisfactory explanation of allosteric effects may be provided by the type of model proposed by Monod *et al.* [72]. These workers suggested that allosteric enzymes consist of aggregates of two or more identical sub-units (protomers). Figure 2.10 shows a model which provides an explanation for homotropic activation, i.e. the substrate functions also as an activator. The model is made up of two protomers, and can exist in two forms A and B which are in equilibrium. Conversion of one form into the other is envisaged as involving a conformational change. In form A, each unit possesses a binding site for the substrate, while in form B these binding sites are distorted so that they have a lower affinity for substrate. When A binds to substrate, it is then unable to rearrange into B; its conformation is thus fixed. The presence of substrate, therefore, upsets the equilibrium between A and B and ensures that more enzyme molecules are in the form more favourable to reaction. The observable effects of such a process would be (a) an activation by substrate manifested in an anomalous kinetic order with respect to substrate, and (b) an apparent net conformational change in the population of molecules. The model can of course be extended to include many protomers.

A similar model can be invoked to describe heterotropic allosteric effects both activating and inhibitory. The enzyme model in Fig. 2.11 exists in two forms in equilibrium, A and B. The molecule contains two sub-units. In A each of these sub-units possesses a binding site for substrate and also a site to accommodate the allosteric activator *AA*. These sites exist in distorted form in B, which therefore has little or no affinity for them, but which does bind the inhibitor *I*. The inhibitor binding site is represented in Fig. 2.11 as being the distorted form of the site for *AA*. *I* does not bind to form A.

Combination of the activator with form A results in an increase in the number of enzyme molecules able to bind the substrate and therefore an increase in the overall activity of the enzyme. The inhibitor *I* functions by holding the enzyme in the conformation unfavourable for substrate binding and therefore unfavourable for enzymic reaction.

The enzyme aspartate carbamoyltransferase (carbamoylphosphate: L-aspartate carbamoyltransferase, 2.1.3.2) is probably subject to essentially the same type of allosteric activation and inhibition. The carbamoylation of aspartate is an important stage in the synthesis of cytosine triphosphate (CTP). The end product of the metabolic sequence, CTP itself, is an inhibitor of the aspartate carbamoyltransferase and thus provides an

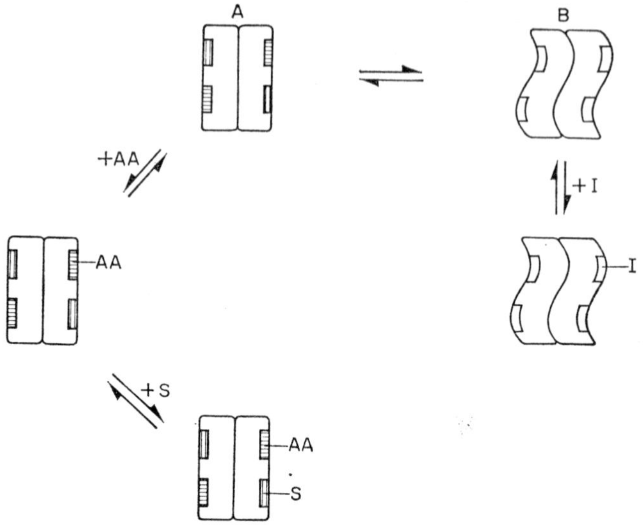


FIG. 2.11. Model to illustrate heterotropic activation and inhibition. A and B represent the active and distorted forms, respectively, of the enzyme which is composed of two protomers. The substrate binding sites are represented by the vertically-hatched areas. Horizontally-hatched areas represent the binding sites for the allosteric activator AA. In B these sites are distorted and ineffective in binding the substrate and the activator. These sites are shown unhatched. The inhibitor binding site in B is represented as being the distorted activator binding site

important feedback control system. ATP, however, is an activator of the enzyme. Thus the biological system possesses a mechanism whereby a balance between the nucleosides is preserved.

## X-RAY CRYSTALLOGRAPHY [73, 74]

Probably the most powerful tool available at the present time for the study of enzymes, and indeed all proteins, is X-ray crystallography. Although in general the task of the X-ray crystallographer is made easier if the amino-acid sequence of the protein is known, in some cases crystallographic studies have contributed to our knowledge of the sequences. Examples are afforded by the case of papain, in which the positions of certain sequences in the primary structure were established, and chymotrypsin, in which the distinction between aspartyl and asparaginyl residues was assisted [75, 76].

In the following chapters the results of several crystallographic studies are discussed in connection with the way in which they have contributed to our knowledge of the mechanisms of enzyme-catalysed reactions. In the case of lysozyme, it was possible to describe in detail not only the groups (of the substrate and the enzyme) involved in enzyme–substrate binding, but also the way in which the enzyme was able to bring about the catalysis of its peculiar reaction [77]. Similarly, in the case of chymotrypsin, the alignments of all the important groups in the enzyme were established and in addition the very important, and previously unsuspected, role of the side-chain carboxyl of aspartyl-102 in the catalytic process [71].

Originally it was thought that the results of X-ray crystallography and their interpretations in terms of enzyme catalysis would be severely limited. First, it must be established that the three-dimensional structure of an enzyme in the crystalline state is the same as that of the enzyme as it functions in solution. Since in several cases it has been shown that crystalline enzymes are active catalytically [78] then it is clear that there is nothing intrinsically different about the conformation in the crystal from that in solution. Another possible limitation, which is related to that just discussed, was that the protein in the crystalline state may be relatively rigid, being able to undergo little more than vibrational movement, while in solution it would be elastic, freely changing from one conformation to another. Thus, it appeared unlikely that X-ray crystallography would be able to make any great contribution to the study of enzymes in which allosteric or induced-fit effects were important. It is interesting that Lipscomb *et al.* [2] were able by the use of X-ray crystallography to demonstrate very clearly the induced-fit effects in carboxypeptidase A. Thus, paradoxically, this technique which deals with rigidly held molecules was able to demonstrate the very flexibility of enzymes.

## REFERENCES

1. Pauling, L., Corey, R. B. and Branson, H. R., *Proc. natn. Acad. Sci. U.S.A.* **37,** 205 (1951); Pauling, L. and Corey, R. B., *Proc. natn. Acad. Sci. U.S.A.* **37,** 235, 241, 251, 256, 261, 272 (1951)
2. Lipscomb, W. N., Hartsuck, J. A., Reeke, G. N., jr., Quiocho, F. A., Bethge, P. H., Ludwig, M. L., Steitz, T. A., Muirhead, H. and Coppola, J. C., *Brookhaven Symposia in Biology*, No. 21 (Brookhaven National Laboratory, Upton, New York, 1968); Reeke, G. N., Hartsuck, J. A., Ludwig, M. L., Quiocho, F. A., Steitz, T. A. and Lipscomb, W. N., *Proc. natn. Acad. Sci. U.S.A.* **58,** 2220 (1967); Lipscomb, W. N., Hartsuck, J. A., Quiocho, F. A. and Reeke, G. N., *Proc. natn. Acad. Sci. U.S.A.* in the press
3. Koshland, D. E., jr., *The Enzymes*, Eds. P. D. Boyer, H. Lardy and K. Myrback, Vol. 1, p. 305 (Academic Press, New York, 1959)
4. Crestfield, A. M., Stein, W. H. and Moore, S., *J. biol. Chem.* **238,** 2413, 2421 (1963)
5. Schoellman, G. and Shaw, E., *Biochemistry* **2,** 252 (1963); *Biochem. biophys. Res. Commun.* **7,** 36 (1962); Ong, E. B., Shaw, E. and Schoellman, G., *J. Am. chem. Soc.* **86,** 1271 (1964); *J. biol. Chem.* **240,** 694 (1965)

6. Baker, B. R., *Design of Active-Site-Directed Irreversible Enzyme Inhibitors. The Organic Chemistry of the Enzymic Active Site* (Wiley, New York, 1967)
7. Whitney, P. L., Folsch, G., Nyman, P. O. and Malmstrom, B. G., *J. biol. Chem.*, **242,** 4206, 4212 (1967)
8. Ingram, J. M. and Hochster, R. M., *Can. J. Biochem.* **45,** 929, 1505 (1967)
9. Westhead, E. W., *Biochemistry* **4,** 2139 (1965)
10. Koshland, D. E., jr., Strumeyer, D. H. and Ray, W. J., jr., *Brookhaven Symposia in Biology*, No. 15, p. 101 (Brookhaven National Laboratory, Upton, New York, 1962); Chaplin, M. F., Gray, C. J. and Kennedy, J. F., *Workshop Conference on Gonadotrophins* (Birmingham, England, 1969) in the press
11. Hoffee, P., Lai, C. Y., Pugh, E. L. and Horecker, B. L., *Proc. natn. Acad. Sci. U.S.A.* **57,** 107 (1967)
12. Kimmel, J. R. and Smith, E. L., *J. biol. Chem.* **207,** 515 (1954); Finkle, B. J. and Smith, E. L., *J. biol. Chem.* **230,** 669 (1958)
13. Brubacher, L. J. and Bender, M. L., *J. Am. chem. Soc.* **88,** 5800 (1966)
14. Husain, S. S. and Lowe, G., *Chem. Commun.* 310 (1968)
15. Katagiri, M., Takemori, S., Nakazawa, K., Suzuki, H. and Agaki, K., *Biochim. biophys. Acta* **139,** 173 (1967)
16. Pfleiderer, G., Holbrook, J. J., Nowicki, L. and Jeckel, R., *Biochem. Z.* **346,** 297 (1966)
17. Eifler, R., Glaesmer, R. and Rapoport, S., *Acta biol. med. Ger.* **17,** 716 (1966) quoted in *Chem. Abst.* **66,** 91831y
18. Chinard, F. P. and Hellerman, L., *Methods of Biochemical Analysis* Vol. 1, p. 1 (Interscience, New York, 1954)
19. Nogare, S. D. *Organic Analysis* Vol. 1, p. 329 (Interscience, New York, 1953)
20. Boyer, P. D., *The Enzymes*, Eds. P. D. Boyer, H. Lardy and K. Myrback, Vol. 1, p. 511 (Academic Press, New York, 1954)
21. Folk, J. E. and Cole, P. W., *J. biol. Chem.* **241,** 5518 (1966)
22. Bocharnikova, I. M. and Petushkova, E. V., *Biokhimiya* **32,** 119 (1967)
23. Yang, S. F., Freer, S. and Benson, A. A., *J. biol. Chem.* **242,** 477 (1967)
24. Kezdy, F. J., Feder, J. and Bender, M. L., *J. Am. chem. Soc.* **89,** 1009 (1967); Hille, M. B. and Koshland, D. E., jr., *J. Am. chem. Soc.* **89,** 5945 (1967)
25. Coleman, R. F., *Biochem. biophys. Res. Commun.* **28,** 222 (1967)
26. Finn, F. M. and Hofmann, K., *J. Am. chem. Soc.* **87,** 645 (1965)
27. Simpson, R. T., Riordan, J. F. and Vallee, B. L., *Biochemistry* **2,** 616 (1963); Riordan, J. F. and Vallee, B. L., *Fedn Proc.* **22,** 594 (1963); Riordan, J. F. and Vallee, B. L., *Biochemistry* **2,** 1460 (1963); Vallee, B. L., *Fedn Proc.* **23,** 8 (1964); Coleman, J. E., Pulido, P. and Vallee, B. L., *Biochemistry* **5,** 2019 (1966)
28. Pugh, E. L. and Horecker, B. L., *Biochem. biophys. Res. Commun.* **26,** 360 (1967)
29. Drechsler, E. R., Boyer, P. D. and Kowalsky, A. G., *J. biol. Chem.* **234,** 2627 (1959); Rutter, W. J., Richards, O. C. and Woodfin, B. M., *J. biol. Chem.* **236,** 3293 (1961)

30. PHILPOTT, J. ST. L. and SMALL, P. A., *Biochem. J.* **32,** 542 (1938); HERRIOTT, R. M., *J. Gen. Physiol.* **19,** 283 (1935); HERRIOTT, R. M., *J. Gen. Physiol.* **31,** 19 (1947); LI, C. H., *J. Am. chem. Soc.* **67,** 1065 (1945)
31. PONTREMOLI, S., GRAZI, E. and ACCORSI, A., *J. biol. Chem.* **242,** 61 (1967); *Biochemistry* **5,** 3072 (1966)
32. GROSS, E. and MORELL, J. L., *J. biol. Chem.* **241,** 3638 (1966); ERLANGER, B. F., VRATSANOS, S. M., WASSERMANN, N. and COOPER, A. G., *J. biol. Chem.* **240,** PC3447 (1965); *Biochem. biophys. Res. Commun.* **23,** 243 (1966)
32a. DELPIERRE, G. R. and FRUTON, J. S., *Proc. natn. Acad. Sci. U.S.A.*, **56,** 1817 (1966)
33. LEGLER, G., *Z. physiol. Chem.* **349,** 767 (1968)
34. HIRS, C. H. W., *Brookhaven Symposia in Biology*, No. 15 (Brookhaven National Laboratory, Upton, New York, 1962); MARFEY, P. S., UZIEL, M. and LITTLE, J., *J. biol. Chem.* **240,** 3270 (1965); MARFEY, P. S., NOWACK, H., UZIEL, M. and YPHANTIS, D. A., *J. biol. Chem.* **240,** 3264 (1965)
35. KANAZAWA, T. and TONOMURA, Y., *J. Biochem., Tokyo* **57,** 604 (1965); KUBO, S., TOKURA, S. and TONOMURA, Y., *J. biol. Chem.* **235,** 2835 (1960)
36. WESTHEIMER, F. H., *Proc. chem. Soc.*, 253 (1963)
37. VENKATARAMAN, R. and RACKER, E., *J. biol. Chem.* **236,** 1883 (1961); PONTREMOLI, S., PRANDINI, B. D., BONSIGNORE, A. and HORECKER, B. L., *Proc. natn. Acad. Sci. U.S.A.* **47,** 1942 (1961); HORECKER, B. L., PONTREMOLI, S., RICCI, C. and CHENG, T., *Proc. natn. Acad. Sci. U.S.A.* **47,** 1947 (1961)
38. GRAZI, T., CHENG, T. and HORECKER, B. L., *Biochem. biophys. Res. Commun.* **7,** 250 (1962)
39. PORTSMOUTH, D., STOOLMILLER, A. C. and ABELES, R. H., *J. biol. Chem.* **242,** 2751 (1967)
40. MORINO, Y. and SNELL, E. E., *J. biol. Chem.* **242,** 2793, 2800 (1967)
41. RICHARDS, F. M. and VITHAYATHIL, P. J., *J. biol. Chem.* **234,** 1459 (1959)
42. KENKARE, U. W. and RICHARDS, F. M., *J. biol. Chem.* **241,** 3197 (1966); VITHAYATHIL, P. J. and RICHARDS, F. M., *J. biol. Chem.* **235,** 2343 (1960)
43. DIXON, M., *Biochem. J.* **55,** 170 (1953)
44. HALDANE, J. B. S., *Enzymes* (Longmans, London, 1930)
45. DALZIEL, K., *Acta chem. Scand.* **11,** 1706 (1957)
46. CLELAND, W. W., *Biochim. biophys. Acta* **67,** 104, 173, 188 (1963)
47. ALBERTY, R. A., *Advances in Enzymol.* **17,** 1 (1956); *The Enzymes* Vol. 1, p. 143 (Academic Press, New York, 1958)
48. KING, E. L. and ALTMAN, C. J., *J. phys. Chem.* **60,** 1375 (1956)
49. BRIGGS, G. E. and HALDANE, J. B. S., *Biochem. J.* **19,** 338 (1925)
50. GUTFREUND, H., *An Introduction to the Study of Enzymes* (Blackwell, Oxford, 1965)
51. DIXON, M. and WEBB, E. C., *Enzymes* 2nd edn (Longmans, London, 1964)
52. WONG, J. T.-F. and HANES, C. S., *Can. J. Biochem. Physiol.* **40,** 763 (1962)
53. MICHAELIS, L. and MENTEN, M. L., *Biochem. Z.* **49,** 333 (1913)
54. LINEWEAVER, H. and BURK, D., *J. Am. chem. Soc.* **56,** 658 (1934)
55. FROMM, H. J. and ZEWE, V., *J. biol. Chem.* **237,** 1661 (1962)

56. Silverstein, R., Voet, J., Reed, D. and Abeles, R. H., *J. biol. Chem.* **242,** 1338 (1967)
57. Hersh, L. B. and Jencks, W. P., *J. biol. Chem.* **242,** 3468 (1967)
58. Fromm, H. J. and Zewe, V., *J. biol. Chem.* **237,** 3027 (1962)
59. Zewe, V. and Fromm, H. J., *J. biol. Chem.* **237,** 1668 (1962)
60. Boyer, P. D., *Archs. Biochem. Biophys.* **82,** 387 (1959); Boyer, P. D. and Silverstein, E., *Acta chem. Scand.* **17,** 5195 (1963)
61. Fromm, H. J., *J. biol. Chem.* **239,** 3645 (1964)
62. Joyce, B. K. and Himes, R. H., *J. biol. Chem.* **241,** 5716, 5725 (1966)
63. Mahler, H. R. and Cordes, E. H., *Biological Chemistry* p. 243 (Harper and Row, New York, 1966)
64. Dalziel, K., *J. biol. Chem.* **238,** 2850 (1963)
65. Theorell, H., *Harvey Lectures, Series 61*, p. 17 (1966)
66. Dixon, M., *Biochem. J.* **55,** 161, 170 (1953)
67. Dodgson, K. S., Spencer, B. and Williams, K., *Biochem. J.* **61,** 374 (1955)
68. Bender, M. L., Gibian, M. J. and Whelan, D. J., *Proc. natn. Acad. Sci. U.S.A.* **56,** 833 (1966); Himoe, A., Parks, P. C. and Hess, G. P., *J. biol. Chem.* **242,** 919 (1967); *Biochem. biophys. Res. Commun.*, **23,** 234 (1965)
69. Brubacher, L. J. and Bender, M. L., *J. Am. chem. Soc.* **88,** 5871 (1968); Bender, M. L. and Brubacher, L. J., *J. Am. chem. Soc.* **88,** 5880 (1968)
70. Gutfreund, H., British Biophysical Society Meeting (London, 1968)
71. Matthews, B. W., Sigler, P. B., Henderson, R. and Blow, D. M., *Nature, Lond.* **214,** 652 (1967)
72. Monod, J., Wyman, J. and Changeux, J.-P., *J. molec. Biol.* **12,** 88 (1965)
73. Crick, F. H. C. and Kendrew, J. C., *Adv. prot. Chem.* **12,** 133 (1957)
74. Low, B. W., *J. polym. Sci.* **49,** 153 (1961)
75. Drenth, J., Jansonius, J. N., Koekoek, R., Swen, H. M. and Wolthers, B. G., *Nature, Lond.*, **218,** 929 (1968)
76. Hartley, B. S., Meeting of the Royal Society (London, Dec. 1968)
77. Blake, C. C. F., Mair, G. A., North, A. C. T., Phillips, D. C. and Sarma, V. R., *Proc. R. Soc.* **167B,** 365 (1967); Blake, C. C. F., Johnson, L. N., Mair, G. A., North, A. C. T., Phillips, D. C. and Sarma, V. R., *Proc. R. Soc.* **167B,** 378 (1967); Blake, C. C. F., Koenig, D. F., Mair, G. A., North, A. C. T., Phillips, D. C. and Sarma, V. R., *Nature, Lond.* **206,** 757 (1965)
78. Quiocho, F. A. and Richards, F. M., *Proc. natn. Acad. Sci. U.S.A.* **52,** 833 (1964)

# CHAPTER 3

# OXIDATION AND REDUCTION

## OXIDATION AND REDUCTION

Oxidation–reduction reactions are of vital importance to all biological systems. These reactions are involved not only in the essential processes of energy storage and interconversion but also in the synthesis and degradation of the various organic compounds which are necessary for the complete functioning of the living cell. The enzymes which catalyse these reactions do so by several different types of mechanism, some of which appear to be understandable in terms of the *in vitro* reactions which are well known in organic chemistry. We shall consider first therefore some aspects of oxidation and reduction as they apply to 'normal' organic chemistry. In some cases, enzyme-catalysed reactions bear at least superficial resemblances to reactions of model compounds, and although such resemblances must by no means be overemphasized, they may throw some light on the processes which are involved.

Although the original meaning of the term oxidation, as the word itself suggests, implied the addition of oxygen to a chemical species, as far as organic chemistry is concerned it is now used to embrace several types of chemical conversions. The original meaning still stands; conversion of acetaldehyde to acetic acid, in which the oxygen content of the molecule is increased by one atom, is considered to be an oxidation, and the reverse process a reduction,

$$CH_3\text{—}CHO \underset{\text{reduction}}{\overset{\text{oxidation}}{\rightleftarrows}} CH_3\text{—}COOH$$

The removal of hydrogen from a molecule is also regarded as an oxidative process. Thus ethanol may be oxidized to acetaldehyde

$$CH_3\text{—}CH_2\text{—}OH \rightarrow CH_3\text{—}CHO$$

Oxidation and reduction reactions therefore may be expressed in terms of transfer processes; transfer of oxygen to a molecule (or of hydrogen from it) results in the oxidation of that molecule. An *oxidizing agent* is a species which effects such a transfer and it is inevitable that in the process of oxidizing a compound the oxidizing agent itself becomes reduced (that is, it loses oxygen or gains hydrogen).

It is possible to arrange classes of organic compounds in series which reflect different oxidation or reduction levels. Two such series are given below:

$$CH_3{-}CH_3;\quad CH_3{-}CH_2{-}OH;\quad CH_3{-}CHO;\quad CH_3{-}COOH$$

$$CH_3{-}CH_3;\quad CH_2{=}CH_2;\quad CH{\equiv}CH.$$

Each compound is said to be at a higher oxidation level than that on its left, and may be derived from it (in terms of its molecular formula at least) either by the removal of hydrogen or by the addition of oxygen.

Of course, not all interconversions in organic chemistry involve oxidation or reduction. For example, addition of the elements of water to a compound is not an oxidative process; thus ethylene and ethanol are said to be at the same oxidation level. Similarly, addition of the elements of a hydrogen halide or of ammonia or some similar H—X compound involves neither oxidation nor reduction. Ethanol for example is at the same level as ethyl chloride or ethylamine, and acetylene is at the same level as acetaldehyde imine, $CH_3CH{=}NH$. Acid–base reactions are also included in this category and we notice therefore that addition of a proton ($H^{\oplus}$) to a compound does not affect its oxidation level whereas addition of a hydrogen atom (H•) involves a reductive process.

The difference between these two species ($H^{\oplus}$ and H•) is one electron, and this leads us to another concept of oxidation and reduction, which is derived mainly from inorganic chemistry. According to this view oxidation is considered in terms of *electron transfer*; for example, removal of one electron from a ferrous ion brings about its oxidation to the ferric state.

$$Fe^{2\oplus} \xrightarrow{-1e^{\ominus}} Fe^{3\oplus}.$$

Conversely, addition of an electron brings about reduction. A formal relationship exists between this concept and those mentioned above, in that the addition of two electrons is equivalent to the addition of two hydrogen atoms or the removal of one oxygen. The relationship may be exemplified as follows. If two electrons are considered to be added to a

$$CH_2{=}CH_2 + 2e^{\ominus} \longrightarrow \overset{\ominus}{C}H_2{-}\overset{\ominus}{C}H_2 \xrightarrow{2H^+} CH_3{-}CH_3$$

molecule of ethylene the resulting dianion may take up two protons to give ethane, the final result being equivalent to adding two hydrogen atoms. Since the second step does not involve reduction, then the change in oxidation level has been produced simply by the addition of the electrons. We may relate compounds of various oxidation levels by the number of electrons (or their equivalent) which are required for their interconversion. Most organic oxidations or reductions involve the equivalent of a two-electron transfer. Transfer of a single electron must lead to the formation of a radical and usually a second electron transfer is required if a stable product is to be obtained.

**Reduction Potentials**

An oxidation–reduction process may be regarded as being made up of two half-reactions involving electrons, just as an acid–base reaction can be viewed as two half-reactions involving protons (See chapter 1, page 13). Thus we may consider reaction (3.1) as being composed of the two half-reactions (3.2 and 3.3), where the subscripts $O$ and $R$ refer to the oxidized and reduced forms of A and B respectively, and we are assuming that a one-electron oxidation is involved.

$$A_R + B_O \rightleftharpoons A_O + B_R \tag{3.1}$$

$$A_R \rightleftharpoons A_O + e \tag{3.2}$$

$$B_O + e \rightleftharpoons B_R \tag{3.3}$$

A measure of the tendency of any species to accept electrons, that is, to become reduced, is provided by its *reduction potential.* This potential, in volts, represents the experimentally-determined electrical potential difference which is observed when a half-cell made up of the oxidized and reduced forms is connected to a standard half-cell. (A detailed account of the theoretical principles involved is given by Glasstone [1].) For the compound B, for example, this potential difference is given by the equation (3.4)

$$E = E_O^B - \frac{RT}{nF} \ln \frac{[B_R]}{[B_O]}, \tag{3.4}$$

where $R$ is the gas constant, $T$ is the absolute temperature, $F$ is the faraday (23,000 cal/absolute volt equivalent), $n$ is the number of electrons involved in the reduction of $B_O$ to $B_R$. $E_O^B$ is the standard reduction potential for $B_R'$ at pH 0 and $E$ is the measured potential difference of the system in which a half-cell containing $B_R$ and $B_O$ is connected to a standard hydrogen half-cell.

$E_O^B$ is numerically equal to the potential difference obtained when $[B_O] = [B_R]$ in the half-cell. In some cases, when the half-reaction under consideration involves hydrogen ions, $E_O$ is dependent on pH, and it is usually more convenient to consider the value of the standard reduction potential at pH 7. Table 3.1 gives a list of some chemical species which are important in biological systems, with corresponding values of $E_O'$, which relate to pH 7.

Let us consider now the reaction (3.1) when it has been allowed to reach equilibrium. The potential of the $B_R/B_O$ system may be expressed by

$$E = E_O^B - \frac{RT}{nF} \ln \frac{[B_R]}{[B_O]}$$

and the potential of the $A_R/A_O$ system by

$$E = E_O^A - \frac{RT}{nF} \ln \frac{[A_R]}{[A_O]}.$$

Table 3.1. *Standard reduction potentials for systems of biochemical importance* [2] †

| System | $E'_0$ (pH 7), volts |
|---|---|
| Oxygen/water | 0·815 |
| Ferric/ferrous | 0·77 |
| Nitrate/nitrite | 0·42 |
| Ferricyanide/ferrocyanide | 0·36 |
| Oxygen/hydrogen peroxide | 0·30 |
| Cytochrome *a*; ferric/ferrous | 0·29 |
| Cytochrome *c*; ferric/ferrous | 0·22 |
| Crotonyl-SCoA/butyryl-SCoA | 0·19 |
| Methemoglobin/hemoglobin | 0·17 |
| Cytochrome $b_2$; ferric/ferrous | 0·12 |
| Ubiquinone; ox/red | 0·10 |
| Dehydroascorbic acid/ascorbic acid | 0·08 |
| Metmyoglobin/myoglobin | 0·046 |
| Fumaric acid/succinic acid | 0·03 |
| Methylene blue, ox/red | 0·01 |
| Yellow enzyme, FMN/$FMNH_2$ | −0·122 |
| Pyruvate + ammonium/alanine | −0·13 |
| α-Ketoglutarate + ammonium/glutamic acid | −0·14 |
| Oxaloacetate/malate | −0·17 |
| Pyruvate/lactate | −0·19 |
| Acetaldehyde/ethanol | −0·20 |
| Riboflavin, ox/red | −0·21 |
| Glutathione, ox/red | −0·23 |
| Acetoacetate/β-hydroxybutyrate | −0·27 |
| Lipoic acid, ox/red | −0·29 |
| $NAD^+$/NADH | −0·32 |
| Pyruvate/malate | −0·33 |
| Uric acid/xanthine | −0·36 |
| Carbon dioxide/formate | −0·42 |
| $H^+/H_2$ | −0·42 |
| Acetate/acetaldehyde | −0·60 |
| Succinate/α-ketoglutarate | −0·67 |
| Acetate + carbon dioxide/pyruvate | −0·70 |

Since equilibrium conditions apply, these two potentials must be equal and therefore

$$E_O^B - \frac{RT}{nF}\ln\frac{[B_R]}{[B_O]} = E_O^A - \frac{RT}{nF}\ln\frac{[A_R]}{[A_O]}.$$

Hence

$$E_O^B - E_O^A = \frac{RT}{nF}\ln\frac{[B_R][A_O]}{[A_R][B_O]} = \frac{RT}{nF}\ln K,$$

where $K$ is the equilibrium constant for the reaction (3.1).

† By courtesy of E. H. Cordes and Harper Rowe.

Therefore, using equation (1.7) we may write

$$\Delta G^0 = -nF(E_O^B - E_O^A),$$

which shows us that the standard free energy change for any oxidation–reduction reaction is a function of the difference in the standard reduction potentials of the two half-reactions which make it up. A corollary to this statement is that the reduction potential of any species is a thermodynamic parameter being related to the free energy change which accompanies reduction of the oxidized form of that species.

Table 3.1 contains some interesting examples of the way in which the reduction potential of a species can be influenced by its immediate environment. The ferric/ferrous system has a reduction potential of +0·77 volts. This refers to aqueous solutions, and the ions are surrounded by water

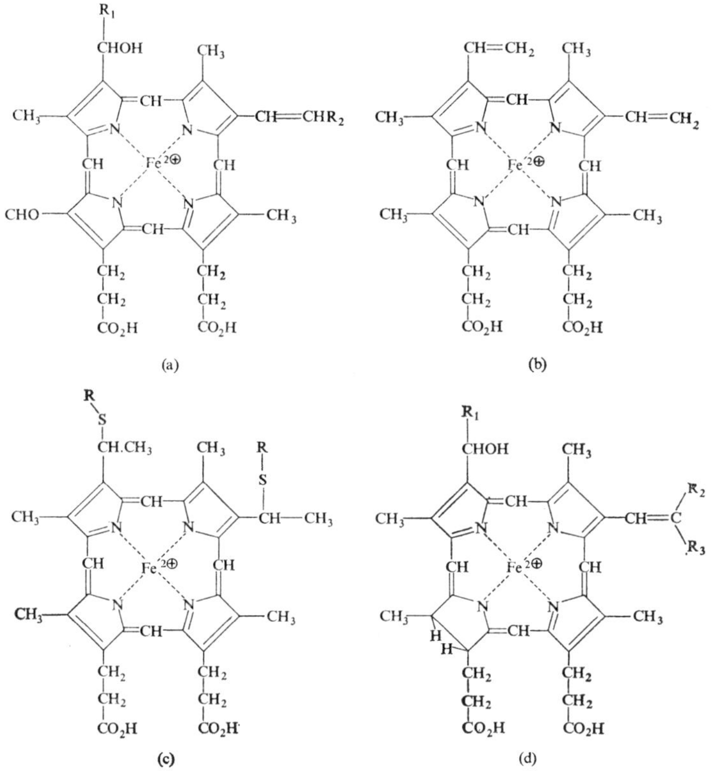


FIG. 3.1. Haem groups of various cytochromes. (a) Haem group containing a formylporphyrin; (b) protoporphyrin; (c) substituted meso-haem group covalently linked to the protein (via R—S—); (d) haem containing a dihydroporphyrin. The four nitrogen atoms of the porphyrin group constitute four ligands to an iron ion. When this iron is in the ferrous state the complex is known as a ferrocytochrome. (From ref. [3]) (By courtesy of M. Dixon and the Longman Group Ltd)

molecules. However, the ferricyanide/ferrocyanide potential is lower, at 0·36 volts. Clearly the presence of six cyanide ion ligands around the metal-ion has a powerful effect on the energy involved in transferring an electron to the iron atom. The potential is reduced even further in the cytochromes *a*, *c*, $b_2$, etc., where the iron atom is surrounded by ligands provided by the haem groups which are represented in Fig. 3.1.

Since changes in the reduction potential of a component in an oxidation–reduction reaction can influence the rate of that reaction (see p. 107), it is probable that the changes in reduction potentials of some metal-ions brought about by the special ligand systems found in enzymes are a major factor in the activity of these enzymes.

The foregoing discussion has been concerned only with oxidation and reduction as formal relationships between molecules, that is, in terms of their molecular formulae. No consideration has been given either to the feasibility in practical terms of any particular reaction or to the mechanism by which such a reaction, if feasible, actually proceeds. A reaction which can quite easily be *formulated* as an addition of two electrons followed by two protons, need not proceed by that mechanism. Some of the mechanisms by which organic oxidations and reductions are thought to proceed are discussed below.

**Hydride Transfer**

The reduction of a carbonyl- to an alcohol-group requires the equivalent of two hydrogen atoms and in many cases the mechanism by which such a reduction is brought about appears to involve addition of a hydride ion ($H^{\ominus}$) and a proton. Metal hydrides are frequently used in the laboratory to bring about the conversion of ketones and aldehydes into alcohols, perhaps the most common being lithium aluminium hydride. The mechanism, as shown, is essentially a nucleophilic attack on the carbonyl

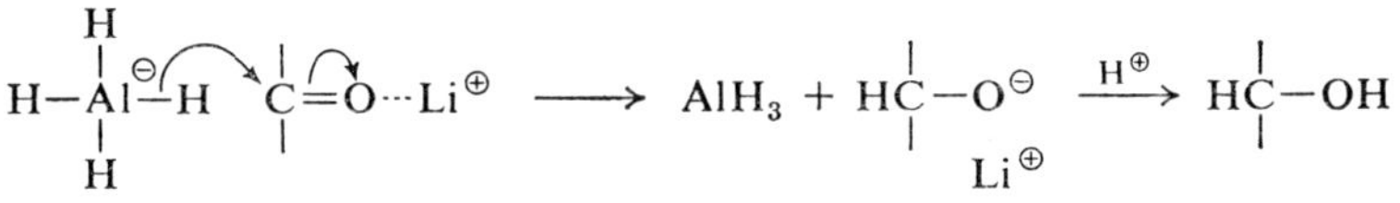

group by the hydrogen atom which is transferred from the aluminium hydride complex with its pair of electrons. Notice that it is not necessary for the hydride ion to exist as an independent entity, and the transfer as envisaged here is direct. Many reactions which appear to involve hydride transfer are performed in potentially acidic solvents and in such solvents a free hydride ion could not exist, since it is an extremely powerful base.

The base-catalysed carbinol–carbonyl equilibrium almost certainly proceeds by a hydride transfer. That the transfer is direct was indicated by deuterium isotope substitution and other studies [4]. The mechanism proposed involves the direct hydride transfer from an alkoxide ion.

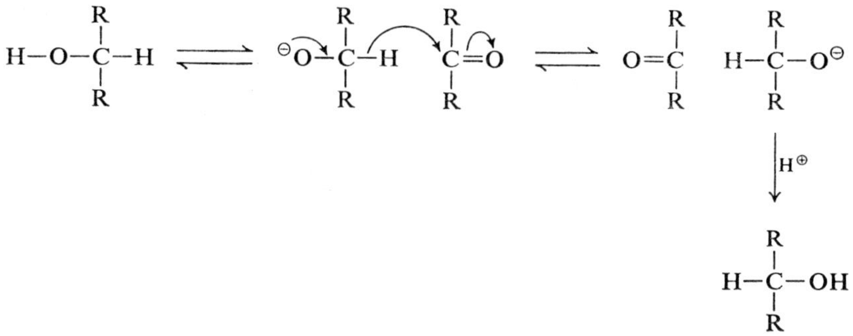

A rather similar mechanism is invoked to explain the Cannizzaro reaction in which benzaldehyde, for example, is converted into a mixture of benzyl alcohol and benzoic acid. Attack of hydroxide ion on the aldehydo-group produces an anionic species (I) from which the hydride departs [5].

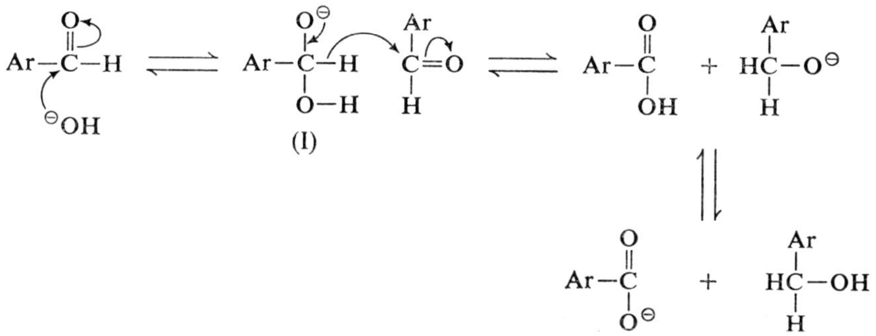

In this case, too, the direct transfer of the hydrogen was confirmed when the reaction was carried out in $D_2O$. The alcohol formed contained no deuterium [6, 7].

An almost identical process is envisaged as taking place intramolecularly during the rearrangement of phenylglyoxal (II) to the mandelic acid ion (III) [8]. In each of these last three reactions the driving force for the

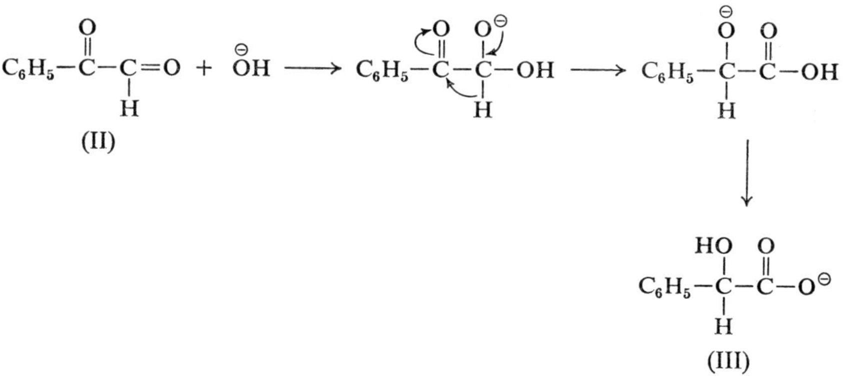

process is provided by the anion, which serves as the electron source. The carbonyl group provides the electron terminus, the resulting negative charge being stabilized by solvation as described in chapter 1. In another example of carbonyl–alcohol equilibrium (the Meerwein–Pondorff–Oppenhauer equilibrium) a cyclic mechanism has been suggested [9] in which the ability of the carbonyl group to accept the hydride transfer is enhanced by the close proximity of a metal atom (IV). The favourable effect of the

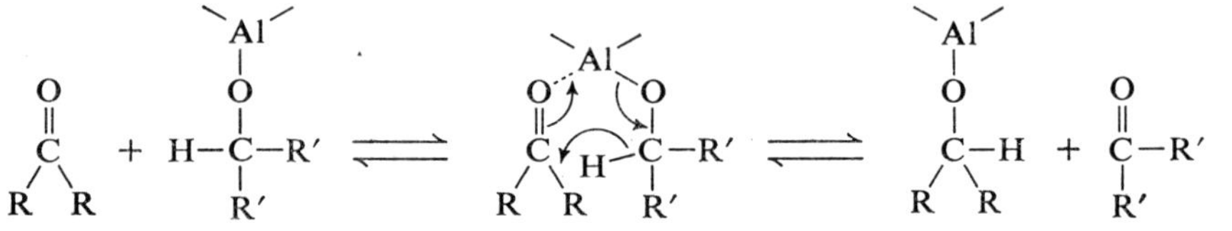

(IV)

presence of a metal ion on a hydride transfer to a carbonyl has been invoked in suggested mechanisms for some enzymic reductions (see page 116).

An intramolecular hydride transfer involving entirely different groups has been proposed to explain the decomposition of the diazonium compound (V) which yields the indole (VII) [10]. The major contribution to

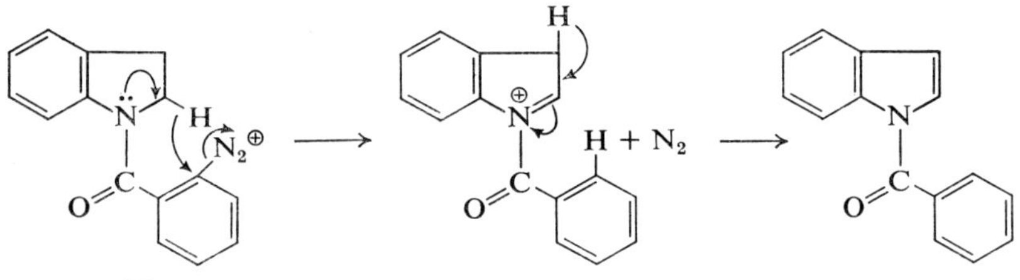

(V) (VI) (VII)

the driving force for this reaction is the extremely powerful, electron-attracting diazonium group which leaves as molecular nitrogen. The process is envisaged as involving two stages; first, the hydrogen transfer, and secondly the loss of a proton to produce the fully-aromatic indole system.

Evidence for the direct, intramolecular nature of the hydrogen transfer was provided in two ways. When the reaction was carried out in $D_2O$, no

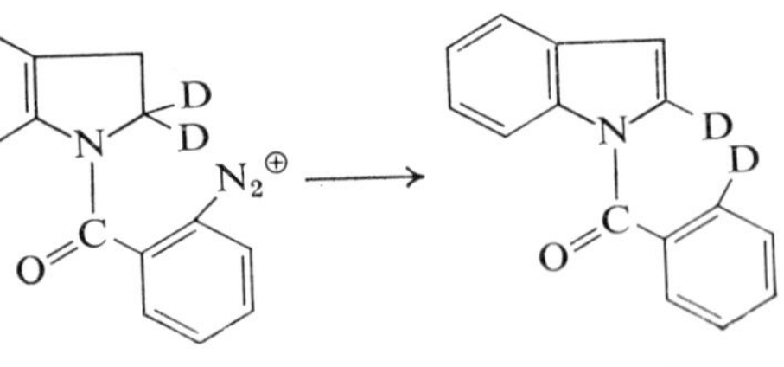

(VIII) (IX)

deuterium was detected in the product molecule. The diazonium compound (VIII) was prepared in which the C-2 position of the indoline ring carried two deuterium atoms. One of these two deuterium atoms was shown to have been transferred to the benzene ring in the product (IX) of the decomposition.

This is an example of a hydride transfer from a hydro-aromatic ring which, by means of the transfer, is able to become more nearly (and eventually fully) aromatic. In practical terms this reaction is not reversible because the nitrogen molecule is completely lost, and also because the positive charge on the nitrogen atom in (VI) is lost when the indole ring becomes fully aromatic. This positive charge would be essential for the reverse process to occur. An example of a hydride transfer in the reverse direction, that is, towards a heterocyclic aromatic ring which possesses a positively charged hetero-atom is provided by the reduction of quinoline methiodide by lithium aluminium hydride [11].

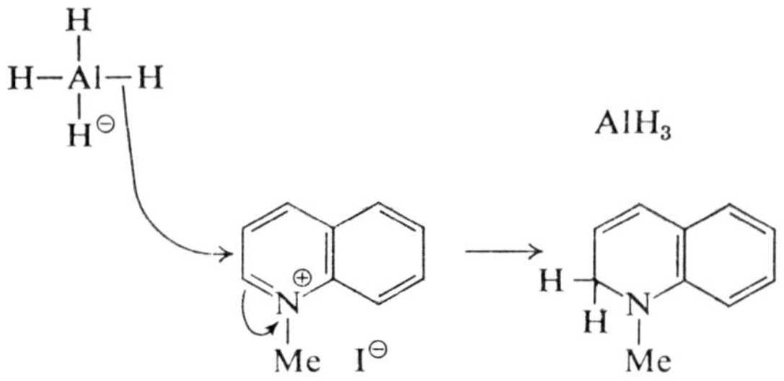


A hydride transfer reaction of great interest was reported by Abeles, Hutton and Westheimer [12]. This involves the reduction of thioketone by a reduced pyridine compound. The direct nature of the hydrogen transfer was demonstrated by deuterium labelling:

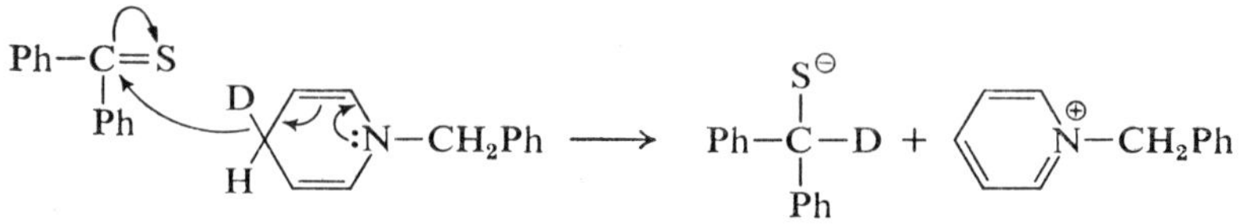


The importance of this reaction lies in that the reducing agent is very similar to the pyridine nucleotide coenzymes which are essential for some enzymic oxidation–reduction reactions.

Many other reactions in organic chemistry, including some which involve Grignard reagents, derivatives of formic acid and some carbonium ions, are generally believed to proceed by similar mechanisms [13, 14, 15].

## Oxidative Elimination

Although removal of a proton from an organic molecule does not in itself constitute an oxidative process, a proton transfer may be involved in an elimination which leads to oxidation. To a certain extent we may relate such oxidations with the electron-transfer concept of oxidation–reduction, in that the leaving group departs having acquired two extra electrons. An example of this type of elimination is the decomposition of benzyl nitrate, the ester of benzyl alcohol and nitric acid, to give benzaldehyde and nitrite. Studies on this reaction indicate that the reaction of benzyl nitrate with ethoxide proceeds via the elimination process shown [16].

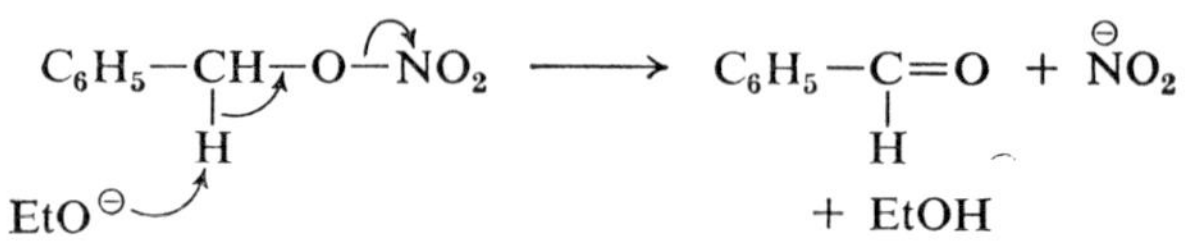

Several types of ester are able to undergo similar eliminations. In general an ester R—$CH_2OZ$ can react in this way if the species $Z^{\ominus}$ is more stable than $OZ^{\ominus}$. An example which finds great application in preparative organic chemistry is the oxidation of alcohols to aldehydes by chromic acid. Westheimer and his group [17] suggested that the oxidation of isopropyl alcohol by this reagent involves formation of a chromate ester followed by an elimination. The resultant Cr(IV) species is able to participate

$$(CH_3)_2CHOH + HCrO_4^{\ominus} + 2H^{\oplus} \longrightarrow (CH_3)CHOCrO_3H_2^{\oplus} + H_2O$$

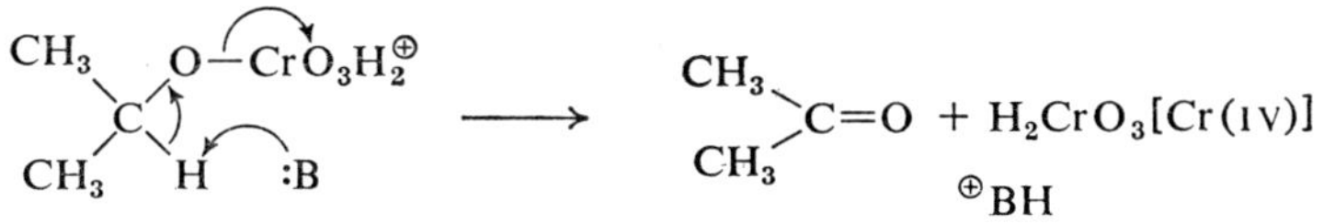

further in the oxidation. Other compounds, such as hypochlorite and bromate esters, are apparently able to undergo similar eliminations [18].

## One- and Two-Electron Transfers

Since reductions may be considered in terms of addition of electrons to a compound, it is at least feasible that certain oxidation–reduction reactions actually do involve direct transfer of electrons. It is, however, extremely difficult to distinguish experimentally between an electron transfer and a transfer of a hydrogen radical, especially for reactions which proceed in

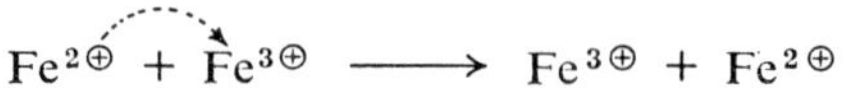

$$Fe^{2\oplus}\cdots OH_2 + H_2O\cdots Fe^{3\oplus} \xrightarrow{\text{H}\cdot\text{transfer}} Fe^{2\oplus}\cdots OH + H_3O\cdots Fe^{3\oplus}$$

$$\Big\downarrow \text{H}^{\oplus}\text{ transfer}$$

$$Fe^{3\oplus}\cdots OH_2 + H_2O\cdots Fe^{2\oplus}$$

certain solvents, including water. This is the case even for the apparently simple exchange between ferrous and ferric ions, for which, in addition to a one-electron transfer, a hydrogen radical transfer involving solvent molecules may be envisaged [19].

Reductions of organic compounds by sodium in liquid ammonia are often considered as involving addition of electrons. Thus, the reduction of naphthalene by this reagent may be formulated as shown. Addition of one electron gives the monoanion radical. As a result of resonance† the

$$Na \longrightarrow Na^{\oplus} + e^{\ominus}$$

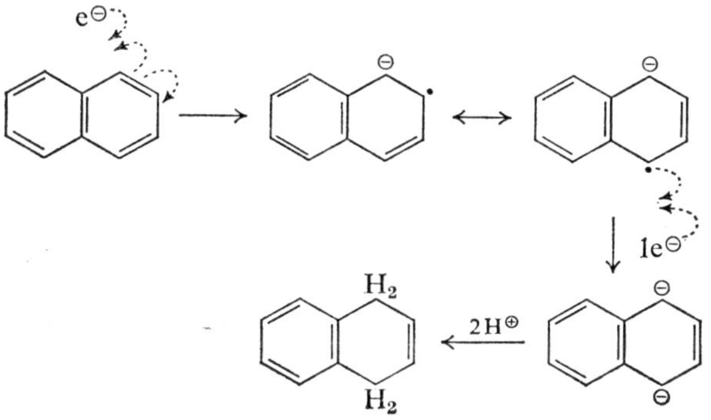


unpaired electron becomes distributed between the ortho- and para-positions. The radical anion takes up a second electron and produces a dianion. Under the appropriate conditions this dianion combines with two protons and gives 1,4-dihydronaphthalene.

Such a process, involving as it does two successive one-electron addition steps, proceeds via an intermediate free radical. The ferricyanide ion, which is a one-electron oxidizing agent, also gives rise to radical intermediates. Thus, the oxidation by ferricyanide of the phenolate anion which leads to dimerization of the phenol involves as a first step the abstraction of one electron. Whether a direct electron transfer is involved is not known.

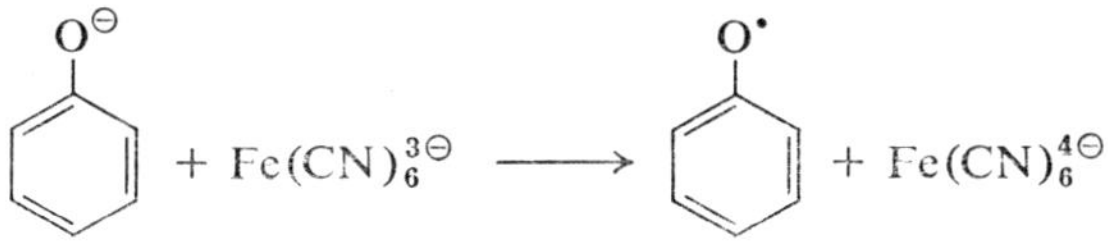


The dimerization may be ortho–ortho-, para–para- or ortho–para [20].

† For a detailed description of resonance see p. 21 of reference [2] from chapter 1. The double-headed arrow used in the diagrams indicates that the true structure of the compound is intermediate between those shown, possessing some of the characteristics of each one.

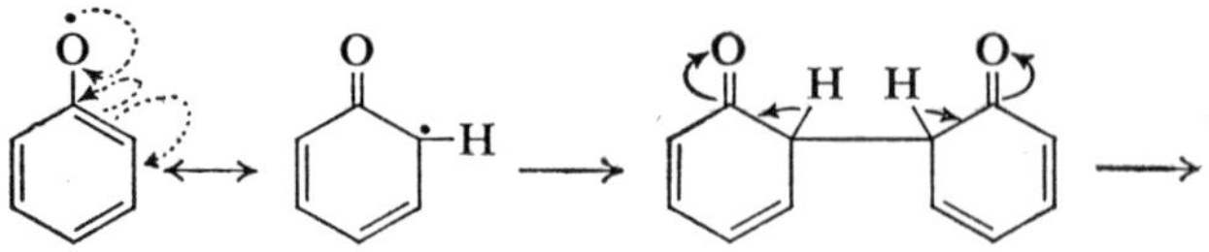


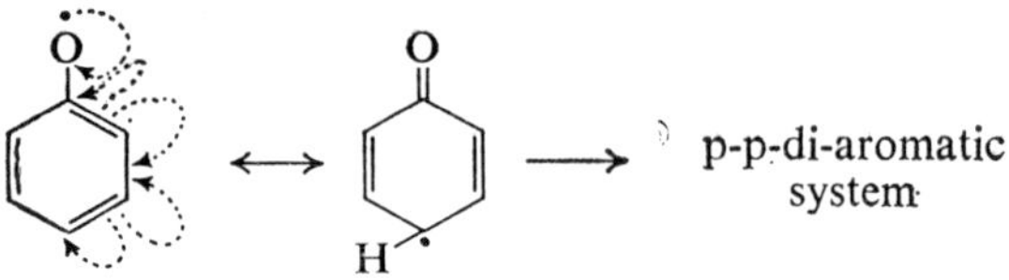


Radicals are also produced when hydroquinone anions react with ferricyanide but in this case the radicals may be relatively stable. Known as semiquinones, such species (X) are the intermediates in the two-electron conversion of hydroquinones into quinones [21].

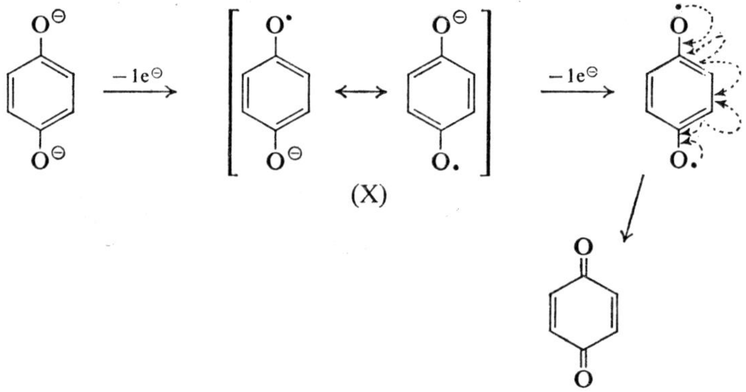

These processes are reversible and therefore we see that semiquinones may function as both acceptors and donors in one-electron processes. This is further shown by the fact that a one-electron transfer may take place between two molecules of semiquinone to produce one molecule each of quinone and hydroquinone [21].

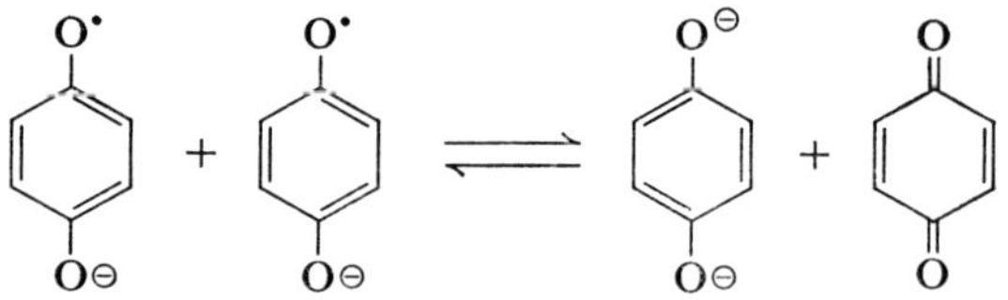

Oxidations involving molecular oxygen itself almost invariably proceed by radical processes. A two-electron reduction of oxygen leads to the production of hydrogen peroxide or its anions, whereas a four-electron reduction is required to produce water. In biological systems the reduction of oxygen to water often requires the two stages and hydrogen peroxide is usually produced at the intermediate stage.

The non-biological reaction of alkanes with oxygen is usually initiated by a process which leads to production of a radical, which then reacts with oxygen. A chain reaction follows when the ROO· radical removes a

$$\text{R—H} \xrightarrow{-\text{H}\cdot} \text{R}\cdot + \text{O}_2 \rightarrow \text{R—O—O}\cdot \xrightarrow{\text{R}-\text{H}} \text{ROOH} + \text{R}\cdot \rightarrow \text{etc.}$$

hydrogen atom from a new R—H molecule [22]. One of the products of this sequence is the hydroperoxide ROOH in which the oxygen has suffered a two-electron reduction. The initiation process may be brought about by various factors including ionizing radiations, high temperatures or the presence of adventitious radicals. The oxygen molecule itself may be able to initiate radical chain processes [23, 24].

Anions of phenols, enols and mercaptans react with molecular oxygen to give radicals in the first instance, but little is known concerning the precise mechanisms [25]. We may formulate the reaction as shown.

$$2\text{Z}^{\ominus} + \text{O}_2 \rightarrow 2\text{Z}\cdot + \text{O}_2{}^{2\ominus} \text{ (peroxide dianion)}$$

In some cases the Z· radicals produced dimerize, although this is not the only possible fate for a radical. The oxidation of the anionic form of ascorbic acid (XI) by oxygen leads to production of a diketone (XII). This reaction is catalysed by the presence of cupric ions, as is also the some-

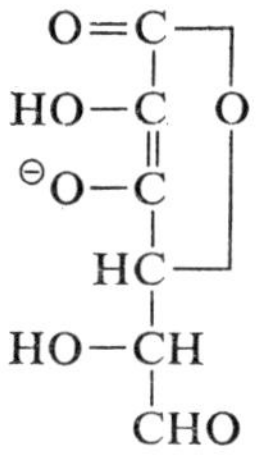

(XI)

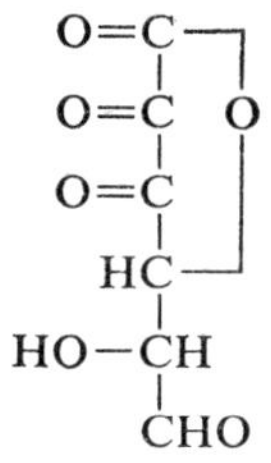

(XII; dehydroascorbic acid)

what similar oxidation of hydroquinone to quinone by oxygen. A suggested mechanism for this catalysis includes the reduction of the cupric to the cuprous ion [26] which then reacts with oxygen.

$$\text{H}_2\text{A} \rightarrow \text{HA}^{\ominus} + \text{Cu}^{2\oplus} \rightarrow \text{HA}\cdot + \text{Cu}^{\oplus}$$

$$2\text{Cu}^{\oplus} + \text{O}_2 + \text{H}^{\oplus} \rightarrow 2\text{Cu}^{2\oplus} + \text{H}_2\text{O}_2$$

Another oxidative process which almost certainly involves radicals and which has some significance in the chemistry of living organisms is the hydroxylation of aromatic rings. The most well-known system for bringing about this type of reaction is Fenton's reagent [27], which consists of ferrous ions and hydrogen peroxide. It is well established that this system functions through the production of hydroxyl radicals produced apparently by a one-electron transfer from the ferrous ion [28].

$$Fe^{2\oplus} + H_2O_2 \rightarrow Fe^{3\oplus} + \cdot OH + OH^{\ominus}$$

The reaction of this radical with the aromatic ring occurs via the formation of an adduct (XIII) which may be converted into the phenol by a one-electron oxidation by either $Fe^{3\oplus}$ or oxygen (produced by the ferrous- or ferric-ion catalysed decomposition of the hydrogen peroxide). The ·OH radical has the properties of an electrophile, and this is reflected in the distribution of isomers obtained when a substituted aromatic ring is hydroxylated.

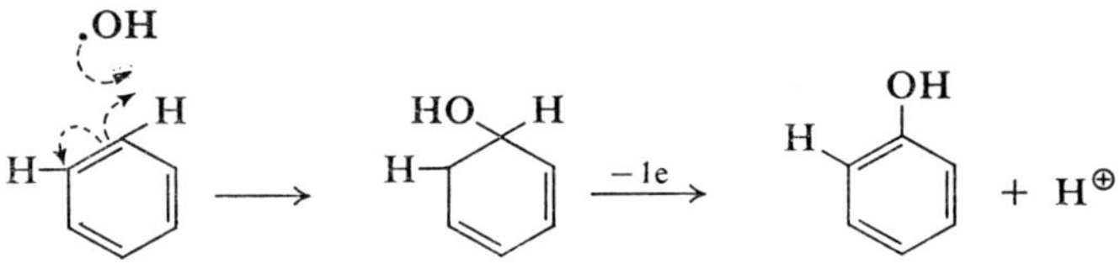


(XIII)

Another system which has received great attention recently is that of Udenfriend *et al.* [29], in which hydroxylation is accomplished by a mixture of ferrous ions, EDTA, ascorbic acid and molecular oxygen. The oxygen can be replaced by hydrogen peroxide and Udenfriend suggested that hydrogen peroxide is an intermediate in the reaction. If this were the case, then the system would be little more than a modification of Fenton's reagent. However, it is unlikely that the reaction involves hydroxyl radicals because it can produce a different distribution of isomers from that produced by Fenton's system [28]. Staudinger *et al.* [30] have suggested that $HO_2\cdot$ radicals are the hydroxylating agents in the Udenfriend system, while Norman and Lindsay Smith [28] favour a mechanism in which a complex of $Fe^{2+}$ and $O_2$ is formed and the hydroxylating agent is an oxygen radical $\overset{\ominus}{O}$—O· which is complexed with a metal ion.

An important feature of the reaction is that an electron donor is required. Staudinger *et al.* [30] have reported that the original ascorbic acid may be replaced by other electron donors.

The Udenfriend system provides an interesting example of the way in which a reagent which changes the redox potential of a single step can influence the overall rate of a reaction. At pH 5·8, the $Fe^{2\oplus}/O_2$ mixture possesses the ability to hydroxylate substrates, but the reaction is extremely slow. On increasing the pH the reaction becomes more rapid; at higher pH values the redox potential of the $Fe^{2\oplus}/Fe^{3\oplus}$ system becomes more negative. When EDTA is added, it complexes with the metal ions and the redox potential drops from +0·77 to +0·14 volts. In addition the rate of hydroxylation increases markedly [30]. Replacement of iron-EDTA with $Cu^{\oplus}$ or $Ti^{3\oplus}$ ($E_0 = +0{\cdot}17$ and $0{\cdot}03$ volts respectively) also gives good hydroxylating activity.

Thus, a change in the oxidation potential of a system can influence the rate of a reaction in which the system is involved. Notice that this implies a relationship between an equilibrium, that is, a thermodynamic, parameter (the redox potential is a measure of the standard free energy change associated with the transfer of an electron from $Fe^{2\oplus}$) and a kinetic parameter (the rate of the reaction reflects the free energy of *activation*). In chapter 1, page 11, we saw that the free energy of activation for a reaction which involves several stages might be a complex function involving the standard free energy changes for all the reaction steps preceding the rate-determining step. A change in the redox potential of a system can therefore be expected to influence the overall energy of activation, and therefore the reaction rate.

## OXIDOREDUCTASES

The Enzyme Commission's classification for enzymes which catalyse oxidations distinguishes the enzymes according to the nature of their substrates and also according to the oxidizing agents or electron acceptors which may be required. A large number of possible electron acceptors are known. Some of the most important are the nicotinamide adenine dinucleotides, flavin mononucleotide and flavin adenine dinucleotide and molecular oxygen. In some cases some synthetic electron acceptors can be used in laboratory studies even though they may not be the physiological acceptors for the enzymes in the living cell.

### Nicotinamide Adenine Dinucleotide (NAD) Enzymes

Many enzyme-catalysed oxidations are absolutely dependent on the presence of a certain co-factor, which was given various names including coenzyme I and diphosphosphopyridine nucleotide (DPN). We shall use

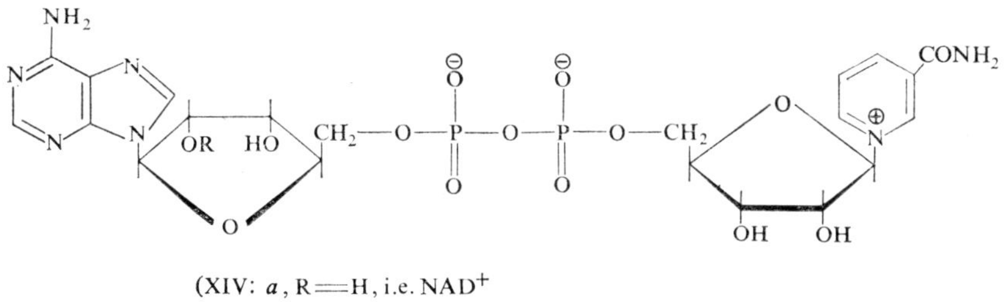


(XIV: *a*, R═H, i.e. $NAD^+$
*b*, R═$PO_3$ i.e. $NADP^+$)

the currently-accepted nomenclature and refer to it by the name in the title. Its structure is shown (XIVa) [31]. During the course of the reaction in which the substrate becomes oxidized, the coenzyme becomes reduced.

In the reduced form only the nicotinamide portion of the molecule has been changed, a dihydropyridine system having been produced (XV) [32]. A closely-related coenzyme is also known (XIVb), nicotinamide adenine

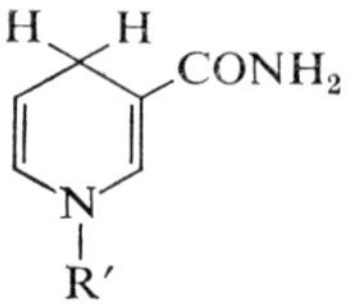

(XV: R′ represents the adenosine diphosphate ribose moiety shown in XIV)

dinucleotide phosphate. It differs from $NAD^+$ in that it carries a phosphate group on the adenosine part of the molecule. We shall refer to the oxidized forms of these two compounds as $NAD^+$ and $NADP^+$, and to their reduced states as NADH and NADPH respectively. Some enzymes are very specific with regard to these coenzymes, requiring either $NAD^+$ or $NADP^+$, while other enzymes are capable of functioning with either compound.

The reduction of the nicotinamide ring in these coenzymes is accompanied by changes in the ultraviolet absorption, the reduced form showing a new absorption at 340 nm. Changes in fluorescence are also observed. These spectroscopic changes may be used to follow kinetically the reactions in which the coenzymes take part.

*Alcohol Dehydrogenases.* There are many enzymes which are capable of catalysing the oxidation of alcohols to carbonyl compounds and many are highly specific with respect to their alcohol substrates. The Enzyme Commission name alcohol: NAD oxidoreductase (alcohol dehydrogenase, 1.1.1.1) actually refers to enzymes from various sources which exhibit fairly broad specificities towards aliphatic primary and secondary alcohols. The enzymes from horse liver and yeast have been studied most intensively and they share many common properties although they are undoubtedly different proteins.

*Stereochemical aspects.* Some extremely important studies have been carried out using isotopically-labelled substrates with yeast alcohol dehydrogenase. When ethanol containing two deuterium atoms on the α-carbon, $CH_3CD_2OH$, was oxidized by $NAD^+$ in the presence of this

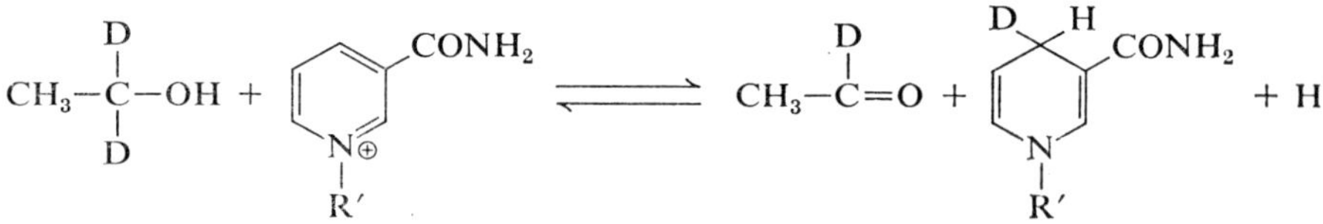

enzyme, it was found that the resulting reduced nicotinamide group contained one atom of deuterium, in the position shown [33]. The reduced deuterated coenzyme was then reoxidized using normal acetaldehyde and the same enzyme. As a result of this reaction the deuterium atom was completely transferred from the coenzyme to the acetaldehyde, giving monodeuteroethanol. This complete transfer of the deuterium atom shows that the enzyme-catalysed reaction is stereospecific with respect to the nicotinamide ring. The two faces of this ring are not sterically equivalent because of the presence of the carboxamide group. The $NAD^+$ becomes bound to the enzyme in such a way that only one face is able to receive the transferred deuterium atom. When this deuterated reduced coenzyme is used for the reverse process, that is, reduction of acetaldehyde, the same face is involved and so the deuterium atom is again transferred. If this were not the case, if the two faces were identical or if the nicotinamide ring were able to rotate and present either face for reaction, then only half an atom of deuterium per molecule would be transferred to the aldehyde.

There are two possible isomers of the reduced deuterated coenzyme. The isomer which is obtained by the above process, that is, by the reduction of $NAD^+$ by deuterated ethanol in the presence of yeast alcohol dehydrogenase, is referred to as 'Form A'. The absolute configuration of this isomer has been elucidated [34] and is shown below (XVI). Thus if the plane of the ring is placed in the plane of the paper with the carboxamide group to the right of the reader, then the deuterium atom lies above the plane and the hydrogen atom on the same carbon lies below it.

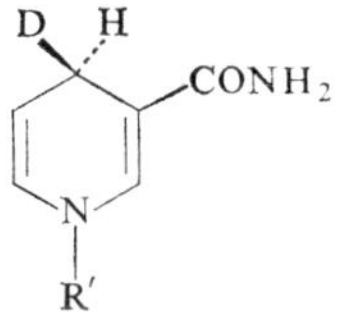


(XVI)

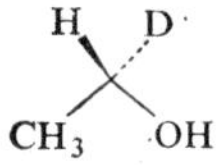


(XVIIa)

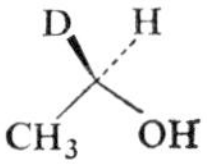


(XVIIb)

The stereospecificity of the reaction is not confined to the nicotinamide ring. It was also demonstrated that the reduction of acetaldehyde to ethanol is subject to strict steric control. Acetaldehyde containing a deuterium atom on the carbonyl carbon was reduced with normal NADH and the monodeuteroethanol (XVIIa) was isolated. Enzymic reoxidation with $NAD^+$ gave acetaldehyde which still possessed the deuterium. Therefore, the same hydrogen atom which is transferred to the aldehyde in the reductive process is removed from it on oxidation. Clearly the enzyme is capable of distinguishing between the two hydrogen atoms on C-1 position of ethanol. Inversion at the C-1 position of the monodeuteroethanol (XVIIa) gave the other isomer (XVIIb) from which the *deuterium* atom was transferred to the $NAD^+$ [35].

It is possible for acetaldehyde (which is planar) to be held on the enzyme so that either of two faces may be presented towards the group from which the transferred hydrogen is donated. These are shown (XVIII and XIX), where the reader views from the direction from which the transferred hydrogen appears. Thus, during the reduction of acetaldehyde by the deuterated reduced coenzyme, the deuterium atom approaches the carbonyl carbon of the acetaldehyde from above the plane of the paper. The isomer (XVIIb) will result only if the molecule is arranged as in (XVIII).

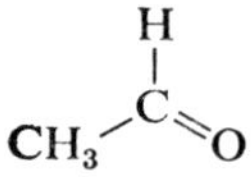

(XVIII)

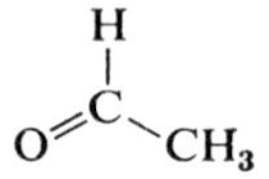

(XIX)

Yeast alcohol dehydrogenase is not alone in exhibiting these types of stereospecificity towards alcohol and coenzyme, demonstrated as described above by deuterium-labelling experiments. The alcohol dehydrogenase from horse liver shares these properties. Thus, reduction of $NAD^+$ with deuteroethanol in the presence of the liver enzyme results in the formation of the same 'A-form' of reduced, deuterated coenzyme [36]. In addition, it is almost certain that the same hydrogen in ethanol is transferred by both the yeast and the liver enzymes [35].

An essential feature of this mechanism is, of course, that the molecules must be so positioned in space that the requirements of the stereospecificity are complied with. The process of catalysis itself probably requires correct alignment of the various interacting molecules. This situation is achieved by means of the binding of the substrate and coenzyme to the enzyme, and a full description of the mechanism of action of these, and indeed all other, enzymes should take this binding into consideration.

*Binding between enzyme, coenzyme and substrate.* Liver alcohol dehydrogenase has a molecular weight of about 80,000 [37] and possesses two binding sites for $NAD^+$ or NADH per molecule. The binding sites are independent of each other and apparently identical, and binding of $NAD^+$ is strictly competitive with that of NADH [38]. The adenosine diphosphate ribose portion of the coenzyme molecule makes an important contribution to the binding of the coenzyme to the enzyme. This is indicated by the fact that adenosine diphosphate ribose is an inhibitor and actually binds to the enzyme more strongly at neutral pH values than does $NAD^+$ itself [39]. In addition, neither nicotinamide methiodide (XX) nor nicotinamide mononucleotide (XXI) is able to function as the coenzyme in the reaction, presumably because they are not bound effectively [40]. The amino-group of the adenine is not essential since deamino-$NAD^+$ functions as well as does $NAD^+$ [41]. Modification of the sugar residue of the adenosine moiety, however, does affect the activity. Thus, $NAD^+$ which contains a deoxyribose unit in place of ribose in the adenosine shows only

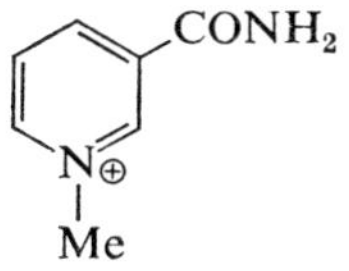


(XX)

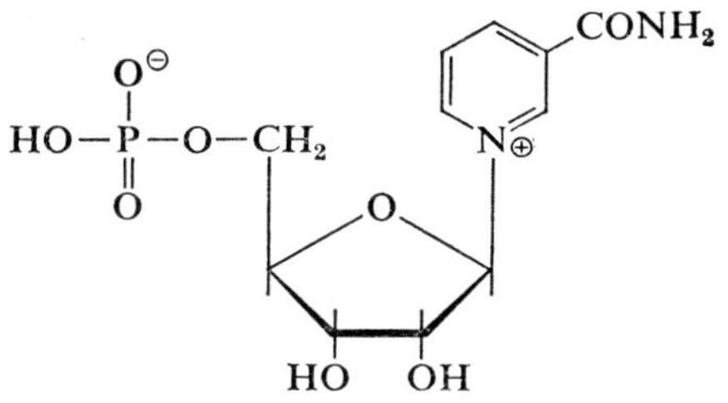


(XXI)

low activity [42], and $NADP^+$ in which that sugar residue is phosphorylated [XIVb], shows only one hundredth of the activity of $NAD^+$ [41]. Since these two modifications are unlikely to influence the ability of the nicotinamide ring to take part in the oxidation–reduction process, they must influence the coenzyme binding.

Since the presence of added phosphate has very little effect on the dissociation constant, it has been deduced that the pyrophosphate bridge does not play any role in the enzyme–coenzyme binding [43].

The steric requirements for the reaction make it probable that the carboxamide group of the nicotinamide moiety becomes bound to the enzyme. This suggestion is supported by evidence from n.m.r. studies of $NAD^+$ and NADH in the presence of liver alcohol dehydrogenase. Hollis [44] reported an apparent intensity decrease in the spectra due to both the nicotinamide and the adenine groups of the reduced and oxidized forms of the coenzyme. This indicates low turnover rates in the association–dissociation equilibria and therefore both groups must be fixed quite strongly to the enzyme.

The ability of the enzyme to distinguish between the two hydrogens on the C-1 of ethanol must mean that the alcohol is held at the active site by at least two positive binding sites. Presumably these interact with the alcohol group itself and with the alkyl side-chain. We may predict therefore that the active site of liver alcohol dehydrogenase includes a hydrophobic region which combines with the alkyl side-chain of the substrate alcohol or aldehyde. This is supported by the fact that fatty acids compete with alcohol for the same binding site on the enzyme, the binding between the fatty acid and the enzyme being stronger as the hydrocarbon chain length of the acid is increased. Similarly if we consider the reverse process, fatty acid amides compete with aldehydes for the same binding region† [43, 45]. Some other aspects of the binding of alcohols are discussed later.

An important feature of this enzyme is that it contains zinc (as $Zn^{2\oplus}$), which is essential for activity. Various estimates of the number of zinc

† An interesting point, which is discussed in some detail by Theorell and Sund [43], is that fatty acids do not compete for the aldehyde binding site, nor do fatty acid amides compete for the alcohol sites.

atoms per molecule of enzyme have been published [46] but the generally accepted view at the present time is that there are four zinc atoms [47]. Vallee *et al.* [48] have suggested that difficulties encountered in the estimation of the zinc content of various preparations were due to the presence of iso-enzymes in different proportions in these preparations. Of the four zinc ions present in the molecule, two are relatively loosely bound and these are essential for catalytic activity. The other two, the more firmly-bound ions, appear to be responsible for holding the enzyme (which is made up of four units of molecular weight 20,000 each) in the required, active conformation [48].

The two zinc ions which are described as being loosely bound and essential for activity are positioned at the active sites, that is, at the sites which combine with the substrate and the coenzyme. That the zinc ions are essential for activity has been demonstrated by the use of two types of metal-ion chelating agents. The very powerful complexing agent sodium diethyldithiocarbamate (XXII) inactivates the enzyme by completely

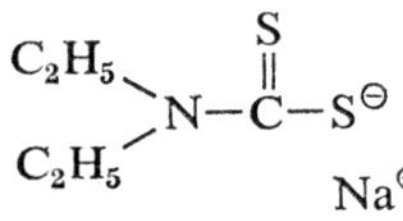


(XXII)

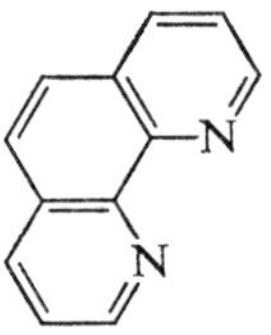


(XXIII)

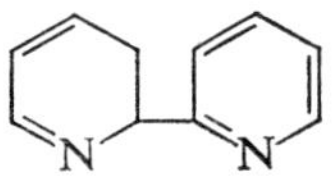


(XXIV)

removing the zinc ions [48]. Weaker complexing agents are able to inhibit the enzyme, not by removing the zinc, but by associating with it and thus occupying the active site. In the resulting inactive complex the zinc ion is bound to both the enzyme and the inhibitor. 1,10-Phenanthroline (XXIII) and o,o-bipyridine (XXIV) are examples of such 'weaker' agents which can form 1 : 1 stoichiometric complexes with each active site of the enzyme [49, 50, 51].

Theorell *et al.* [50] studied kinetically the inhibition of the alcohol dehydrogenase by phenanthroline. They found, in agreement with Coombs and Vallee [49], that inhibition by phenanthroline competed with $NAD^+$ and NADH, but did not compete with acetaldehyde. Some mixed inhibition was observed with alcohol and this suggested that the alcohol was bound at or near the zinc atom in the active site. Support for the suggestion that the alcohol might be very close to the zinc atom, if not actually bound to it, was provided by Sigman [51], who studied the reaction of the enzyme with bipyridine. Formation of the enzyme–bipyridine complex leads to an increase in ultraviolet absorption at 308 nm. In the presence of substrates (alcohols and aldehydes) this absorption was diminished indicating competition of the substrate with the bipyridine for the zinc ion (bipyridine is a weaker complexing agent than is phenanthroline).

The binding of the substrates to the enzyme, as determined by their effect on the absorption of the enzyme–bipyridine complex, increased as the hydrocarbon-chain length was increased, a result which corresponds to a certain extent with the substrate specificity of the enzyme. Mercaptans were found to be bound to the enzyme more strongly than were alcohols, and since it is known that thiols are bound more strongly to metal-ions than are alcohols, this suggests that in the case of this enzyme the alcohol is bound to the zinc ion.

Adenosine diphosphate ribose does not compete with either bipyridine or phenanthroline for the zinc ion and this eliminates the possibility that the corresponding section of the coenzyme is bound to the enzyme via a zinc atom bridge [51, 52]. As far as $NAD^+$ is concerned, the nicotinamide portion of the coenzyme cannot be coordinated to the zinc ion; Sigman has shown that the carboxamide group is not involved with the zinc [51], Kosower states that interaction of the zinc ion with the carboxamide group is incompatible with spectral data [53], and the positively charged pyridine ring is more likely to be repelled by the metal-ion than attracted by it [39, 43]. However, the nicotinamide ring must be quite close to the zinc ion in the enzyme–coenzyme complex and this may explain the competitive inhibition of coenzyme binding by the complexing agents referred to above. It may also serve to explain spectra data reported by Kosower [53].

In summary then, these results appear to indicate (a) that the coenzyme is bound directly to the enzyme via the adenosine diphosphate ribose and carboxamide moieties but is not coordinated to the zinc ion, (b) that the aliphatic chain of an alcohol substrate is located in a complementary site on the enzyme, and (c) that the hydroxyl group of the alcohol is close to the zinc ion, and probably coordinated to it. The carbonyl group of an aldehyde substrate is probably also coordinated to the zinc. The natures of the groups in the enzyme which are involved in the binding of the zinc ion have not yet been established.

*Kinetic studies*. The kinetic scheme originally proposed by Theorell and Chance [54] for the liver enzyme-catalysed oxidation of alcohol has been investigated thoroughly, and is now well established. The scheme is shown in reactions 3.5 to 3.7,

$$E + O \underset{k_2'}{\overset{k_1'}{\rightleftharpoons}} EO \text{ (equilibrium constant } K_{EO}) \tag{3.5}$$

$$EO + Alc \underset{k_3}{\overset{k_3'}{\rightleftharpoons}} ER + Ald + \mathrm{H}^{\oplus} \tag{3.6}$$

$$ER \underset{k_1}{\overset{k_2}{\rightleftharpoons}} E + R \text{ (equilibrium constant } K_{ER}) \tag{3.7}$$

where $E$ = liver alcohol dehydrogenase and $O$, $R$, $Alc$ and $Ald$ refer to oxidized and reduced coenzyme, alcohol and aldehyde respectively. From equation (3.8) which is obtained from the above scheme by means of a steady-state treatment, it is possible to obtain experimental values for individual rate constants. In equation (3.8), $e$ represents the total

$$\frac{e}{v} = \frac{1}{k_2} + \frac{1}{k_1'[\mathrm{O}]} + \frac{1}{k_3'[\mathrm{S}']} + \frac{k_2'}{k_1'[\mathrm{O}]k_3'[\mathrm{S}']} \tag{3.8}$$

enzyme concentration, $S'$ alcohol and $v$ the reaction velocity. Dalziel has used a modification of this equation [55].

In the Theorell–Chance mechanism for the oxidation of alcohols the rate-determining step is the final one, the dissociation of the enzyme-reduced coenzyme complex ($k_2$). Thus, when different alcohols are oxidized, the reactions should all have the same maximal velocities (at high substrate and coenzyme concentrations). Theorell and Bonnichsen [38] confirmed this experimentally for n-propanol, n-butanol and allyl alcohol. Similarly, even when maximal velocity conditions do not apply, the experimentally-determined values for $k_1$, $k_1'$, $k_2$ and $k_2'$ should be independent of the nature of the alcohol or aldehyde; a point which Dalziel has confirmed for ethanol, acetaldehyde, butanol and butyraldehyde [56]. Further support was obtained for the scheme when the equilibrium constants $K_{EO}$ and $K_{ER}$ as obtained by equilibrium studies were shown to be in good agreement with values obtained from measurements of individual rate constants [57, 58, 38].

The mechanism requires a compulsory order in the construction of the ternary complex. The coenzyme binds to the enzyme and then the substrate is incorporated. This does not mean that an enzyme–alcohol complex may not form, but if it does it probably does not lead to reaction. Recent work by Theorell *et al.* has suggested a reason for this compulsory order at least for the reverse reaction [59]. The intensity of the fluorescence of NADH is increased when the enzyme–NADH complex is formed. It was found that the formation of this complex does not involve simply a single-stage process, but that an intermediate, $ER_0$, is produced first (3.9). In

$$\mathrm{E} + \mathrm{R} \underset{k_{-1}}{\overset{k_{+1}}{\rightleftharpoons}} \mathrm{ER}_0 \underset{k_{-1}}{\overset{k_{+2}}{\rightleftharpoons}} \mathrm{ER} \tag{3.9}$$

this intermediate the fluorescence of the NADH is unaltered. It appears that $ER_0$ undergoes a conformational change which brings the reduced nicotinamide group into a new position so that the fluorescence is enhanced. In this position the nicotinamide group can take part in the hydrogen transfer process. This conformational change is thought to produce the true active site with which the substrate may combine, thus explaining the compulsory order. The coenzyme therefore appears to function not only as a reactant but also as a type of allosteric activator. This type of conformational change provides an explanation for the earlier observation by Theorell, Chance and Yonetani [60] that the appearance of fluorescence following the combination of NADH with the *crystalline* enzyme was much slower than that when the enzyme was in solution. The $ER_0$ complex in the crystalline state may be prevented by the rigidity of the system from undergoing the conformation change which results in the increased fluorescence.

The Theorell–Chance mechanism appears to be at least compatible with all the data available concerning liver alcohol dehydrogenase. It does not apply, however, to the yeast enzyme for which a different kinetic scheme is required. For example, there is no compulsory order for the binding of the substrate and coenzyme to the enzyme. A rapid equilibrium is set up with the formation of all the possible complexes, and the rate-determining step appears to be the interconversion of the two ternary complexes. Detailed studies of the kinetic scheme (3.10) have been made [61].

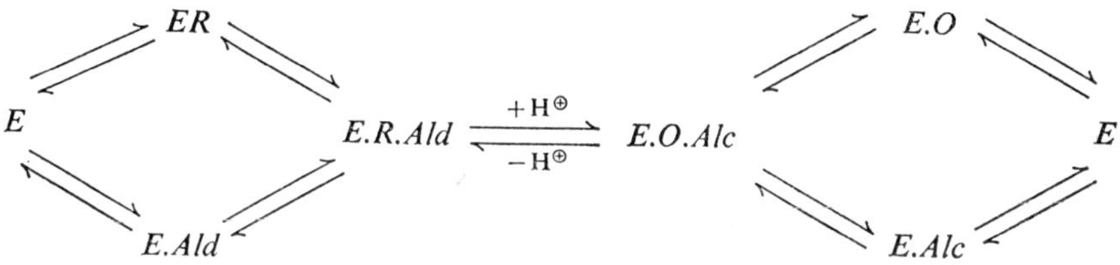


(3.10)

*The mechanism of the reaction.* The deuterium transfer experiments which have been described appeared to indicate that the transfer of hydrogen between $NAD^+$ and alcohol, and that between NADH and aldehyde, was direct and probably involved a hydride species. The similarity between these reactions and the hydride transfers described earlier in the chapter is noticeable. A direct hydride transfer forms the basis of most suggested mechanisms for liver alcohol dehydrogenase [43], and on the whole these are similar to the sequence of events which is summarized as follows. First, an enzyme–coenzyme complex is formed through the binding of the ADPR portion and the carboxamide group to the enzyme. Because of the latter, the nicotinamide ring is held so that the correct face is presented to the alcohol molecule from which it will receive the hydride. In this position the pyridine ring may be quite close to the zinc ion. Next, the alcohol becomes bound to the enzyme–coenzyme complex through its hydroxyl group which probably becomes a ligand to the metal-ion. At this stage a proton may be lost from the hydroxyl group. The alcohol is also bound through its aliphatic side-chain (methyl group) which has an affinity for a hydrophobic region in the enzyme. The alcohol, too, is held in the correct steric position. When the ternary complex has been formed, the hydride transfer takes place (3.11). The aldehyde then dissociates away from the complex, leaving the species *ER* in which the reduced coenzyme is bound strongly to the enzyme. The conformational change giving $ER_0$ then occurs and eventually the reduced coenzyme, too, dissociates away. The role of the zinc ion in this mechanism may be simply to provide a binding site for the alcohol. In the reverse direction the presence of the cation would certainly be expected to favour the reaction, just as the aluminium ion is considered to favour the Meerwein–Ponndorf–Oppenhauer reaction (see page 100).

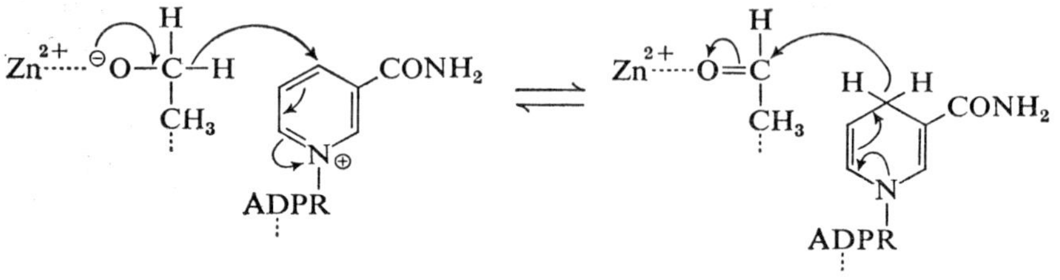

(3.11)

Most of the mechanisms proposed for the NAD-enzymes are in essence similar to that summarized in equation (3.11). However, ideas that a single direct transfer was involved were challenged at least for the yeast enzyme when Schellenberg reported that during the enzyme-catalysed oxidation of tritium-containing ethanol some tritium became incorporated into the protein itself [62]. That this incorporation was related to the enzymic activity was suggested by experiments in which (a) enzyme inactivated by boiling was used, (b) the alcohol used was labelled with tritium in the non-transferred position (see page 109), (c) $NAD^+$ was absent. In each of these three cases there was insignificant incorporation of tritium into the enzyme.

Schellenberg was able to show that the incorporated tritium was carried by the methylene group in the side-chain of a tryptophanyl residue, and on this basis proposed that the transfer of hydride between substrate and coenzyme was in fact indirect, proceeding via a tryptophanyl residue. Schellenberg's proposed mechanism was essentially as shown in (XXV to XXVII).

Thus, the hydride transfer to the $NAD^+$ is envisaged as being brought about by a flow of electrons which originates at the nitrogen of the indole ring of tryptophan (XXV) and terminates at the positively-charged nitrogen in the nicotinamide group. An indolenine (XXVI) is thereby produced. The reduction of this indolenine is brought about by a second hydride transfer, this time from the alcohol (XXVII). It is this process which leads to the labelling of the tryptophan in the protein. It is known that the relevant position on indolenines is susceptible to nucleophilic attack [63]. The hydride transfer shown in (XXV) is somewhat similar to that shown in (V) (see page 100), the tryptophan molecule possessing an extra double bond and therefore representing a more conjugated system.

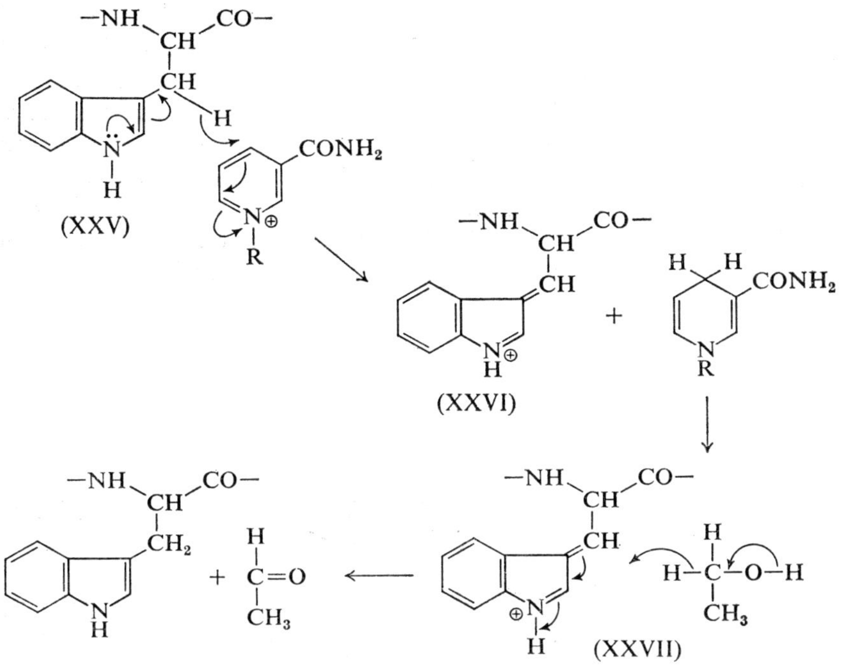

(XXV) (XXVI) (XXVII)

More recent reports, however [62a], have suggested that the incorporation of tritium into the side-chain of a tryptophanyl residue may not necessarily be associated with the enzymic process itself, but may in fact be a typical reaction of tryptophanyl residues in any non-specific protein.

In view of these reports we shall continue to regard the reaction as involving a direct transfer between substrate and coenzyme until more evidence is available.

*Substrate specificity.* Although both yeast and liver alcohol dehydrogenases catalyse the oxidation of wide ranges of alcohols, the yeast enzyme exhibits a narrower specificity. This is clearly shown when the actions of the two enzymes with the stereo-isomers of butan-2-ol (XXVIII and XXIX) are compared. Both isomers of this alcohol are oxidized by the $NAD^+$ in the presence of the liver enzyme, but the yeast enzyme is specific for the (+)-enantiomer. The lack of stereospecificity demonstrated by the enzyme from liver does not, however, prevent it from distinguishing between the two hydrogens on the C-1 position of ethanol. Dickinson and Dalziel [35] explained this phenomenon on the basis of the structures of the active sites of the two enzymes. Thus, there are two binding sites for an alcohol substrate in the active site of the liver enzyme. These accommodate the methyl and the hydroxyl group in ethanol and are represented in diagram (XXX) as (1) and (2), respectively. The deuterium atom which is transferred selectively from mono-deuteroethanol is shown. The other group carried by the central carbon atom (a hydrogen in the case of ethanol) occupies a space which is designated as (3). A primary alcohol will always take up this position on the enzyme because (1) is a positive binding site for aliphatic side-chains. In the case of the secondary alcohol butan-2-ol two possibilities arise. For (+)-butan-2-ol the ethyl group is bound to site (1) and the methyl group occupies space (3) (XXVIII). The reverse situation applies for (—)-butan-2-ol (XXIX). Since both of these isomers are oxidized in the presence of the liver enzyme it is clear that the region (3) is able to accommodate the ethyl group. This is also indicated by the

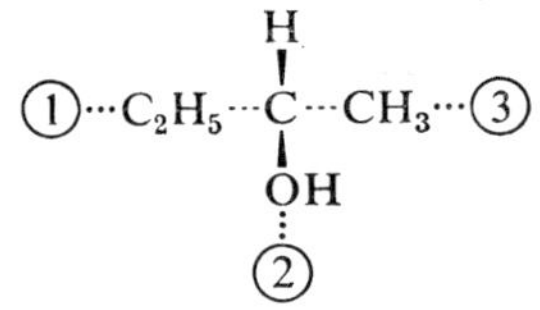

(XXVIII)
(+)-butan-2-ol

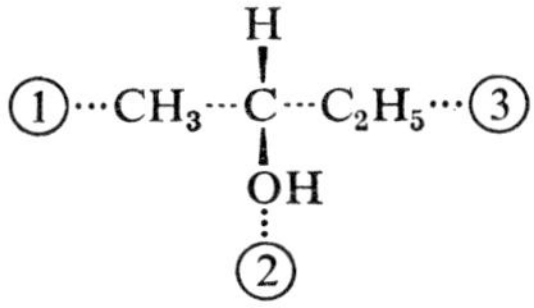

(XXIX)
(—)-butan-2-ol

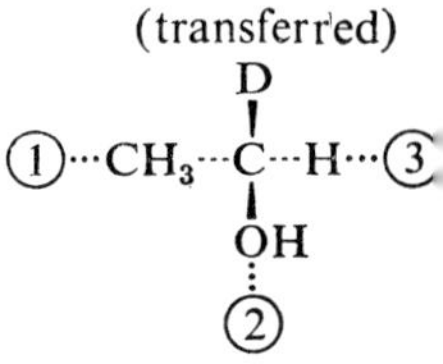


(XXX)

fact that cyclohexanol is a substrate for this enzyme [64, 65]. However, the *yeast* enzyme cannot catalyse the oxidation of the (—)-enantiomer and this must mean that the region (3) in the active site of this enzyme cannot accommodate the ethyl group. It presumably can accept no group larger than a methyl group. This explanation would predict that yeast alcohol dehydrogenase cannot catalyse the oxidation of any secondary alcohol in which both alkyl groups are larger than methyl. Neither cyclohexanol [64] nor hexan-3-ol [66] are, apparently, substrates for this enzyme while they can function as such for the liver enzyme [67].

The term 'alcohol dehydrogenase' is used to refer to those enzymes which catalyse the oxidation of a fairly broad range of primary and secondary alcohols. However, most $NAD^+$- or $NADP^+$-requiring enzymes which bring about the oxidation of hydroxy- to carbonyl-compounds are very specific. Thus, in addition to the alcohol dehydrogenases (alcohol: NAD oxidoreductase, 1.1.1.1; alcohol: NADP oxidoreductase, 1.1.1.2) the Enzyme-Commission lists sixty-six specific enzymes according to the hydroxy compounds which serve as their substrates. In comparison with normal organic chemistry some of the specificities manifested by these enzymes are indeed remarkable. Thus, two lactate dehydrogenases are known, one (1.1.1.27) oxidizes only L-lactate to pyruvic acid and the other (1.1.1.28) oxidizes only D-lactate to the same product. Again, the oxidation of 3-α-hydroxysteroids is catalysed by one enzyme (1.1.1.50) while 3-β-hydroxysteroids are oxidized in the presence of another (1.1.1.51). Of the five hydroxyl groups in ribitol only one, that on C-2, becomes oxidized by the action of ribitol: NAD oxidoreductase (1.1.1.56). These remarkable specificities must arise from the unique nature of the active site of each enzyme, which is able to bind in a specific way the appropriate substrate. The stereospecificities demonstrated must be due to the presence of asymmetric binding sites and to the very strong enzyme–substrate binding which prevents the substrate presenting anything but a particular side to the group which is responsible for the hydrogen transfer.

Apart from these specificities it is very likely that the mechanisms of action of these enzymes are essentially similar. They all involve primary or secondary alcohols and nicotinamide coenzymes. If a deuterated substrate is used a deuterated reduced coenzyme is obtained, and this is usually stereochemically pure, although not all enzymes give the 'A-form' (see Table 3.2). All of the NAD-dependent enzymes appear to require the presence of metal-ions and of thiol groups in the protein in their reduced state.

Table 3.2. *Some examples of stereospecificity in the reduction of NAD and NADP*†

| |
|---|
| Enzymes producing Form 'A' |
| Alcohol dehydrogenase (1.1.1.1) from liver or yeast (NAD) |
| L-Lactate dehydrogenase (1.1.1.27) from heart (NAD) |
| D-Lactate dehydrogenase (1.1.1.28) from bacteria (NAD) |
| Glycerate dehydrogenase (1.1.1.29) from spinach (NAD) |
| Malate dehydrogenase (1.1.1.37) from heart or wheat germ (NAD) |
| Dihydroorotate dehydrogenase (1.3.3.1) from bacteria (NAD) |
| Aldehyde dehydrogenase (1.2.1.3) from liver (NAD) |
| Isocitrate dehydrogenase (1.1.1.42) from heart (NADP) |
| Enzymes producing Form 'B' |
| Glycerolphosphate dehydrogenase (1.1.1.8) from muscle (NAD) |
| Glucose-6-phosphate dehydrogenase (1.1.1.49) from yeast (NADP) |
| Glucose dehydrogenase (1.1.1.47) from liver (NAD) |
| 3-α-Hydroxysteroid dehydrogenase (1.1.1.50) from bacteria (NAD) |
| Phosphogluconate dehydrogenase (1.1.1.44) from yeast (NADP) |
| Glyceraldehydephosphate dehydrogenase (1.2.1.12) from yeast or muscle (NAD) |

† Taken from ref. [3].

*Oxidative decarboxylations.* In some cases, an enzyme may catalyse more than just an oxidative process. For example, while L-malate: NAD oxidoreductase (1.1.1.37) simply brings about the oxidation of L-malate to oxaloacetate, another enzyme (malate dehydrogenase, decarboxylating, 1.1.1.38) brings about its conversion into pyruvate and carbon dioxide, in other words a decarboxylation is also involved. Similarly, isocitrate serves as substrate for an enzyme (isocitrate dehydrogenase, 1.1.1.41) which effects both oxidation and decarboxylation. Notice that decarboxylation involves neither oxidation nor reduction.

Two possibilities exist for the reaction pathway in such a case. The hydroxy-compound may be oxidized and the resulting $\beta$-ketoacid may subsequently undergo decarboxylation. Alternatively, the two processes, decarboxylation and oxidation, occur simultaneously. As far as malate dehydrogenase (1.1.1.38) is concerned the two-stage mechanism cannot be ruled out because it has been shown that the decarboxylation of oxalo-acetate, which is the product of the oxidation step, is catalysed by the enzyme [68]. That no free oxaloacetate has been detected as an intermediate does not disprove the two-step mechanism since the oxaloacetate produced may at all times be very tightly bound to the enzyme [69, 70]. The enzyme requires a metal-ion ($Mn^{2\oplus}$ or $Mg^{2\oplus}$) for activity and a role for this metal-ion in the mechanism of the decarboxylation may be suggested (see chapter 1).†

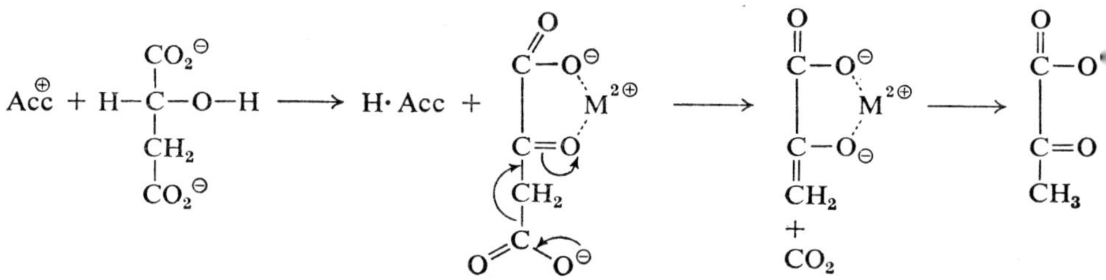

Another decarboxylating malate dehydrogenase, however, has been shown *not* to catalyse the decarboxylation of oxaloacetate (1.1.1.39) [71]. In addition, this enzyme did not catalyse the incorporation of $^{14}CO_2$ into oxaloacetate (the reverse direction) although incorporation into malate was detected. For this enzyme, therefore, the oxidative decarboxylation of malate probably proceeds through a one-step process, and the simultaneous loss of $CO_2$ and transfer of hydride may follow the mechanism shown below:

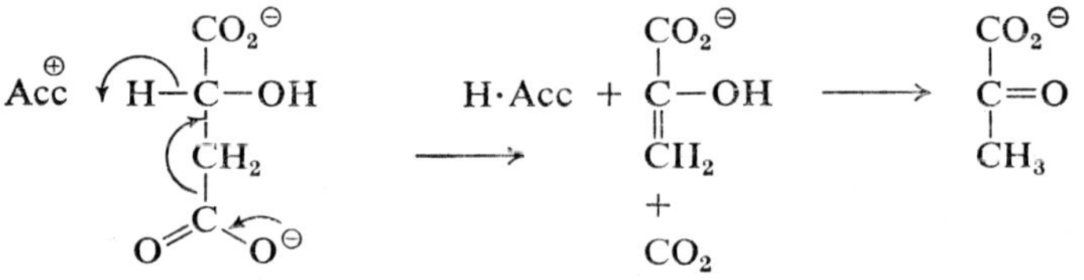

A similar situation may be envisaged for an isocitrate dehydrogenase (decarboxylating) from yeast (1.1.1.41) which has been reported not to catalyse the decarboxylation of oxalosuccinate [72].

† In representing the mechanisms by which these reactions take place we shall use the symbol $Acc^{\oplus}$ to refer to the group which accepts a hydride from the species which suffers oxidation. $Acc^{\oplus}$ therefore probably represents $NAD^+$ but in view of Schellenberg's work it is not impossible that in some cases at least $Acc^{\oplus}$ represents an indolenine species.

*Oxidation of aldehydes.* In addition to the oxidation of alcohols, liver alcohol dehydrogenase also brings about the oxidation of formaldehyde to formic acid [73]. In the presence of methanol, methyl formate is formed. The ability of this enzyme to catalyse the oxidation of two apparently different groups, alcohol and aldehyde, is not necessarily an example of great versatility on its part. The reactions are essentially very similar because in aqueous solution formaldehyde exists predominantly in its hydrated form which is, in effect, a primary alcohol. In the presence of methanol, the hemiacetal is formed. The oxidation proceeds as for a normal alcohol. Similarly, the oxidation of D-glucose-6-phosphate to

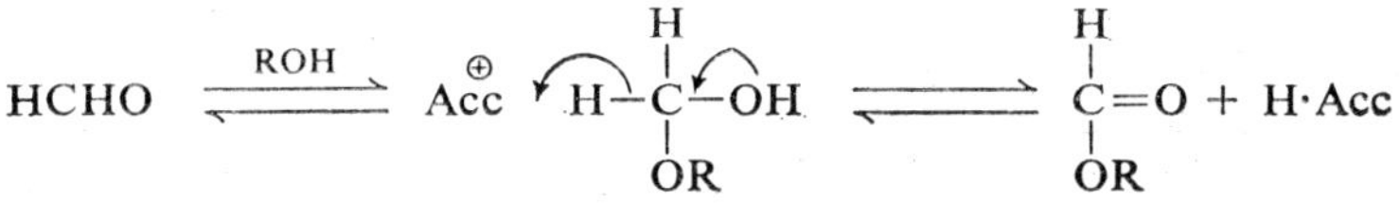


6-phospho-gluconolactone catalysed by D-glucose-6-phosphate dehydrogenase (1.1.1.49) is the equivalent of an aldehyde–carboxylic acid oxidation, but it appears to proceed via the pyranose form (XXXI) directly to the lactone [74]. It is therefore the oxidation of a hydroxyl to a carbonyl group and as such is classified in the first group of oxidoreductases.

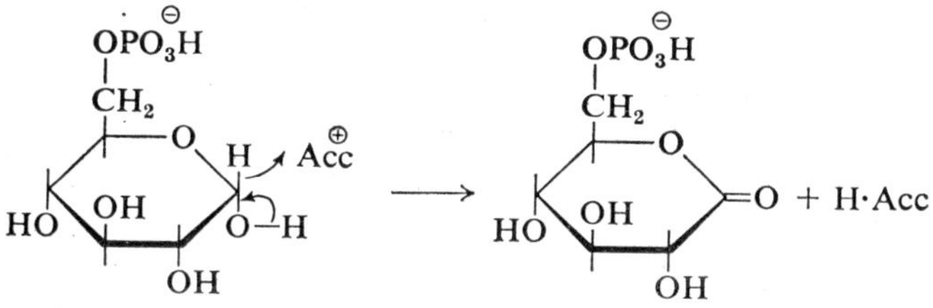


(XXXI)

It is probable that, for the NAD$^+$-requiring enzymes which oxidize aldehydes, the NAD$^+$ is involved in the reaction in exactly the same way as it is for the alcohol dehydrogenases. Deuterium transfer to the coenzyme appears to be general and the resultant reduced coenzyme may be either A-form or B-form. In Table 3.2 are listed several NAD$^+$- and NADP$^+$-requiring enzymes showing which stereochemical form of the coenzyme is produced in the enzyme-catalysed deuterium transfer reactions.

Of special interest in this group is the enzyme D-glyceraldehyde-3-phosphate: NAD oxidoreductase (phosphorylating) (1.2.1.12), for which an acceptable mechanism of action has been established [75]. This enzyme catalyses the oxidation, in the presence of phosphate ions, of D-glyceraldehyde-3-phosphate to 1,3-diphospho-D-glyceric acid (XXXII), which possesses a carboxylic-phosphate anhydride system in place of the aldehyde group. The enzyme also possesses some phosphate-ester hydrolase activity which is explicable in terms of the mechanism proposed. Reaction (3.12) is brought about by the enzyme in two stages.

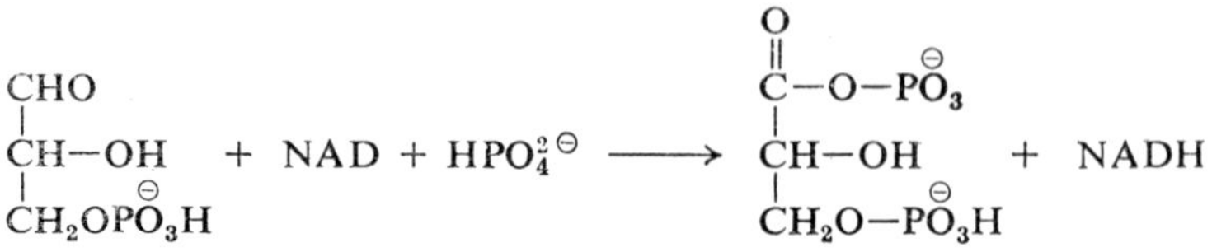

(XXXII)

(3.12)

In the absence of phosphate the reduction of $NAD^+$ to NADH takes place to an extent determined completely by the quantity of enzyme present. The enzyme acts as a reagent in this situation and not as a catalyst [76, 77] and a relatively stable acyl derivative of the protein is produced. When phosphate is added the acyl-enzyme is broken down, the free enzyme is liberated and the reduction of the $NAD^+$ proceeds. Regeneration of the free enzyme allows it to fulfil its catalytic function.

This is the first time we have come across a covalently-bound enzyme–substrate complex. Indeed, it was the first enzyme for which such a complex could be formulated on the basis of direct evidence. We shall see that many covalent complexes involving various groups have been implicated in enzyme reaction.

It is possible to prepare the acetyl derivative of the protein by treating it with acetaldehyde in the presence of NADH but in the absence of phosphate [78, 79]. Enzymic digestion of acetyl-glyceraldehyde-3-phosphate dehydrogenase has afforded peptides whose structures could be elucidated [81]. Such studies have shown that the acetyl group is carried on a cysteine residue via a thio-ester. The amino-acid sequence around this cysteine is shown (XXXIII). The acetyl-enzyme may also be prepared by treatment

$COCH_3$ (on Cys)

Lys.Ile.Val.Ser.Asn.Ala.Ser.Cys.Thr.Thr.Asn.Cys.Leu.Ala.Pro.-
Leu.Ala.Lys.

(XXXIII)

with acetyl-phosphate or p-nitrophenylacetate [78, 80, 83] and the same cysteine residue is substituted in each case. Moreover, this is the same cysteine residue which is acylated during the oxidation of glyceraldehyde-3-phosphate itself. Carboxymethylation of this cysteine with iodoacetate causes complete loss of enzymic activity [82].

The preparation of the acetyl-derivative of the enzyme using acetyl-phosphate is an indication that the whole process is freely reversible. In the presence of NADH, the acetyl-enzyme will yield acetaldehyde [79].

On the basis of the information available a mechanistic scheme for the reaction has been proposed in the following terms. The enzyme, coenzyme and substrate associate to form a ternary complex in which the desired three-dimensional arrangement is achieved. Within this complex the reaction proceeds by nucleophilic attack by a thiol group on the carbonyl of the aldehyde group. Mathew *et al.* [78] suggest that for the reverse reaction the nucleophilicity of the thiol group may be enhanced by basic catalysis by a histidine side-chain; it may be that such an enhancement is

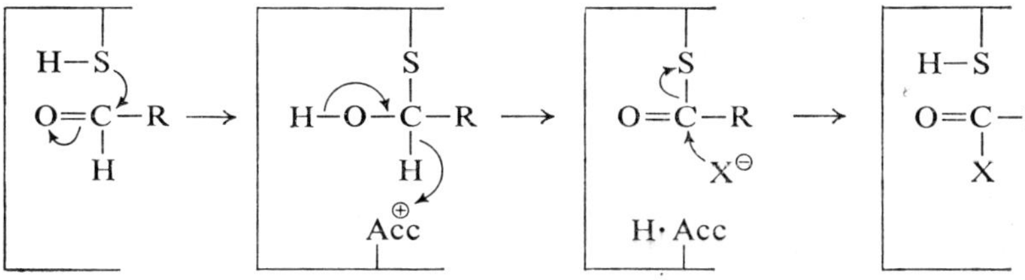

involved in the forward process also. A thio-hemiacetal is produced, which breaks down by means of a hydride transfer giving the acyl-enzyme which involves a thio-ester linkage. This process is essentially similar to that described above for the oxidation of hydrates and hemiacetals by alcohol dehydrogenases. In the absence of phosphate the acyl-enzyme may be hydrolysed slowly. Phosphate itself may react giving the acyl-phosphate and regenerating the thiol group and hence the free enzyme ($X = HPO_4^{2\ominus}$). If a thiol replaces the phosphate then a thiolester can be synthesized.

Some of the important reactions of D-glyceraldehyde-3-phosphate dehydrogenase are summarized in Fig. 3.2.

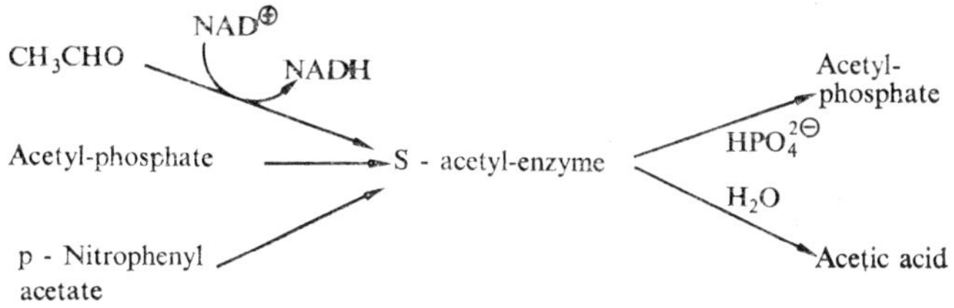


FIG. 3.2. Some reactions of glyceraldehyde-3-phosphate dehydrogenase. (From ref. [78])

The role of the coenzyme in contributing to the conformational structure of the enzyme has been emphasized by Bolotina *et al.* [84], who reported that binding of the $NAD^+$ with the enzyme caused a change in conformation leading to a structure with more $\alpha$-helix content. Such a structure is probably more rigid. These results are in agreement with those of Furfine *et al.* [85].

From the reaction scheme shown (Fig. 3.2) we can see that in the presence of glyceraldehyde-3-phosphate dehydrogenase, and in the absence of phosphate, acetylphosphate will be hydrolysed to acetic acid via the acetyl enzyme. This provides an explanation for the phosphatase activity which has been observed.

Jakoby, in a review [86], has emphasized the essential nature of the thiol group for all dehydrogenases which catalyse the oxidation of aldehydes by $NAD^+$ or $NADP^+$. These enzymes may differ not only in their substrate specificities but also in the nature of the final product. Some produce free acids, some phosphates, some thiol-esters, etc. Jakoby suggests that all these enzymes function by mechanisms which are essentially similar to that described above for glyceraldehyde-3-phosphate dehydrogenase.

### Flavoenzymes

Certain derivatives of isoalloxazine, known as flavins, have long been known to be essential for some biological oxidations. These include riboflavin (XXXIV) in which a 6,7-dimethylisoalloxazine carries on the 9-position a group derived from ribitol. Riboflavin-5′-phosphate (XXXV) is usually known as flavin mononucleotide (FMN) while the third important flavin is flavin adenine dinucleotide (XXXVI, FAD).

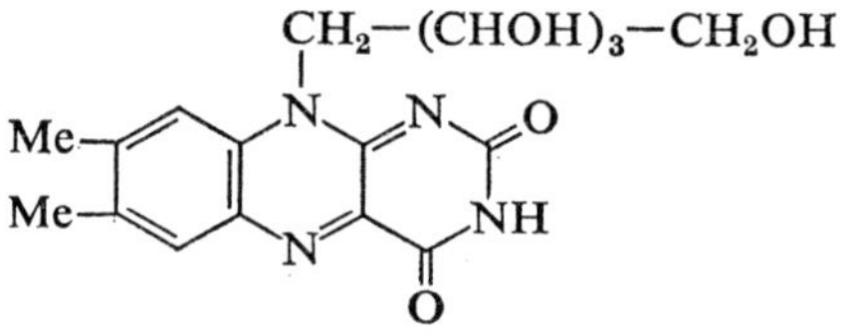


(XXXIV)
Riboflavin

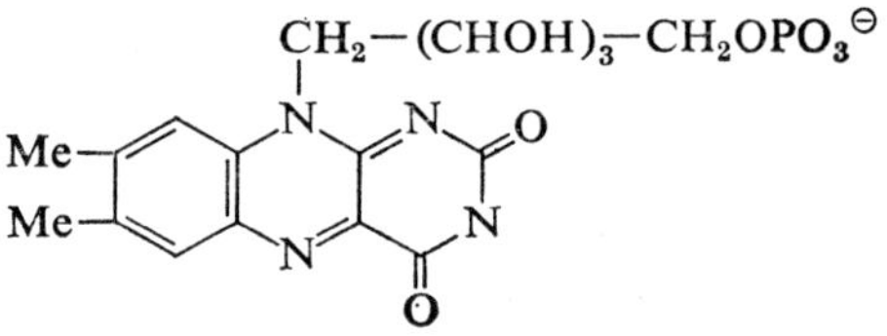


(XXXV)
FMN

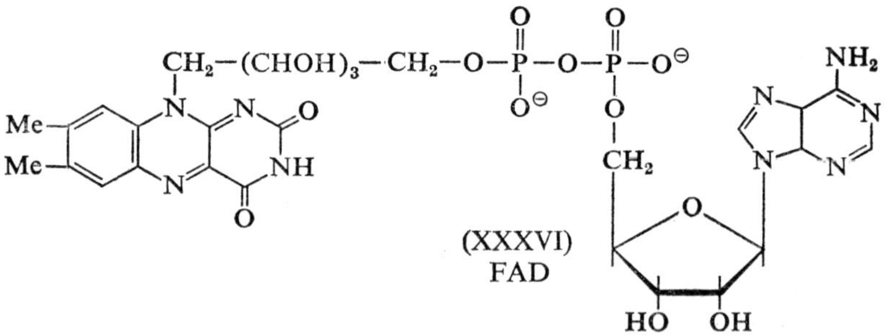


(XXXVI)
FAD

FMN and FAD function as coenzymes to a large number of enzymes which catalyse oxidation–reduction processes. Their roles, however, differ from those of the pyridine nucleotides. Whereas the pyridine nucleotides function as final electron donors or acceptors in the reactions and may dissociate from the enzymes in the manner of a normal substrate at the completion of the reaction, the flavin coenzymes are usually bound very strongly to their proteins and do not dissociate away. In addition, they do not function as terminal electron acceptors or donors, that is, as substrates, but as we shall see they appear to act as intermediates in the electron transfer chain.

The ultraviolet spectra of flavins and flavoproteins are very characteristic and are important not only for the identification of flavoproteins and the determination of the flavin content of any system, but also, where the oxidized and reduced forms are considered, for the study of the kinetics and mechanisms of the reactions. The ultraviolet spectra of riboflavin and FMN in water are almost identical, possessing bands at 220, 265, 375 and 447 nm. The spectra are highly dependent on the nature of the surrounding medium. Certain changes occur in the spectra of these flavins when molecular complexes with aromatic compounds are formed. These changes are also reflected in the spectrum of FAD and it is assumed that the adenine group of this molecule is very close to (probably coplanar with) the flavin moiety. The ability of flavins to form complexes with certain aromatic rings may account, at least in part, for the binding of these coenzymes to the proteins. From ultraviolet absorption data, Penzer and Radda have concluded that in flavoproteins the flavin chromophore may be surrounded by an atmosphere of non-polar groups while at the same time interacting at a particular point with a polar species [87].

Reduction of the flavin nucleus can occur either by a two-electron process or by two successive one-electron reactions. Thus, at normal pH

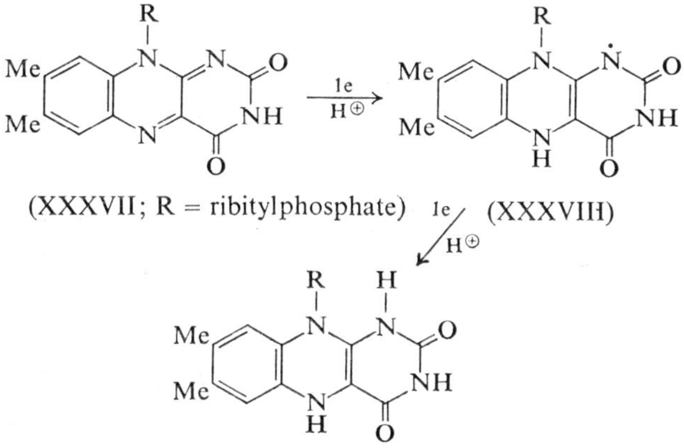

(XXXVII; R = ribitylphosphate) (XXXVIII)

(XXXIX)

values, one-electron reduction of FMN (XXXVII) gives the radical FMNH· (XXXVIII) and a second electron transfer yields the fully reduced form $FMNH_2$ (XXXIX). The radical FMNH· is stabilized by resonance and the situation is comparable to that which applies to the quinone–hydroquinone system (page 104). Indeed the FMNH· form is often referred to as a flavin semiquinone. The similarity is emphasized by the reaction in which the oxidized and reduced flavins undergo electron transfer to produce the radical.

$$FMNH_2 + FMN \rightleftharpoons 2FMNH\cdot \qquad (3.13)$$

The semiquinone form is often detected during flavoenzyme reactions by virtue of its characteristic e.s.r., ultraviolet and visible absorptions (it shows a distinctive red colour).

Being a radical the flavin semiquinone reacts with oxygen, hydrogen peroxide being formed by the oxidation of two moles of semiquinone by one mole of oxygen [88]. Fully-reduced flavins are also oxidized rapidly

$$2\ FMNH\cdot + O_2 \rightleftharpoons 2FMN + H_2O_2$$

by oxygen by a two-electron process which involves a flavin peroxide (XL). Equilibration between the FMN produced and the $FMNH_2$

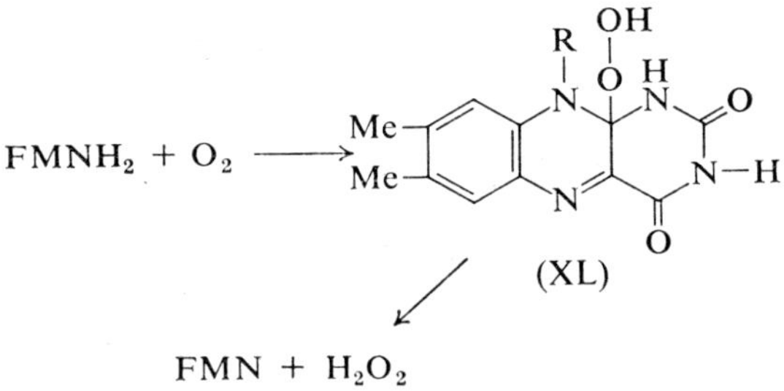


(XL)

(3.14)

(equation 3.13) can lead to the production of a semiquinone which can react with oxygen and therefore the air oxidation of $FMNH_2$ can proceed by two simultaneous competitive processes [88].

*Reduced Pyridine Nucleotide Dehydrogenases.* These enzymes catalyse reactions corresponding to the general equation (3.15). Different enzymes

$$NAD(P)H + \text{acceptor} \rightleftharpoons NAD(P)^+ + \text{reduced acceptor} \qquad (3.15)$$

in this group may be specific for different electron acceptors. Some typical acceptors are: oxidized pyridine nucleotides (NAD(P) transhydrogenase; 1.6.1.1), cytochrome $b_5$ (cytochrome $b_5$ reductase; 1.6.2.2) in which a ferric ion becomes reduced to the ferrous state, disulphides such as cystine (cystine reductase; 1.6.4.1) or lipoamide (lipoamide dehydrogenase; 1.6.4.3), ascorbate (oxidized ascorbate reductase, 1.6.5.4) and the iron-containing protein ferredoxin (ferredoxin-NADP reductase 1.6.99.4).

The overall reaction (3.15) occurs in at least two stages. The first process is the reaction between the NADH (or NADPH) and the flavoprotein which yields a reduced flavoprotein. During the subsequent stage or stages electron transfer from the reduced flavin to the acceptor molecule is effected.

(a) NADH + $H^{\oplus}$ + flavin ⟶ $NAD^{+}$ + reduced flavin

(b) reduced flavin + acceptor ⟶ flavin + reduced acceptor.

Studies have been made of the comparable non-enzymic processes, that is, the reaction between NADH and FMN [89] and the reduction of oxygen or cytochrome *c* by NADH as catalysed by various flavins [90]. Flavin radicals are produced when NADH reacts with FMN. However, Fox and Tollin [89] showed that the radicals are not produced directly by a reaction such as (3.16), but rather that an initial two-electron reduction

$$\text{NADH} + \text{FMN} \rightarrow \text{FMNH}\cdot \ldots \qquad (3.16)$$

takes place and that the radicals are produced during some subsequent reaction. Suelter and Metzler [91] had found that when deuterated analogues of NADH were used, the reaction manifested an isotope effect of (3.16). The suggestion was therefore that the reaction between NADH and FMN involves a direct hydrogen transfer accompanied by two electrons (a hydride transfer). It is difficult to prove the direct nature of the hydride transfer because, in contrast to the case of NAD (page 108), the transferred hydrogen on a reduced flavin rapidly exchanges with solvent. The reduction of a flavin by a reduced pyridine nucleotide may therefore be described mechanistically as shown.

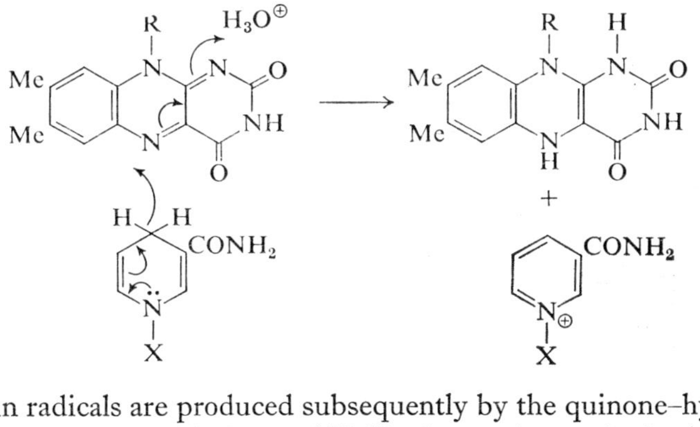


The flavin radicals are produced subsequently by the quinone–hydroquinone type of process (3.13). Fox and Tollin detected a species in the reaction system which they identify as the $FMNH_2$–FMN complex.

It is not unreasonable to suggest that the enzymic reduction of flavins by reduced pyridine nucleotides proceeds by a mechanism similar to the non-enzymic process, that is, by a hydride transfer. The situation with regard to the second phase of the overall reaction (the reoxidation of the flavin) is more complicated. For the different enzymes there is a variety of electron acceptors some of which function by two-electron and others by one-electron reactions. In the case of NAD(P) transhydrogenase, where the ultimate electron acceptor is an oxidized pyridine nucleotide, symmetry would suggest that the reduced flavin becomes reoxidized by a hydride transfer.

Gascoigne and Radda [92] reported a study of the nonenzymic reduction of FMN with dihydrolipoic acid (XLI) which indicated that the reaction

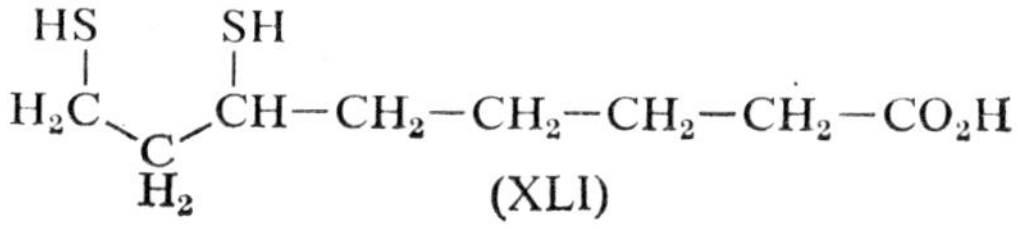

(XLI)

between these compounds involves a hydride transfer. However, the comparable enzyme system appears to be more complicated. Inhibitor studies on lipoamide dehydrogenase have led to the suggestion that two neighbouring thiol groups in the enzyme are closely implicated in the reaction [93]. According to Ide *et al.* [94] the electron-transfer sequence during the oxidation of NADH by lipoamide in the presence of this enzyme is as shown.

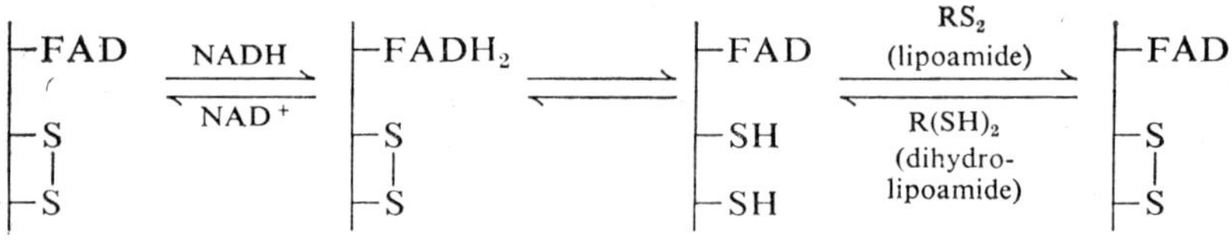


Thus, the reoxidation of the reduced flavin proceeds via the dithiol system. In some non-enzymic systems, the oxidation of thiols to disulphides may involve radicals, and Massey and Veeger [95] have detected a flavin free radical during the enzymic reaction. They suggest that this radical may be stabilized by a thiol group and that a diradical is an intermediate in this reaction.

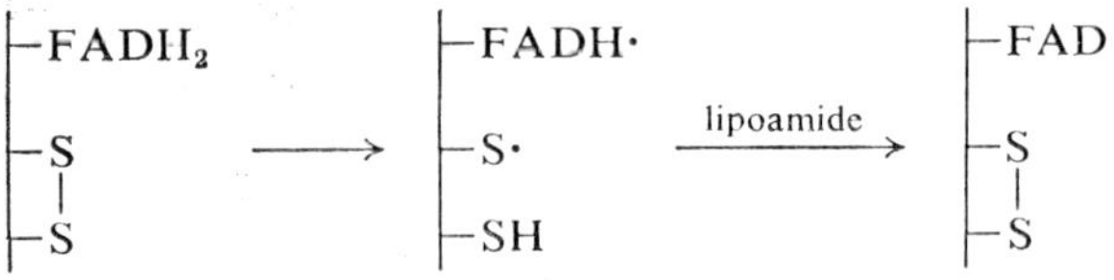


Some electron acceptors are essentially one-electron reagents. Fox and Tollin used a model (non-enzyme) system in which a complexed ferric ion acted as electron acceptor. The reduction of the ferric ion by $FMNH_2$ proceeded by a one-electron transfer which gave the flavin radical, the radicals could then disproportionate (equation 3.13) and regenerate fully-reduced flavin to continue the process. Masters *et al.* [96] proposed a

$$FMNH_2 + Fe^{3\oplus} \rightarrow FMNH\cdot + Fe^{2\oplus} + H^{\oplus}$$

similar mechanism for NADPH–cytochrome *c* reductase in which a complexed iron functions as electron acceptor. This envisages the flavin moiety existing either as a semiquinone or as the fully-reduced form. Such a mechanism can only apply if the enzymes contain two flavin groups per active site, sufficiently close together to interact in this way. In other cases the reduced flavin may be reoxidized simply by two consecutive one-electron steps.

In summary, we envisage the NAD(P) H-dehydrogenases as functioning by an initial hydride transfer from the reduced nicotinamide to the protein-bound flavin. The reduced flavin may become reoxidized by a two-electron process or by two one-electron processes according to the nature of the ultimate acceptor and the individual enzyme concerned.

*Glucose Oxidase* (1.1.3.4). This flavoprotein catalyses the oxidation of D-glucose to D-glucono-$\delta$-lactone (XLII) by molecular oxygen (equation 3.17). The product is therefore similar to that obtained when glucose-6-phosphate is oxidized with dehydrogenase (page 121). The reaction occurs

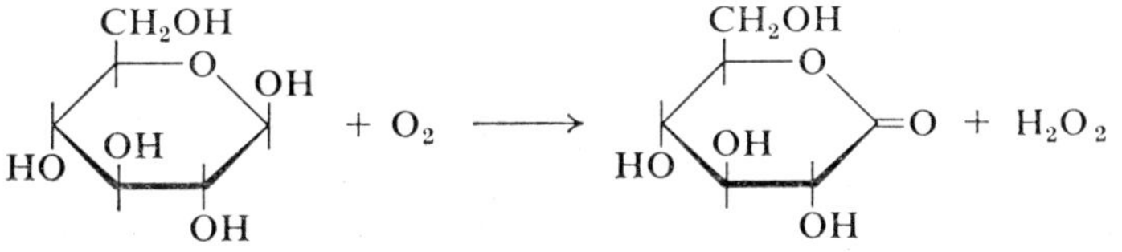


(XLII)

(3.17)

in two stages. In the first stage glucose is oxidized to the product and the reduced flavin is formed. The gluconolactone dissociates away from the reduced enzyme before the reoxidation of the flavin by oxygen takes place [97]. The kinetic scheme (3.18) therefore applies and this has been confirmed by rate and inhibition studies by Bright and Gibson [98],

$$E_{ox} + S \underset{}{\overset{k_{+1}}{\rightleftharpoons}} E_{ox}\cdots S \overset{k_{+2}}{\rightleftharpoons} E_{red}\cdots P_1 \overset{k_{+3}}{\rightleftharpoons} E_{red} + P_1$$

$$E_{red} + O_2 \overset{k_{+4}}{\rightleftharpoons} E_{ox}\cdots P_2 \overset{k_{+5}}{\rightleftharpoons} E_{ox} + P_2, \qquad (3.18)$$

where S is D-glucose, $P_1$ is gluconolactone, $P_2$ is $H_2O_2$, $E_{ox}$ is the enzyme with oxidized flavin and $E_{red}$ contains reduced flavin.

In the absence of oxygen the reaction proceeds as far as the $k_{+3}$ stage. When glucose-1-$^2$H was used, $k_{+2}$ was decreased significantly [98]. This kinetic isotope effect suggests that the reaction between glucose and the flavin group includes a hydrogen transfer. If this is a hydride transfer we may formulate the reaction thus:

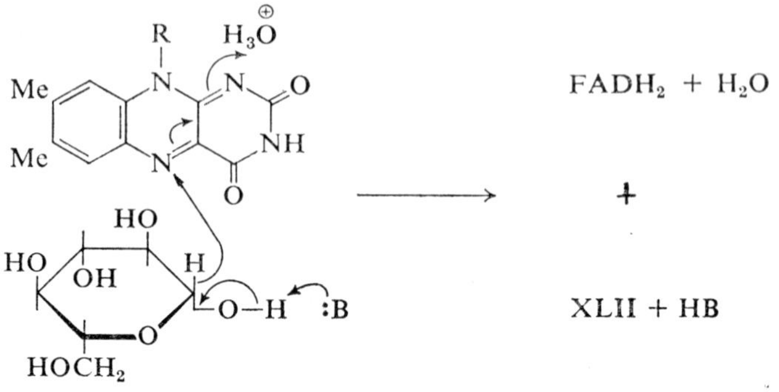


The available spectrophotometric evidence indicates that the enzyme exists in only two states, an oxidized and a reduced form [99]. If the reduction of the oxidized enzyme is effected by a two-electron process then the reduced form must be the $FADH_2$ state and the semiquinone takes no part in the process. This is supported by the observations that the semiquinone form, as produced by irradiation, does not react with glucose [100] and that no e.s.r. signals are detected during the reaction. If this is the case, then the reoxidation of the reduced flavin by the oxygen must also be a two-electron process, possibly similar to that described for the non-enzymic reaction of $FMNH_2$ with oxygen on page 126.

Bright and Gibson point out that there is an alternative mechanism for the two-electron reduction of the flavin by glucose which is also consistent with the data. This mechanism is suggested by the non-enzymic oxidation of glucose by bromine, which yields the same product. The established mechanism for this reaction is an example of the oxidative eliminations referred to on page 101.

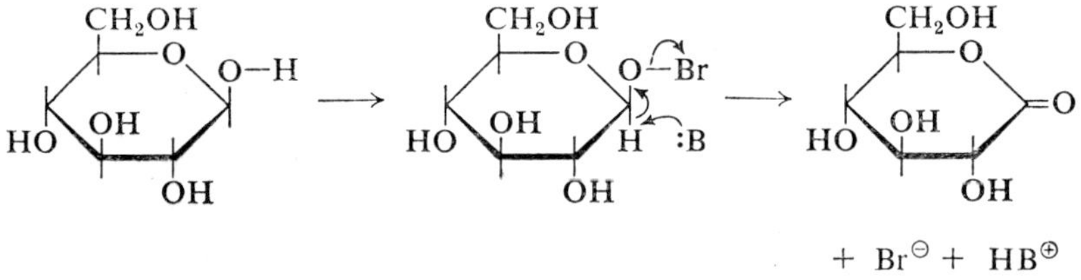


An important feature of this reaction is that the $\beta$-isomer of D-glucose (as shown) reacts more rapidly than does the $\alpha$-form. This is usually considered to be due to the presence of the hydroxyl group on position-2. Similar considerations would imply that a hydride-transfer mechanism such as the one proposed above would favour the $\alpha$-isomer. Now glucose oxidase reacts with the $\beta$-isomer far more rapidly than with the $\alpha$-isomer. Therefore Bright and Gibson proposed a mechanism in which the flavin group becomes covalently bound to the glucose in a hemiacetal and becomes reduced through an elimination. The reaction involves the cleavage of the C—H bond at position-1 and hence could be subject to a kinetic isotope effect.

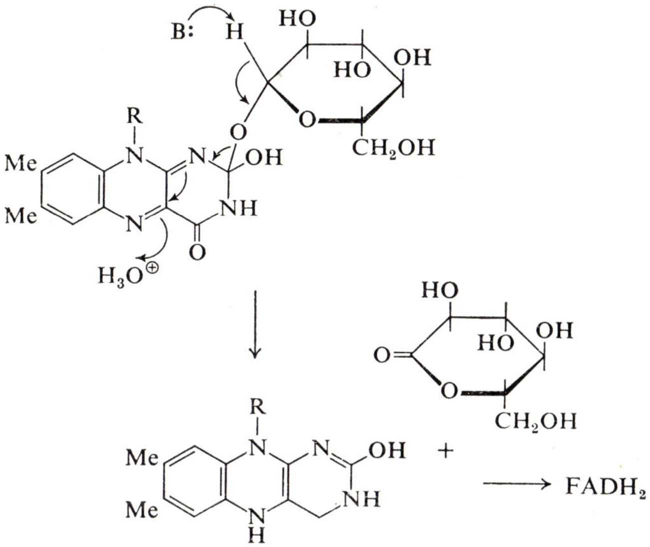

However, it must be stated that the $\beta$-stereospecificity of the enzyme does not eliminate the possibility of the hydride-transfer mechanism. The asymmetric nature of the active site and the correct positioning of the groups involved (including the basic group :B) can account for the stereospecificity, as in the case of most other enzymes.

*Amino-Acid Oxidases.* These flavoenzymes catalyse the deaminative oxidation of amino-acids, the products of the reactions being ketoacids. There is little doubt that the an amino-acid molecule undergoes at least two reactions; first oxidation to an imino-acid (XLIII) and then hydrolysis to give the ketoacid product and ammonia.

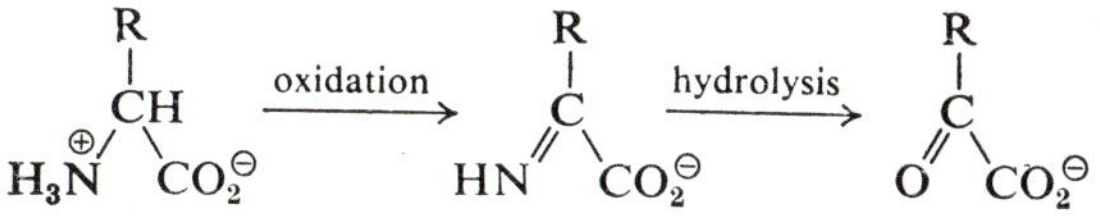


(XLIII)

Some amino-acid oxidases are also capable of bringing about the oxidation of $\alpha$-hydroxy-acids. The same active sites are involved and it is very likely that the mechanisms of the oxidations are similar [101].

Early suggestions concerning the mechanism were that a hydride transfer from the amino-acid takes place. A similar mechanism can be envisaged for the oxidation of amino-acids to imino-acids by $NAD^+$-requiring enzymes. Clearly, α-hydroxy-acids could be oxidized by such a mechanism.

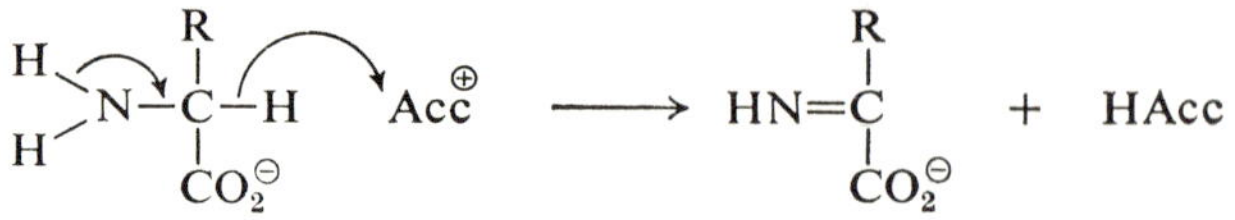


Massey *et al.* [102] studied pig kidney D-amino-acid oxidase (1.4.3.3) and proposed a different mechanism for the oxidation of the amino-acid. In the absence of oxygen the amino-acid substrate can reduce the flavin groups in the enzyme (FAD) fully. An intermediate is produced in this reaction which is suggested to be a complex of the enzyme, flavin semiquinone and the amino-acid radical (XLIV). Thus, Massey and his group envisaged that the oxidation of substrate to imino-acid proceeds by two successive one-electron steps, as shown below where AA represents the D-amino-acid, and IA the imino-acid.

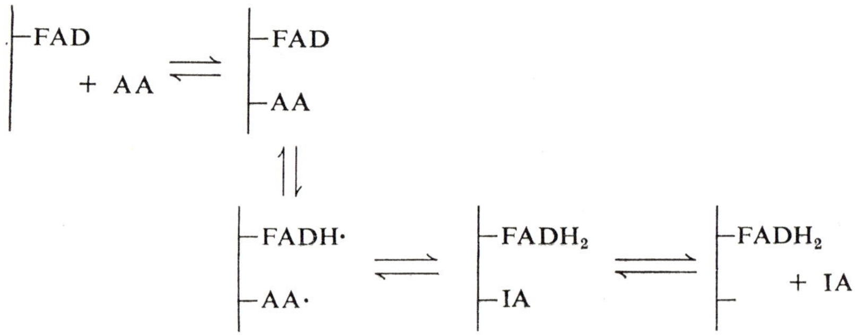


(XLIV)

The complete reduction of the flavin is, it was found, far too slow for the fully-reduced flavin species to be an intermediate in the aerobic reaction. The aerobic process, however, does involve (XLIV) as an intermediate.

The nature of this 'biradical' (XLIV) is not clear. It is not simply the half-reduced enzyme (in which all the flavin groups are in the semiquinone form). The half-reduced enzyme can be prepared in other ways and it is certainly different from the intermediate which is detected. Its most important property, perhaps, is that the ultraviolet spectrum varies according to the amino-acid substrate used. The amino-acid, or some derivative of it, must therefore be intimately involved in the complex, presumably in the very near vicinity of the flavin moiety which is the chromophore.

At present, the description of the intermediate as a 'biradical' (XLIV) must suffice. Massey points out that it may in fact be a charge transfer complex between the reduced flavin and the imino-acid species.

The overall, aerobic reaction may be described as shown:

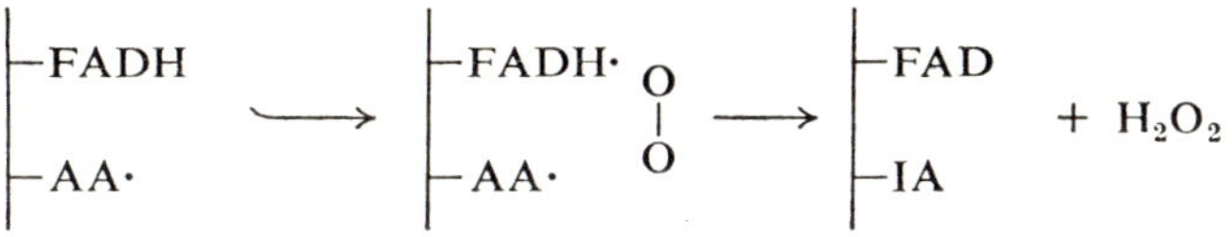

(XLV)

The rate-determining step is the final one, that is, the dissociation of the imino-acid from the complex.

The overall reaction involves a ternary complex between enzyme, amino-acid and oxygen; this is in agreement with kinetic studies. Very similar results were obtained by Massey and Curti [103] for the L-amino-acid oxidase (1.4.3.2) from *Crotalus adamanteus* venom. In this case too, an intermediate similar to that suggested for the D-amino-acid oxidase appears to be involved. Again, evidence was presented that the intermediate was not simply the half-reduced enzyme and that it was characteristic of the amino-acid substrate involved.

An interesting result, obtained by Hellerman and Coffey [104] using mammalian kidney D-amino-acid oxidase, may eventually throw some light on the structures of the intermediates in these enzymes. They treated the enzyme-FAD–alanine-$^{14}C$ complex with sodium borohydride and obtained a labelled protein.

The ε-amino-group of a lysine residue was involved, since ε-N-(1-carboxyethyl)-lysine (XLVI) was identified in the product after hydrolysis of the protein. This indicates that the alanine had been bound as an imine to the protein.

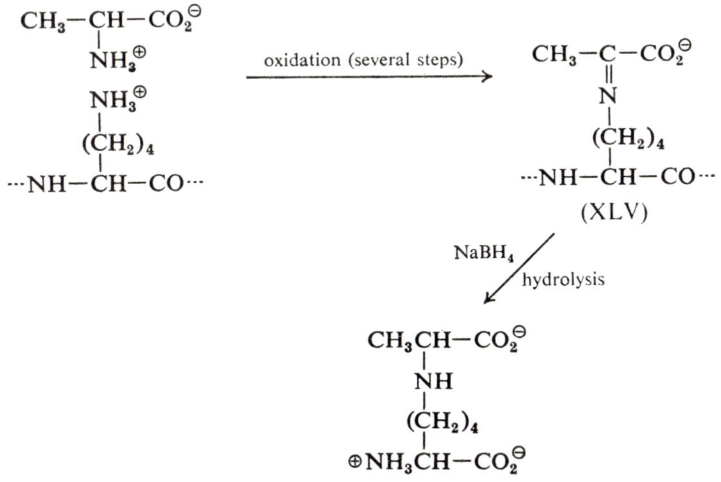


(XLVI)

The formation of this imine is probably an essential step in the catalytic process since (a) the extent of protein labelling was greatest under conditions which gave maximum enzymic activity, and (b) the labelling process was stereospecific in that the D-amino-acid oxidase required a D-amino-acid for labelling whereas a snake venom L-amino-acid oxidase became labelled only when an L-amino-acid substrate was used. In addition, 0·9 mole of labelled alanine was incorporated per mole of FAD in the enzyme.

It is possible then that the substrate-characteristic intermediate (XLIV) described in Massey's mechanism for the amino-acid oxidases involves the ketimine structure (XLV). Such a system is of course at the same oxidation level as the imino-acid itself.

*Metalloflavoenzymes.* If an enzyme contains, in addition to the flavin prosthetic group, one or more metal ions, then the number of possible steps in the electron-transfer process is increased and the situation becomes more complicated. For example, L-4,5-dihydroorotate; oxygen oxidoreductase (dihydroorotate dehydrogenase, 1.3.3.1) contains 2 FAD, 2FMN and 4Fe groups for each molecule, assuming a molecular weight of 115,000. It is probable that each active site contains 1FAD, 1FMN and 2Fe. The enzyme catalyses the oxidation of L-4,5-dihydroorotate (XLVII) to orotate (XLVIII) by molecular oxygen.

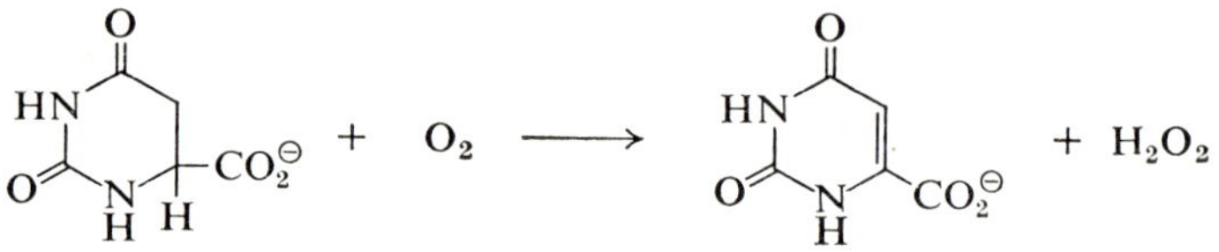


(XLVII) (XLVIII)

As in the case of other flavin enzymes, the enzyme becomes reduced by the substrate and is subsequently reoxidized by the oxygen.

An interesting feature of this enzyme is its versatility. Thus, it can use $NAD^+$ in place of oxygen in the oxidation of dihydroorotate. In addition,

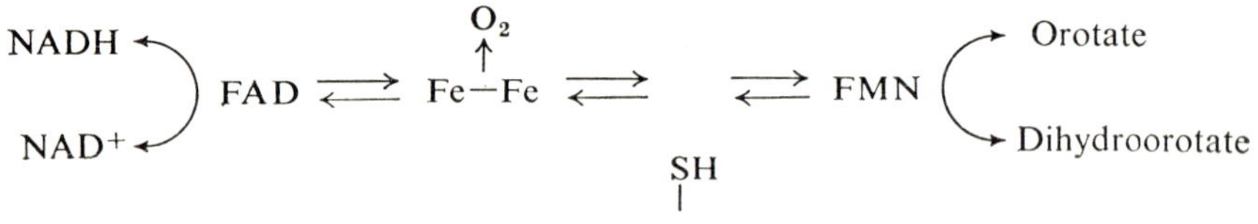


(3.19)

it also catalyses the oxidation of NADH by oxygen. This means that both both NADH and dihydroorotate can reduce the enzyme and that $NAD^+$ and $O_2$ can oxidize its reduced form. Aleman-Aleman *et al.* [105] presented the electron flow path (3.19) to explain the various aspects of the enzymic process.

Notice that this scheme shows that the reduction of the enzyme can be effected by both dihydroorotate and NADH, but by different mechanisms.

The electron flow path was based on the following results. Anaerobic reduction of the enzyme by NADH (one mole) leads to the production of e.s.r. signals at $g = 2{\cdot}00$ and $1{\cdot}94$. The signal at $g = 2{\cdot}00$ is attributed to a flavin semiquinone species and that at $g = 1{\cdot}94$ to a reduced iron system [105]. Kinetic studies have indicated that the initial event is a full reduction (that is, two electrons) of a flavin group by the NADH, and that this is followed by a partial reoxidation involving the iron atoms. Thus the final state consists of a flavin semiquinone and reduced iron.

Addition of more than one mole of NADH can cause reduction of the other flavin group in the active site.

Absorption spectroscopy suggests that the dihydroorotate and orotate bind to the enzyme near a flavin group and fluorescence quenching indicates that it may be the FMN group [106]. If this is the case, then presumably reduction of the enzyme by dihydroorotate involves the FMN group and reduction by NADH affects the FAD.

The anaerobic reduction of the enzyme by one mole of dihydroorotate results in complete reduction of one of the flavins (presumably FMN) and the semiquinone is not produced [105]. This two-electron reduction may proceed by a hydride transfer mechanism also. In connection with this aspect of the enzymic reaction, activators such as cysteine play an important role. The enzyme, as normally isolated, usually requires activation by cysteine or some other thiol for maximum activity in the oxidation of dihydroorotate. However, this activation is not needed for the reduction of the enzyme by NADH, nor is it required for the aerobic oxidation of NADH. In addition treatment of the enzyme with one mole of p-chloromercuribenzoate inhibits the oxidation of dihydroorotate but has no effect on the aerobic oxidation of NADH. It appears, therefore, that the reactions involving the dihydroorotate require the presence of some thiol group (probably a cysteine residue in the protein) which must be in its reduced form. When this thiol group, which Handler [107] refers to as a 'switch', is in its reduced state, electron flow in all the directions indicated in the scheme (3.19) can proceed unimpeded. However, when this thiol is oxidized or blocked by the mercurial derivative, electron transfer at the point marked —SH is prevented although the reactions in the other parts of the system remain unaffected. Even when the 'switch' is in its reduced condition, the first pair of electrons to enter the oxidized enzyme remain on the side of the switch at which it entered. Thus, two electrons from NADH can distribute themselves over the FAD and Fe–Fe systems but not on the FMN group. Two electrons from one mole of dihydroorotate can reduce the FMN to $FMNH_2$ but may proceed no further. In either case, when a second mole of reduced substrate is introduced into the system, the electrons succeed in crossing the barrier [107].

Aleman-Aleman *et al.* [105] have shown that, during the oxidation of the reduced form of the enzyme by molecular oxygen, oxygen radicals ($O_2^{\ominus}$) are formed. Several criteria for the involvement of such radicals were established; these were initiation of sulphite autoxidation, oxygen-dependent reduction of cytochrome *c* and induction of chemiluminescence in the presence of luminol. Dihydroorotate dehydrogenase was found to yield $O_2^{\ominus}$ radicals by all these criteria. This is a property shown by several iron-containing flavoenzymes (in particular, xanthine oxidase and aldehyde oxidase). Iron-free flavoenzymes, however, have not been shown to produce these radicals, and this leads to the suggestion that it is interaction between the reduced iron system in the enzyme and the molecular oxygen which gives the radicals.

According to the present state of our knowledge, therefore, we can summarize the mechanism of action of dihydroorotate dehydrogenase as follows. NADH reduces the FAD group in the enzyme to $FADH_2$, probably by a hydride-transfer process. A one-electron transfer from this reduced flavin to the $Fe^{3+}$–$Fe^{3+}$ system gives the flavin semiquinone and (possibly) a $Fe^{2+}$–$Fe^{3+}$ complex. This and/or some similar reduced iron complex interacts with molecular oxygen transferring first one electron (giving the oxygen radical) and then another (giving the peroxide oxidation level).

Similarly, dihydroorotate reacts with the FMN group, possibly by a hydride transfer, to give $FMNH_2$. Electrons may be transferred via the intermediacy of some thiol species from the $FMNH_2$ to the Fe–Fe system apparently in such a way that the FMNH semiquinone is not obtained. Miller and Massey [108] have proposed a mechanism in which specific roles for the thiol groups necessary are formulated.

An even more complicated situation appears to exist in the case of xanthine oxidase (1.2.3.2) and aldehyde oxidase (1.2.3.1), two enzymes which are very similar to each other. Xanthine oxidase (cow's milk) has a molecular weight of 275,000 and contains 8 iron, 2 FAD and 2 molybdenum groups per molecule [109]. It is assumed that each active site contains 1 FAD group, 1 molybdenum and 4 iron atoms, only one of which is actually involved in the reaction. It catalyses the oxidation of a variety of substrates including purines (such as xanthine and hypoxanthine), pteridines and other heterocyclic molecules, aldehydes and also of NADH to $NAD^+$. A variety of electron acceptors can function with the enzyme and these include $NAD^+$, $O_2$, methylene blue, quinones, ferricyanide and nitrate [110]. Liver aldehyde oxidase is very similar both in composition and function, although its range of substrates is rather more restricted.

The conversion of xanthine (XLIX) into uric acid (L) is typical of the reactions catalysed by these enzymes. With the exception of the oxidation of NADH to $NAD^+$, all the reactions can be regarded as hydroxylations, that is, reactions in which a hydrogen atom is replaced by a hydroxyl group, as shown below for xanthine and for an aldehyde [111].

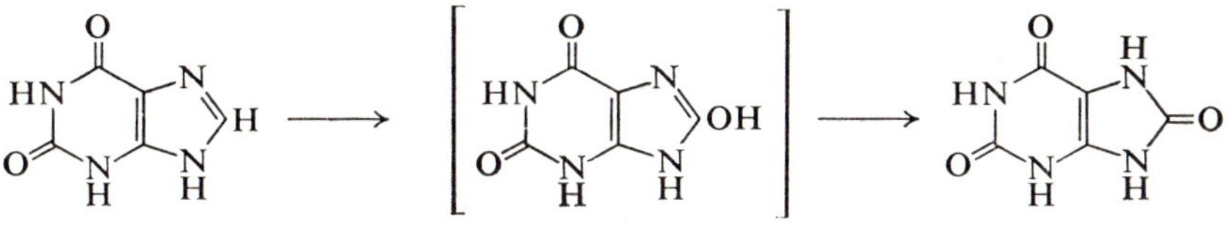

(XLIX) (L)

$$\mathrm{R{-}C({=}O){-}H \longrightarrow R{-}C({=}O){-}OH}$$

The use of isotopes has established that the oxygen atom incorporated into the molecule is derived from the solvent and not from the molecular oxygen which may function as the electron acceptor [112].

The electron-transfer sequence in xanthine oxidase has been established mainly by the e.s.r. studies of Bray *et al.* [110]. The resting, oxidized enzyme gives no e.s.r. signal, but when reduced by xanthine it shows several, at $g = 2{\cdot}00$, $g = 1{\cdot}97$ and $g = 1{\cdot}9$, the last being a broad double signal which is temperature-sensitive. The first signal is attributed to the flavin semiquinone (as in the case of dihydroorotate dehydrogenase) and the third is assumed to be due to a reduced iron system. The 1·97 $g$ signal is quite complex and the $Mo^{5\oplus}$ ion which gives rise to it exists in the enzyme in two forms distinguishable by e.s.r. These two forms have been designated $Mo_{\alpha\beta}$ and $Mo_{\gamma\delta}$. A rapid reaction technique showed that when xanthine oxidase is treated with xanthine in the presence of oxygen these signals appear in the following order. The $Mo_{\gamma\delta}$ signal reaches its maximum at 15 msec, then $Mo_{\alpha\beta}$ at 40 msec followed by the flavin radical at 45 msec and finally the Fe signal at about 100 msec. Therefore, all the components appear to be involved in the electron transfer, the molybdenum ion reacts first and the iron last.

The role of the rather unusual molybdenum ion in the reaction is of great interest. Since it is involved very early on in the electron-transfer sequence it is probably very close to, if not part of, the substrate-binding region of the enzyme. The e.s.r. spectrum is compatible with the presence of a sulphur group functioning as a ligand to a molybdenum ion [113], and the presence of a thiol group at the binding site has been indicated by the fact that the thiol group can be protected from thiol reagents by the presence of substrate [114]. On the basis of these results Aleman-Aleman *et al.* [105] concluded that a $Mo^{6\oplus}$–S complex might be present in the enzyme and that reduction of this complex by substrate might be the initial event in the sequence of reactions.

In the presence of substrates, the enzymes are inhibited by methanol. The normal signal at 1·97 is increased markedly in the methanol-inhibited enzyme and the nature of the signal suggests that at least one methanol molecule has replaced a water molecule as a ligand to the molybdenum ion [105]. This reaction of methanol with the enzyme requires the presence of a substrate (i.e. the reduced enzyme reacts) and it is most likely that the methanol complexes with the molybdenum in the $Mo^{5\oplus}$ form and 'freezes' the enzyme in this state so that the reduction of the flavin and iron groups cannot take place. In the methanol-inhibited enzyme no reduction of flavin or iron was observed and thus the suggestion that molybdenum is the first link in the electron-transport chain is strengthened.

Both enzymes are inhibited by cyanide ion and it is assumed that this species also inhibits by replacing one of the molybdenum ion ligands.

The oxidation of NADH must proceed by a mechanism different from that which applies to normal substrates, because methanol does not inhibit the enzymes with respect to the oxidation of this coenzyme [107]. The NADH oxidation seems to by-pass the Mo-system altogether; spectral studies on chicken liver xanthine oxidase indicate that the nicotinamide coenzyme binds near the flavin group [115]. Therefore, to the electron-transfer pathway proposed by Bray *et al.* [110] involving substrate, molybdenum, flavin, iron and oxygen in that order we may add a branch for NADH.

$$\text{Substrate} \longrightarrow \text{Mo} \longrightarrow \underset{\underset{\text{NADH}}{\uparrow}}{\text{Flavin}} \longrightarrow \text{Fe} \longrightarrow O_2$$

Bray *et al.* [110] point out that the sequence from substrate to oxygen must involve single-electron steps and that therefore substrate and oxygen radicals must be produced. The production of oxygen radicals has been indicated by chemiluminescence [116].

On the basis of the evidence which has been obtained a mechanism may be proposed for the oxidation of xanthine by oxygen. An enzyme–substrate complex is first formed in which the xanthine is very near to the $Mo^{6\oplus}$ complex. The substrate becomes oxidized and the molybdenum becomes reduced to the $Mo^{5+}$ state. A one-electron transfer results in reoxidation of the molybdenum to $Mo^{6+}$ and production of the flavin radical. Transfer of one electron from the flavin semiquinone to oxygen via the ferric–ferrous system can give an oxygen radical. The oxidation of xanthine to uric acid, or of an aldehyde to a carboxylic acid, involves two electrons and therefore the above sequence must be carried out twice for each molecule of substrate. At the 'half-way stage' substrate and oxygen radicals must be present. A possible mechanism for the hydroxylation of the substrate at the molybdenum site is presented in Fig. 3.3. Replacement of the

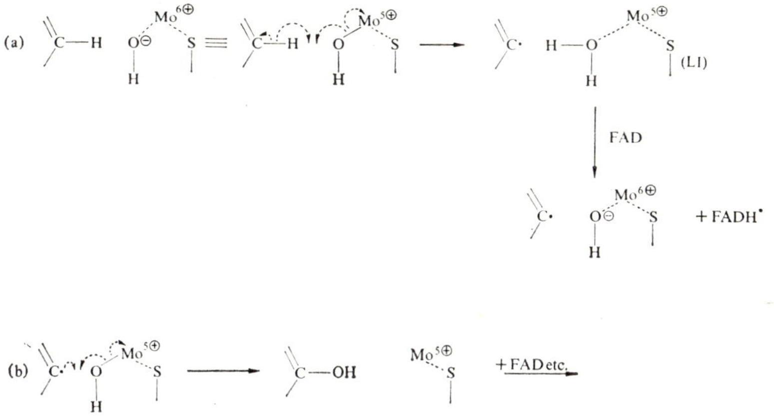

FIG. 3.3. Possible mechanism for hydroxylation of a substrate at the molybdenum site of xanthine oxidase or aldehyde oxidase

water molecule in (LI) by a methanol may block further reaction and this could explain the methanol inhibition. The molybdenum ion is envisaged as having a hydroxyl-ion ligand at the start of each one-electron sequence.

## Hydroxylations and Ring Cleavage

In the hydroxylations referred to above for the xanthine and aldehyde oxidases, the oxygen atom incorporated into the substrate originates in the solvent and not in the molecular oxygen. However, a different situation obtains for certain other enzymes which bring about hydroxylation. These enzymes, referred to as *hydroxylases*, catalyse reactions which may be generalized in reaction (3.20)

$$AH + DH_2 + O_2^* \rightarrow A\overset{*}{O}H + D + H_2O^*, \qquad (3.20)$$

where AH is the substrate which becomes hydroxylated and $DH_2$ is an electron donor. The overall reaction is a four-electron process. Isotope studies have shown that the oxygen atom incorporated into the substrate is derived from the molecular oxygen, the other atom appearing in the water [117]. Enzymes which catalyse such reactions are often referred to as 'mixed function oxidases' [118]. An essential feature of the reactions is the requirement for an electron donor, $DH_2$. This means, in effect, that in addition to the oxygen, the enzyme is concerned with two substrates, that which becomes hydroxylated and the electron donor, which is sometimes called a co-substrate.

The hydroxylases all appear to contain metal-ions (iron is very common) and superficially the reactions resemble the hydroxylations brought about by Udenfriend's system. Not surprisingly, some mechanisms which have been proposed for the enzymic reactions are similar to those suggested for the Udenfriend reaction. For example, Goldstein *et al.* [119] have suggested a reaction scheme for the hydroxylation of 3,4-dihydroxyphenylethylamine (dopamine) as catalysed by the copper-enzyme dopamine hydroxylase (1.14.2.1). In this scheme the cupric ion is first reduced to cuprous. Then a complex of the cuprous ion with the molecular oxygen is formed, and this complex is involved in the hydroxylation. It will be recalled that Norman and Lindsay-Smith [28] suggested that a complex of ferrous ion and molecular oxygen participated in non-enzymic hydroxylations (page 106).

In contrast with the hydroxylases, the true oxygenases catalyse reactions in which both atoms of the oxygen become incorporated into the substrate. These are usually reactions in which ring systems are opened. For example, catechol: oxygen 2,3-oxidoreductase (catechol 2,3-oxygenase 1.13.1.2) brings about the conversion of catechol (LII) into 2-hydroxy-

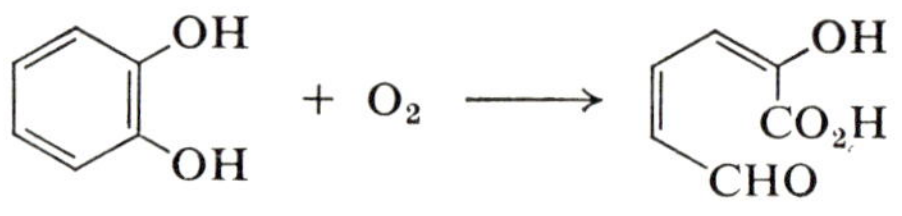


(LII) (LIII)

muconate semialdehyde (LIII). If $^{18}O$-labelled oxygen is used all of the label appears in the product. Similarly the indole ring of tryptophan is

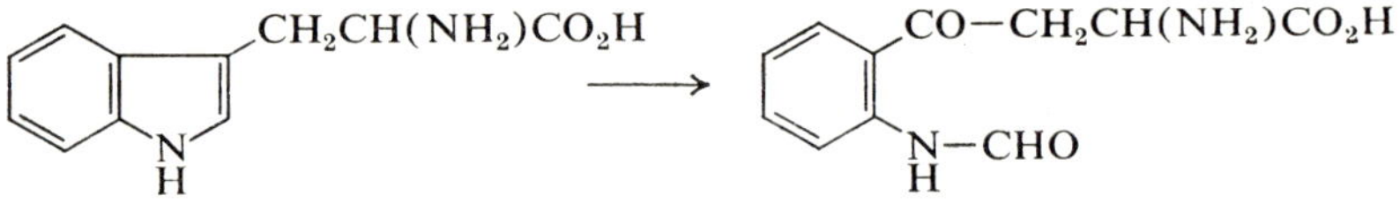


(LIV)

cleaved in the presence of tryptophan oxygenase (1.13.1.12) giving L-formylkynurenine (LIV). For the catechol 2,3-oxygenase reaction,

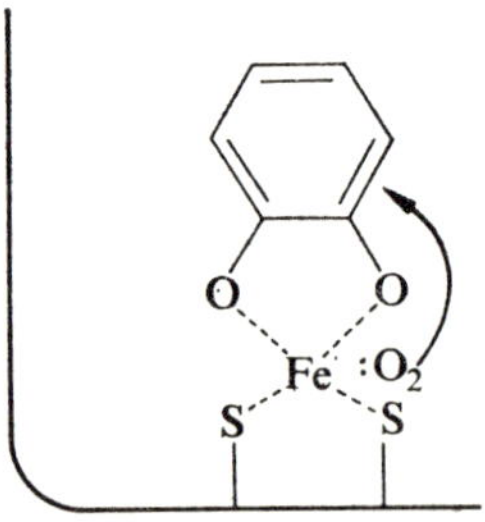


FIG. 3.4. Proposed arrangement of iron atom, substrate, oxygen molecule and thiol groups in catechol 2,3-oxygenase (By courtesy of O. Hayaishi and John Wiley & Sons Ltd)

Hayaishi [120] proposed a mechanism in which oxygen becomes bound to the ferrous ion in the enzyme (Fig. 3.4) and in which the aromatic ring is attacked by what is essentially an oxygen radical ($O_2^{\ominus}$) bound to a metal-ion (Fig. 3.5). In a suggested mechanism for the tryptophan oxygenase

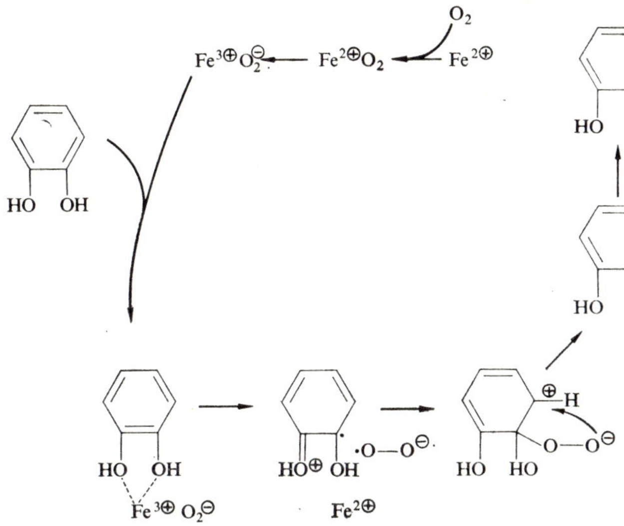


FIG. 3.5. Proposed mechanism for ring cleavage in the presence of catechol 2,3-oxygenase. (By courtesy of O. Hayaishi and John Wiley & Sons Ltd)

reaction [121] Hayaishi again suggests that the attacking species is an oxygen radical bound to a metal-ion (Fig. 3.6).

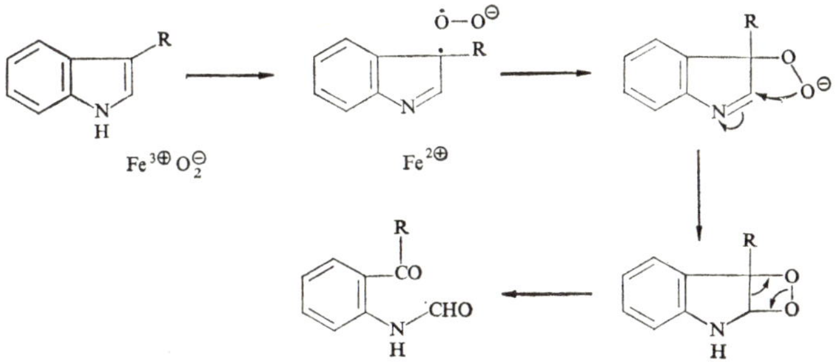


FIG. 3.6. Proposed mechanism for the tryptophan oxygenase reaction. (By courtesy of O. Hayaishi and John Wiley & Sons Ltd)

## Molecular Oxygen as an Electron Acceptor

In general we find that those enzymes which can utilize molecular oxygen as an electron acceptor contain either flavins or metal-ions (or both). Handler [107] reports evidence that suggests that oxygen radicals are formed only by metal-containing enzymes. If this is the case then the reduction of molecular oxygen by metal-free flavoenzymes must involve one-stage two-electron processes such as that suggested on page 126 for the non-enzymic reaction of $FMNH_2$ with oxygen, or that suggested for the reduction of oxygen in the case of the amino-acid oxidases (page 133). In metal-containing enzymes, the oxygen appears to form an association with the reduced form of the metal. A one-electron transfer then gives the oxygen radical and the oxidized form of the metal-ion. In some cases, such as dihydroorotate dehydrogenase, this oxygen radical, which may be detected by the various chemical methods available, takes up a second electron, eventually to yield hydrogen peroxide. In other cases, the metal-bound oxygen radical may attack a substrate and either hydroxylation or ring cleavage can result.

## REFERENCES


1. Glasstone, S., *Textbook of Physical Chemistry* (Macmillan, London, 1962)
2. Mahler, H. R. and Cordes, E. H., *Biological Chemistry* p. 207 (Harper and Row, New York, 1966)
3. Dixon, M. and Webb, E. C., *Enzymes*, 2nd edn. (Longmans, London, 1964)
4. Doering, W. von E. and Aschner, T. C., *J. Am. chem. Soc.* **75,** 393 (1953)
5. Molt, E. L., *Recl Trav. chim.* **56,** 233 (1937); Eitel, A. and Lock, G., *Monatsh.* **72,** 392 (1939)
6. Tommila, E., *Ann. Acad. Sci. Fennicae* **59A,** 8 (1942); Hammett, L. P., *Physical Organic Chemistry* p. 350 (McGraw-Hill, New York, 1940)
7. Fredenhagen, H. and Bonhoeffer, K. F., *Z. physik. Chem.* **181A**, 379 (1938)
8. Alexander, E. R., *J. Am. chem. Soc.* **69,** 289 (1947)
9. Woodward, R. B., Wendler, N. L. and Brutschy, F. J., *J. Am. chem. Soc.* **67,** 1425 (1945)
10. Forbes, E. J. and Gray, C. J., *Tetrahedron* **24,** 6223 (1968)
11. Schmid, H. and Karrer, P., *Helv. chim. Acta* **32,** 960 (1949)
12. Abeles, R. H., Hutton, R. F. and Westheimer, F. H., *J. Am. chem. Soc.* **79,** 712 (1957)
13. Dunn, G. E. and Warkentin, J., *Can. J. Chem.* **34,** 75 (1956)
14. Moore, M. L., *Organic Reactions*, Eds. R. Adams, W. E. Bachman, A. H. Blatt, L. F. Fieser, J. R. Johnson and H. R. Snyder, Vol. 5, p. 301 (Wiley, New York)
15. Leonard, N. J. and Sauers, R. R., *J. Am. chem. Soc.* **79,** 6210 (1957)
16. Baker, J. W. and Heggs, T. G., *J. chem. Soc.* 616 (1955); Buncel, E. and Bourns, A. N., *Can. J. Chem.* **38,** 2457 (1960)
17. Leo, A. and Westheimer, F. H., *J. Am. chem. Soc.* **74**, 4383 (1952); Westheimer, F. H., *Chem. Rev.* **45,** 419 (1949)
18. Barker, I. R. L., Overend, W. G. and Rees, C. W., *Chemy Ind.* 1297, 1298 (1960); Bentley, R., *Nature, Lond.* **176,** 870 (1955); Bentley, R., *The Enzymes* Vol. 7, p. 567 (Academic Press, New York, 1963)

19. HUDIS, J. and DODSON, R. W., *J. Am. chem. Soc.* **78,** 911 (1956); HALPERN, J., *Q. Rev. chem. Soc.* **15,** 207 (1961)
20. BARTON, D. H. R. and COHEN, T., *Festschrift A. Stoll* p. 117 (Birkhauser, Basel, 1957)
21. MICHAELIS, M., *Currents in Biochemical Research*, Ed. D. E. Green, p. 213 (Interscience, New York, 1946)
22. BATEMAN, L., *Q. Rev. chem. Soc.* **8,** 147 (1954)
23. KAHN, N. A., *J. chem. Phys.* **22,** 2090 (1954)
24. BATEMAN, L., HUGHES, H. and MORRIS, A. L., *Diss. Faraday Soc.* **14,** 190 (1953)
25. GERSMANN, H. R. and BICKEL, A. F., *J. chem. Soc.* 2356 (1962)
26. LEVITSKI, A., PECHT, I. and ANBAR, M., *Nature, Lond.* **207,** 1386 (1965)
27. FENTON, H. J. H., *J. chem. Soc.* **65,** 899 (1894); FENTON, H. J. H. and JONES, H. O., *J. chem. Soc.* **77,** 69 (1900)
28. NORMAN, R. O. C. and LINDSAY SMITH, J. R., *Oxidases and Related Redox Systems*, Eds. T. E. King, H. S. Mason and M. Morrison, Vol. 1, p. 131 (Wiley, New York, 1965)
29. UDENFRIEND, S., CLARK, C. T., AXELROD, J. and BRODIE, B. B., *J. biol. Chem.* **208,** 731 (1954); BRODIE, B. B., AXELROD, J., SHORE, P. A. and UDENFRIEND, S., *J. biol. Chem.* **208,** 741 (1954)
30. STAUDINGER, H., KEREKJARTO, B., ULLRICH, V. and ZUBRZYICKI, Z., *Oxidases and Related Redox Systems*, Vol. 2, p. 815 (Wiley, New York, 1965)
31. KAPLAN, N. O., *The Enzymes* Vol. 3, p. 105 (Academic Press, New York, 1959)
32. WARBURG, O. and CHRISTIAN, W., *Biochem. Z.* **286,** 81, 142 (1936)
33. VENNESLAND, B. and WESTHEIMER, F. H., *The Mechanism of Enzyme Action*, Ed. W. D. McElroy and B. Glass, p. 357 (Johns Hopkins Press, Baltimore, 1954)
34. CORNFORTH, J. W. and RYBACK. G., *Biochem. biophys. Res. Commun.* **9,** 371 (1962)
35. DICKINSON, F. M. and DALZIEL, K., *Biochem. J.* **104,** 165 (1967)
36. LEVY, H. R. and VENNESLAND, B., *J. biol. Chem.* **228,** 85 (1957)
37. EHRENBERG, A. and DALZIEL, K., *Acta chem. scand.* **12,** 465 (1958); EHRENBERG, A., *Acta chem. scand.* **11,** 1257 (1957)
38. THEORELL, H. and BONNICHSEN, R., *Acta chem. scand.* **5,** 1105 (1951); THEORELL, H. and WINER, A. D., *Archs biochem. Biophys.* **38,** 291 (1959)
39. THEORELL, H., *Harvey Lectures, Series 61* 17 (1966)
40. DIXON, M. and WEBB, E. C., *Enzymes*, 2nd edn p. 368 (Longmans, London, 1964)
41. PULLMAN, M. E., COLOWICK, S. P. and KAPLAN, N. O., *J. biol. Chem.* **194,** 593 (1952)
42. FAWCETT, C. P. and KAPLAN, N. O., *J. biol. Chem.* **237,** 1709 (1962)
43. SUND, H. and THEORELL, H., *The Enzymes* Vol. 7, p. 25 (Academic Press, New York, 1963)
44. HOLLIS, D. P., *Biochemistry* **6,** 2080 (1967)
45. WINER, A. D. and THEORELL, H., *Acta chem. scand.* **14,** 1729 (1960)

46. VALLEE, B. L. and FOCH, F. L., *J. biol. Chem.* **225,** 185 (1957); THEORELL, H., NYGAARD, A. P. and BONNICHSEN, R., *Acta chem. scand.* **9,** 1148 (1955)
47. AKESON, A., *Biochem. biophys. Res. Commun.* **17,** 211 (1964); OPPENHEIMER, H. L. and McKAY, R. H., *Fedn Proc.* **25,** 585 (1966)
48. DRUM, D. E., HARRISON, J. H., LI, T.-K., BETHUNE, J. L. and VALLEE, B. L., *Proc. natn Acad. Sci. U.S.A.* **57,** 1434 (1967)
49. VALLEE, B. L. and COOMBS, T. L., *J. biol. Chem.* **234,** 2615 (1959)
50. PLANE, R. A. and THEORELL, H., *Acta chem. scand.* **15,** 1866 (1961)
51. SIGMAN, D. S., *J. biol. Chem.* **242,** 3815 (1967)
52. YONETANI, T. and THEORELL, H., *Archs biochem. Biophys.* **106,** 243 (1964)
53. KOSOWER, E. M., *Molecular Biochemistry* p. 217 (McGraw-Hill, New York, 1962)
54. THEORELL, H. and CHANCE, B., *Acta chem. scand.* **5,** 1127 (1951)
55. DALZIEL, K., *Acta chem. scand.* **11,** 1706 (1957)
56. DALZIEL, K., *Biochem. J.* **84,** 244 (1962)
57. THEORELL, H. and McKINLEY-McKEE, J. S., *Acta chem. scand.* **15,** 1797 (1961)
58. BOYER, P. D. and THEORELL, H., *Acta chem. scand.* **10,** 447 (1956)
59. THEORELL, H., EHRENBERG, A. and DE ZALENSKI, C., *Biochem. biophys. Res. Commun.* **27,** 309 (1967)
60. THEORELL, H., CHANCE, B. and YONETANI, T., *J. molec. Biol.* **17,** 513 (1966)
61. NYGAARD, A. P. and THEORELL, H., *Acta chem. scand.* **9,** 1300 (1955); MAHLER, H. R. and Douglas, J., *J. Am. chem. Soc.* **79,** 1159 (1957); HAYES, J. E. and VELICK, S. F., *J. biol. Chem.* **207,** 225 (1954)
62. SCHELLENBERG, K. A., *J. biol. Chem.* **240,** 1165 (1965); *J. biol. Chem.* **241,** 2446 (1966)
62a. GUTFREUND, H. and RABIN, B. R., Communicated to the British Biophysical Society Meeting (London, Dec. 1968)
63. BURR, G. O. and GORTNER, R. A., *J. Am. chem. Soc.* **46,** 1224 (1924); THESING, J. and MAYER, H., *Chem. Ber.* **87,** 1084 (1954)
64. MERRITT, A. D. and THOMKINS, G. M., *J. biol. Chem.* **234,** 2778 (1958)
65. DALZIEL, K. and DICKINSON, F. M., *Biochem. J.* **100,** 491 (1966)
66. VAN EYS, J. and KAPLAN, N. O., *J. Am. chem. Soc.* **79,** 2782 (1957)
67. WINER, A. D., *Acta chem. scand.* **12,** 1695 (1958)
68. VEIGA SALLES, J. B. and OCHOA, S., *J. biol. Chem.* **187,** 849 (1950)
69. KUN, E., *The Enzymes* Vol. 7, p. 149 (Academic Press, New York, 1963)
70. BOYER, P. D., *Ann. Rev. Biochem.* **29,** 15 (1960)
71. SAZ, H. J. and HUBBARD, J. A., *J. biol. Chem.* **225,** 921 (1957)
72. KORNBERG, A. and PRICER, W. E., *J. biol. Chem.* **189,** 123 (1951)
73. ABELES, R. H. and LEE, H. A., JR., *J. biol. Chem.* **235,** 1499 (1960)
74. NOLTMANN, E. A. and KUBY, S. A., *The Enzymes* Vol. 7, p. 223 (Academic Press, New York, 1963)
75. WALEY, S. G., *Q. Rev. chem. Soc.* **21,** 379 (1967)
76. VELICK, S. F. and FURFINE, C., *The Enzymes* Vol. 7, p. 243 (Academic Press, New York, 1963)
77. VELICK, S. F. and HAYES, J. E., *J. biol. Chem.* **203,** 545 (1953)

78. MATHEW, E., Meriweather, B. P. and PARK, J. H., *J. biol. Chem.* **242,** 5024 (1967)
79. HARTING, J., *Fedn Proc.* **10,** 195 (1951); HARTING, J. and VELICK, S. F., *J. biol. Chem.* **207,** 857 (1954)
80. PARK, J. H., MERIWEATHER, B. P., CLODFELDER, P. and CUNNINGHAM, L. W. *J. biol, Chem.* **236,** 136 (1961)
81. HARRIS, J. I., MERIWEATHER, B. P. and PARK, J. H., *Nature, Lond.* **198,** 154 (1963); PERHAM, R. N., *Biochem. J.* **99,** 14C (1966)
82. HARRIS, J. I. and POLGER, L., *J. molec. Biol.* **14,** 630 (1965)
83. PARK, J. H., AGNELLO, C. F. and MATHEW, E., *J. biol. Chem.* **241,** 769 (1966)
84. BOLOTINA, I. A., MARKOWICH. D. S., VOLKENSTEIN, M. V. and ZAVODSKY, P., *Biochim. biophys. Acta* **132,** 260 (1967)
85. LISTOWSKY, I., FURFINE, C., BETHEIL, I. and ENGLARD, S., *J. biol. Chem.* **240,** 4253 (1965)
86. JAKOBY, W. B., *The Enzymes* Vol. 7, p. 203 (Academic Press, New York, 1963)
87. PENZER, G. R. and RADDA, G. K., *Q. Rev. chem. Soc.* **21,** 43 (1967)
88. GIBSON, Q. H. and HASTINGS, J. W., *Biochem. J.* **83,** 368 (1962)
89. FOX, J. L. and TOLLIN, G., *Biochemistry* **5,** 3865, 3873 (1966)
90. SINGER, T. P. and KEARNEY, E. B., *J. biol. Chem.* **183,** 409 (1950)
91. SUELTER, C. H. and METZLER, D. E., *Biochim. biophys. Acta* **44,** 23 (1960)
92. GASCOIGNE, I. M. and RADDA, G. K., *Chem. Commun.* 211 (1965)
93. SEARLS, R. L., PETERS, J. M. and SAMADI, D. R., *J. biol. Chem.* **236,** 2317 (1961); MATTHEW, J. and REED, L. J., *J. biol. Chem.* **238,** 1869 (1963); WILLIAMS, C. H., JR., *J. biol. Chem.* **240,** 4793 (1965)
94. IDE, S., HAYAKAWA, T., OKABE, K. and KOIKE, M., *J. biol. Chem.* **242,** 54 (1967)
95. MASSEY, V. and VEEGER, C., *Biochim. biophys. Acta* **48,** 33 (1961)
96. MASTERS, B. S. S., BILIMORIA, M. H. and KAMINS, H., *J. biol. Chem.* **240,** 4081 (1965)
97. GIBSON, Q. H., SWOBODA, B. E. P. and MASSEY, V., *J. biol. Chem.* **239,** 3927 (1964)
98. BRIGHT, H. J. and GIBSON, Q. H., *J. biol. Chem.* **242,** 994 (1967)
99. GIBSON, Q. H., SWOBODA, B. E. P. and MASSEY, V., *J. biol. Chem.* **239,** 3927 (1964)
100. MASSEY, V. and GIBSON, Q. H., *Fedn Proc.* **23,** 18 (1964)
101. MEISTER, A. and WELLNER, D., *The Enzymes* Vol. 7, p. 609 (Academic Press, New York, 1963)
102. MASSEY, V., GANTHER, H., BRUMBY, P. E. and CURTI, P., *Oxidases and Related Redox Systems* Vol. 1, p. 335 (Wiley, New York, 1965)
103. MASSEY, V. and CURTI, P., *J. biol. Chem.* **242,** 1259 (1967)
104. HELLERMAN, L. and COFFEY, D. S., *J. biol. Chem.* **242,** 582 (1967)
105. ALEMAN-ALEMAN, V., RAJAGOPALAN, K. V., HANDLER, P., BEINERT, H. and PALMER, G., *Oxidases and Related Redox Systems* Vol. 1, p. 380 (Wiley, New York, 1965)

106. Aleman, V. and Handler, P., *J. biol. Chem.* **242,** 4087 (1967)
107. Handler, P., *Oxidases and Related Redox Systems* Vol. 1, p. 399 (Wiley, New York, 1965)
108. Miller, R. W. and Massey, V., *J. biol. Chem.* **240,** 1453 (1965)
109. Bray, R. C., *The Enzymes* Vol. 7, p. 533 (Academic Press, New York, 1963)
110. Bray, R. C., Palmer, G. and Beinert, H., *Oxidases and Related Redox Systems* Vol. 1, p. 359 (Wiley, New York, 1965)
111. Rajagopalan, K. V. and Handler, P., *J. biol. Chem.* **239,** 2027 (1964)
112. Handler, P., *Oxidases and Related Redox Systems* Vol. 1, p. 400 (Wiley, New York, 1965)
113. Meriweather, L. S., Marzluff, W. F. and Hodgson, W. G., *Nature, Lond.* **212,** 465 (1966); Bray, R. C. and Meriweather, L. S., *Nature, Lond.* **212,** 469 (1966)
114. Davidson, A. J. and Wainio, W. W., *Fedn Proc.* **23,** 323 (1964)
115. Rajagopalan, K. V. and Handler, P., *J. biol. Chem.* **242,** 4097 (1967)
116. Trotter, J. R., de Dugros, E. C. and Riviero, C., *J. biol. Chem.* **235,** 1839 (1960); Stauff, J., Schmidkunz, H. and Hartman, G., *Nature, Lond.* **198,** 281 (1963)
117. Mason, H. S., Foulks, W. L. and Peterson, E., *J. Am. chem. Soc.* **77,** 2914 (1957)
118. Mason, H. S., *Adv. Enzymol.* **19,** 79 (1957); *Science* **125,** 1185 (1957)
119. Goldstein, M., Lauber, E., Blumberg, W. E. and Peisach, J., *Fedn Proc.* **24,** 604 (1965)
120. Hayaishi, O., *Oxidases and Related Redox Systems* Vol. 1, p. 286 (Wiley, New York, 1965)
121. Hayaishi, O., *Oxidases and Related Redox Systems* Vol. 1, p. 527 (Wiley, New York, 1965)

# CHAPTER 4

# HYDROLYSIS I

It has already been pointed out that the Enzyme Commission classification for enzymes is, of necessity, based on the nature of the reaction catalysed and not on the type of mechanism involved. Since we are concerned primarily with mechanism the order in which we shall consider the enzymes may deviate from that set out by the Commission. This is especially the case for the hydrolytic enzymes where enzymes which catalyse similar reactions, for example various peptidases, may do so by different mechanisms. In addition, some peptidases function by a mechanism similar to that utilized by some esterases.

In this chapter we consider the enzymes which catalyse the hydrolysis of carboxylic acid derivatives, that is, of esters, amides and peptides. We have already seen in chapter 1 how the hydrolysis of these derivatives may be catalysed by acids, bases and metal-ions, and also by certain nucleophiles such as imidazole or acetate which provide alternative pathways. There is good reason to believe that the catalysis of hydrolytic reactions by enzymes may be explained to some extent in terms of these processes.

For our purposes, it is convenient to classify these enzymes into the following groups. Enzymes which belong to the first group probably function by very similar mechanisms; the same may be said of the second group. At present we know very much less about the mechanism of action of any enzymes in the last two groups.

1. Peptide- and carboxylic ester-hydrolases which possess active serine residues.
2. Those which possess active cysteine residues.
3. Metal-ion dependent hydrolases.
4. Peptide hydrolases which function at very low pH values.

## HYDROLASES WHICH POSSESS ACTIVE SERINE RESIDUES

This group includes certain peptide hydrolases such as chymotrypsin A (3.4.4.5), trypsin (3.4.4.4), thrombin (3.4.4.13), plasmin (3.4.4.14) and subtilopeptidase A (3.4.4.16) and certain esterases such as carboxylesterase (3.1.1.1), acetylcholinesterase (3.1.1.7) and cholinesterase (3.1.1.8).

Their most important common property is that each is inhibited by organic phosphorus compounds like di-isopropylfluorophosphate (I) and that a phosphate ester is produced by reaction of the reagent with the hydroxymethyl group of a serine residue in the protein chain [1]. Of these enzymes

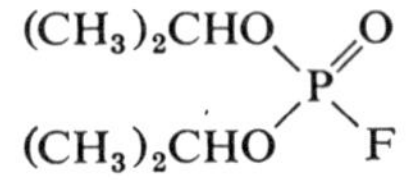

(I)

chymotrypsin A has been studied in by far the greatest detail and it is now possible to formulate a mechanism for catalysis by this enzyme. It may be assumed that the other enzymes in the group catalyse by essentially similar mechanisms, and indeed in some cases there is good evidence for this assumption, although there may be minor differences.

**Chymotrypsin A (3.4.4.5)**

Chymotrypsin A, like the closely-related enzyme trypsin and the even more closely-related chymotrypsin B (3.4.4.6), occurs in the pancreatic juice of mammals as an inactive precursor which is converted into the active enzyme in the duodenum. It is a proteolytic enzyme and its biological function is almost certainly protein degradation in digestion.

The inactive precursor, chymotrypsinogen A (the Enzyme Commission prefers the term pre-chymotrypsin A), is a single polypeptide-chain protein of molecular weight 25,000. Five intramolecular disulphide bridges help to hold the molecule in its natural conformation. The amino-acid sequence is shown in Fig. 4.1. Conversion of pre-chymotrypsin into the various active forms of chymotrypsin involves a complicated sequence of events. Essentially the process includes conversion by trypsin into $\pi$-chymotrypsin by cleavage of the arginyl-15—isoleucine-16 bond followed by autocatalytic cleavage of the leucyl-13—serine-14 bond giving $\delta$-chymotrypsin. Two additional processes take place in which the tyrosyl-146—threonine-147 and the asparaginyl-148—alanine-149 bonds are broken. The final result is $\alpha$-chymotrypsin [3]. The overall process therefore involves breakage of four peptide bonds and the loss of two dipeptides. $\alpha$-Chymotrypsin as a consequence consists of three polypeptide chains, two of the disulphide bridges having become *inter-chain*. Two new C-terminal residues (leucine-13 and tyrosine-146) and two new N-terminal residues (isoleucine-16 and alanine-149) are produced. Presumably the loss of the two peptides allows the protein to take up a different conformation in which the active site of the enzyme is generated. That a conformational change occurs is indicated by the observation that activation of pre-chymotrypsin causes a decrease in optical rotation, which suggests that the protein becomes more compactly coiled [4]. The ultraviolet spectrum also changes [5]. However, since the pre-enzyme, although inactive, possesses the ability to bind with certain potential substrates [6], the conformational change cannot be a drastic one.

Chymotrypsin most readily effects the cleavage of tyrosyl, tryptophany

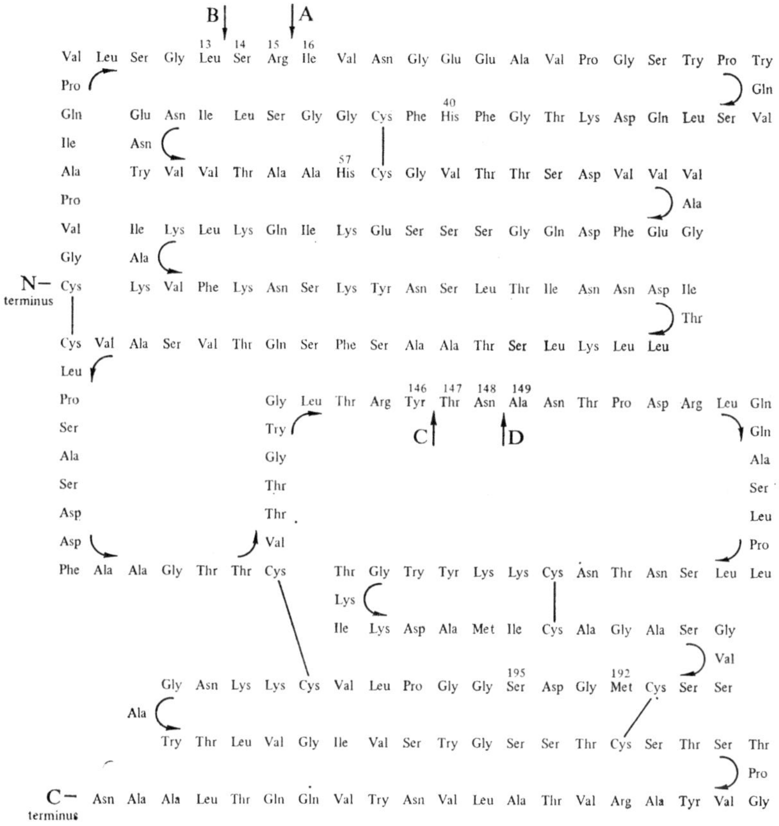


FIG. 4.1. Amino-acid sequence of α-chymotrypsinogen (pre-chymotrypsin) [2]. The curved arrows indicate the sequence from the N-terminus to the C-terminus. A, B, C and D show bonds broken during conversion to α-chymotrypsin. (By courtesy of B. S. Hartley and the Editorial Board of the Biochemical Journal)

and phenylalanyl peptides, amides and esters [7]. Its main specificity therefore is said to be directed towards derivatives of the aromatic amino-acids. However, derivatives of other amino-acids such as leucine, methionine, asparagine and glutamine are also hydrolysed efficiently and this indicates that the specificity is not a narrow one. Indeed, esters such as p-nitrophenyl acetate may be hydrolysed at appreciable rates.

*Chemical Studies with Chymotrypsin.* One of the most important chemical processes which involve the activity of chymotrypsin is the activation process itself. In Fig. 4.1 the four peptide bonds which are cleaved are labelled A, B, C and D. Cleavage of bond A leads to $\pi$-chymotrypsin and subsequent cleavage of B affords the $\delta$-form. This is converted into $\alpha$-chymotrypsin by breakage at C and D. Under other conditions a different intermediate is formed; this is neo-chymotrypsin in which bonds C and D have been broken but A and B are intact. Neo-chymotrypsin is inactive whereas the other forms, $\alpha$-, $\pi$- and $\delta$-, are active. Quite clearly it is cleavage of bond A which confers activity on the protein. Cleavage of A produces the new N-terminal residue, isoleucine-16.

More information about the importance of this N-terminal isoleucine residue was obtained by Hess and his co-workers [8]. When pre-chymotrypsin was treated with acetic anhydride all the amino-groups were acetylated. Treatment of the product with trypsin afforded an active chymotrypsin which was able to catalyse the hydrolysis of N-acetyl-L-tyrosine ethyl ester and which reacted with di-isopropylfluorophosphate (DFP) (I). A single N-terminal amino-acid residue was demonstrated in this active material; this was isoleucine-16. Further acetylation, which blocked the isoleucine amino-group, caused complete loss of activity. The dependence of the catalytic activity of chymotrypsin on pH suggests that a group with $pK_a$ of about 9 is involved in some way and that for activity this group must be in its conjugate acid form [6, 9]. It is probable that this is the amino-group of the isoleucine-16 residue.

Chemical reaction of other amino-acid residues affects the catalytic activity of the enzyme; one is the methionine at position-192. Knowles [10] has shown that oxidation of this methionine side-chain causes a decrease in activity but not a total loss. It is likely that this oxidation affects the binding of some substrates rather than the catalytic process itself. This is

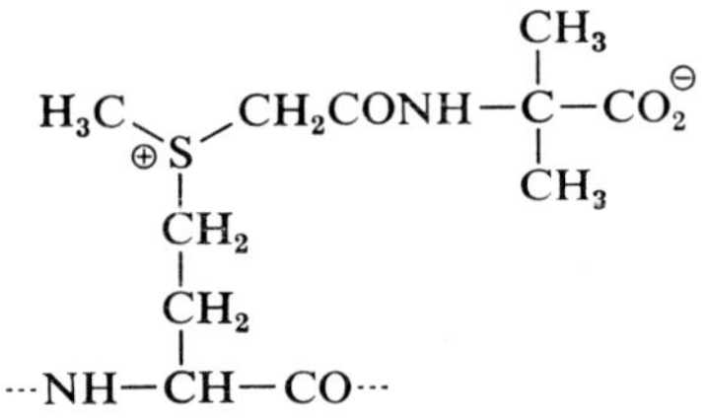

(II)

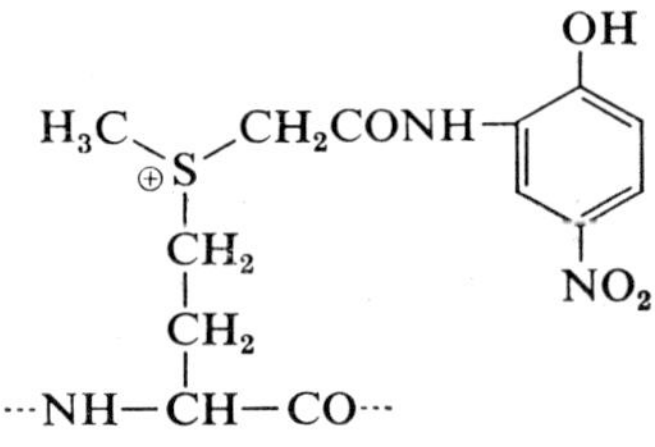

(III)

supported by the observation that alkylation of the methionine residue with the N-(1,1-dimethyl-1-carboxymethyl)-acetamido group which affords the methionine derivative (II) also causes a decrease in activity by interfering with the binding of some substrates [11]. A similar result was obtained by Hille and Koshland, who attached a 'reporter-group' (III) to the methionine residue [12]. Again, the product showed decreased activity. These and other results suggest that the methionine residue at position-192 is in the vicinity of the active site but not involved in the catalytic process. It may make a contribution to the binding of the substrate to the enzyme.

As a result of various oxidation studies it has been suggested that a tryptophan residue is present at or near the active site [13, 14].

The critical importance of one or more histidine residues to the activity of chymotrypsin has been demonstrated many times. Photo-oxidation studies have been reported [13, 15]. In particular, Koshland, Strumeyer and Ray showed that under certain conditions photo-oxidation resulted in the destruction of one histidine residue and the oxidation of one methionine (at position-192) to its sulphoxide. Using their 'all-or-nothing' assay procedure these workers were able to show that destruction of the histidine residue caused total inactivation, whereas oxidation of the methionine caused only partial loss [15]. The histidine involved in this reaction is that at position-57.

Whitaker and Jandorf treated chymotrypsin with dinitrofluorobenzene at pH 10·7 under which conditions only the imidazole groups of histidine residues react with the reagent [16]. They showed that reaction of one histidine residue per molecule could cause complete loss of activity. Similarly, Schramm [17] showed that the inactivation of chymotrypsin by 1,4-dibromo-2-phenylacetoin was caused by the reaction of this reagent with the histidine residue at position-57.

The involvement of one or more histidine residues in the catalytic process has been assumed generally, because of the way in which the enzyme's activity varied with pH [18], although conclusions drawn from this type of evidence must be treated with some caution.

The use by Schoellman and Shaw [19] of chloroketones related to enzyme substrates has afforded the most striking evidence for the involvement of a histidine residue at the active site of chymotrypsin. Esters of

$$\begin{array}{c} \phantom{CH_3-C_6H_4-SO_2-NH-}C_6H_5\phantom{-CO_2R} \\ \phantom{CH_3-C_6H_4-SO_2-NH-}|\phantom{-CO_2R} \\ \phantom{CH_3-C_6H_4-SO_2-NH-}CH_2\phantom{-CO_2R} \\ \phantom{CH_3-C_6H_4-SO_2-NH-}|\phantom{-CO_2R} \\ CH_3-C_6H_4-SO_2-NH-CH-CO_2R \end{array}$$

(IV)

$$\begin{array}{c} \phantom{CH_3-C_6H_4-SO_2-NH-}C_6H_5\phantom{-COCH_2Cl} \\ \phantom{CH_3-C_6H_4-SO_2-NH-}|\phantom{-COCH_2Cl} \\ \phantom{CH_3-C_6H_4-SO_2-NH-}CH_2\phantom{-COCH_2Cl} \\ \phantom{CH_3-C_6H_4-SO_2-NH-}|\phantom{-COCH_2Cl} \\ CH_3-C_6H_4-SO_2-NH-CH-COCH_2Cl \end{array}$$

(V)

N-toluenesulphonyl-L-phenylalanine (IV) are good substrates for the enzyme, an important point being that they are strongly bound to the active site. Schoellman and Shaw treated chymotrypsin with the chloro-compound (V) which is obviously related to the ester (IV) and which is usually known by the trivial name of tosyl-phenylalanine chloroketone (TPCK). The product, in which histidine-57 had become alkylated, was inactive [20]. Trypsin was not affected by this reagent. Its specificity is not for aromatic amino-acid derivatives but for derivatives of the basic amino-acids, lysine and arginine. Trypsin is completely inhibited by a similar derivative of lysine, tosyl-lysine chloroketone (TLCK) (VI) which, however, has no effect on chymotrypsin.

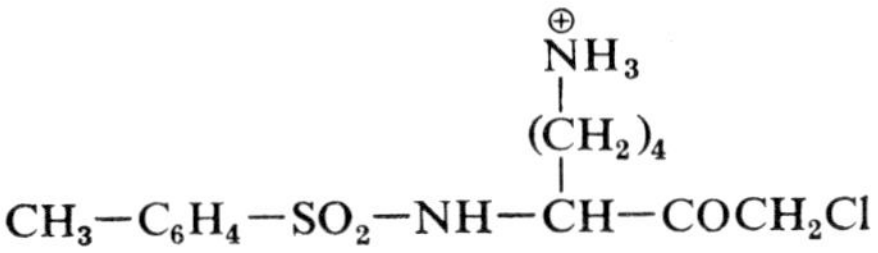


(VI)

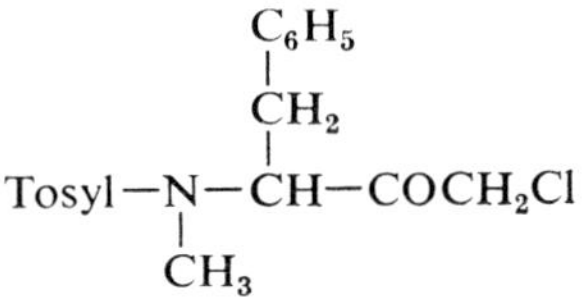


(VII)

Thus inhibition by these chloroketones is subject to the same specificities as those which apply to hydrolysis of ester substrates. This point is emphasized by the fact that N-methyl-TPCK (VII) is not an inhibitor, nor is the D-isomer of TPCK [21]. The corresponding derivatives of phenylalanine esters are not substrates. That TPCK reacts at the active site of the chymotrypsin is shown by the fact that the product is inactive and supported by evidence that common inhibitors of chymotrypsin catalysis such as $\beta$-phenylpropionate and benzamide decrease the rate of the inactivation [21]. Furthermore the alkylation of methionine-192 which interferes with enzyme–substrate binding also inhibits the reaction between TPCK and the enzyme [11].

Many important chemical modifications of chymotrypsin have involved the hydroxymethyl group of a serine residue. The inhibition of the enzyme by dialkylfluorophosphates has already been mentioned. As a result of this inhibition onc dialkylphosphoryl residue was found per molecule in the inactive product, and on acid hydrolysis phosphoserine was isolated [22]. Enzymic digestion of the dialkylphosphoryl-chymotrypsin gave a peptide which carried the phosphoryl group attached to a serine; sequence analysis of this peptide showed that the serine is at position-195 [23]. Similarly, if p-nitrophenyl acetate reacts with chymotrypsin under acid conditions an intermediate may be isolated in which the acetyl group is covalently bound to the enzyme. Analysis of peptides obtained by enzymic digestion of this inactive acetyl-chymotrypsin showed again that serine-195 was involved [24, 25]. Many other acyl-chymotrypsins, prepared in a similar way, may be isolated and in each case there appears to be an ester link between serine-195 and the acyl-residue [26].

The absolute requirement of the hydroxyl group of the serine-195 residue for the catalytic process has been clearly demonstrated by Koshland and his colleagues [27]. The mono-toluenesulphonyl derivative of chymotrypsin (tosyl-chymotrypsin), which contains a sulphonyl-serine ester link, is inactive. This might have been considered to be due to the size of the tosyl group. If the serine residue is at or near the active site the presence of

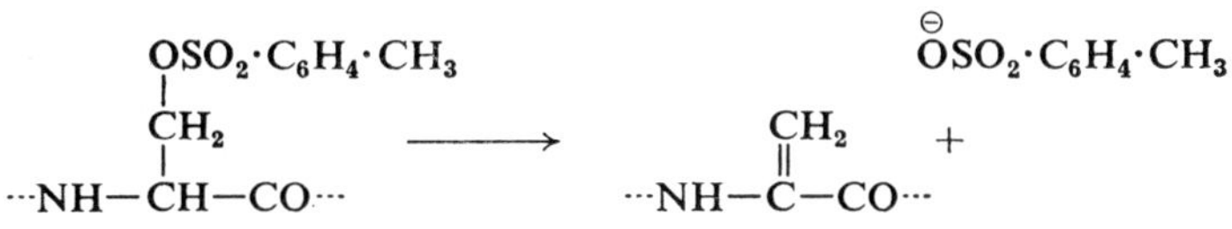

(VIII)

a bulky substituent could prevent the binding of the substrate. However, elimination of toluenesulphonic acid results in the conversion of the original serine into a dehydroalanine residue (VIII). The product, dehydro-chymotrypsin, is also inactive. Interestingly, dehydro-chymotrypsin, like phosphoryl-chymotrypsin, does not react with the inhibitor TPCK [28].

*The Active Site Environment.* As in the case of other enzymes the active region of chymotrypsin constitutes a peculiar environment in which the reactions of the substrates may take place. From specificity considerations alone it may be supposed that a hydrophobic or non-polar region is present. This enables the enzyme to form a 'hydrophobic link' with the aromatic ring of the specific substrates. Here a direct contrast with trypsin is observed. Trypsin, whose specificity is essentially directed towards molecules which possess positively-charged side-chains, would be expected to contain a region which is either negatively charged or sufficiently polar to accommodate the charged group.

However, the most important consequence of the peculiar nature of the active site environment is that the reactivity of certain groups is greatly enhanced. For example the hydrolysis of acetyl-chymotrypsin, which yields acetic acid and the free enzyme, proceeds far more rapidly than would be expected for an ordinary O-acetyl derivative. Acetyl-chymotrypsin also reacts rapidly with hydroxylamine to give acethydroxamic acid. In a medium which contains 8M urea, however, under which conditions the protein is denatured and the active site destroyed, these reactions of the acetyl-enzyme proceed no faster than would be expected for an ordinary ester [29]. Bender, Schonbaum and Zerner have reported similar results for another acyl-enzyme, *trans*-cinnamoyl-chymotrypsin [30]. The rate of the non-enzymic hydrolysis of this compound in a solution of high urea concentration is almost identical to that of O-cinnamoyl-N-acetylserinamide. On dialysis of urea-containing solutions of acetyl-chymotrypsins, the active site is re-formed to such an extent that the high reactivity is regenerated.

The fact that this serine residue becomes acylated at all is an indication that the nucleophilicity of its hydroxyl group has been enhanced. There are twenty-nine serine residues in the whole molecule and only one becomes phosphorylated during reaction with DFP or acylated during reaction with carboxyl derivatives. Under normal conditions a serine hydroxyl would not react with these reagents. Indeed, the fact that a histidine residue is known to be involved at the active site has led to suggestions that two stages are involved in the formation of the final acyl-enzyme. According to these suggestions the imidazole group of a histidine residue attacks the reagent to form either a phosphoryl- or an acyl-imidazole. Subsequently the phosphoryl- or acyl-group is transferred to the serine. Such a mechanism is supported by the fact that imidazole under normal conditions can react with DFP to give a phospho-imidazole (IX), which is an effective

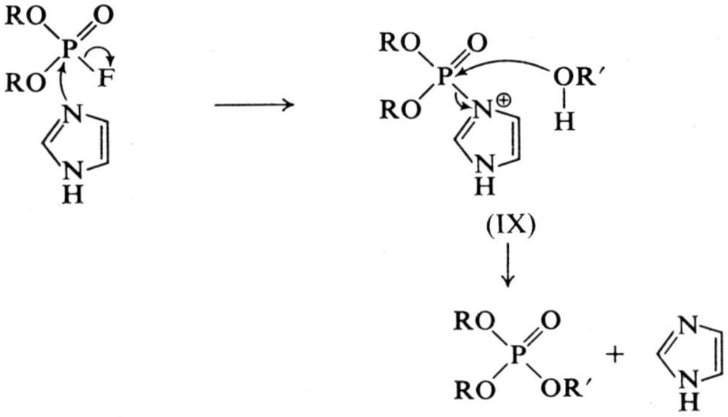

phosphorylating agent for alcohols [31]. In chapter 1 we saw that imidazoles may react with esters to give acyl-imidazoles, which are efficient acylating agents.

Several attempts have been made to detect acyl-imidazole intermediates in the chymotrypsin reaction using techniques suitable for fast reactions [18, 32]. At present it is clear that if an acyl-imidazole species is formed, it is not detectable with the techniques available; the first detectable intermediate is the O-acyl-serine species. When we discuss the mechanism of action of chymotrypsin (see later) we shall assume that an acyl-imidazole is not formed.

The special nature of the active site of chymotrypsin has been explored in several ways. One of these is the spectroscopic study by Kallos and Avatis [33]. These workers showed that the ultraviolet absorption spectra of p-nitrobenzenesulphonyl (nisyl) and toluenesulphonyl (tosyl) esters were different and that the extent of the difference depended on the polarity of the environment of the two groups. From the difference spectrum of nisyl- and tosyl-chymotrypsin they deduced that the aromatic substituents are buried in a strongly hydrophobic region. The dielectric constant of this region was estimated to be similar to that of cyclohexane. This region is quite inaccessible to water molecules. Disruption of the three-dimensional structure of the protein resulted in the displacement of the substituents into the aqueous medium with a consequent change in the spectrum. This portion of the active site is presumably involved in the binding of the aromatic groups of the substrates.

Using their 'reporter-group', a p-nitrophenol attached to methionine-192 via an acetamido bridge (see III), Hille and Koshland examined another region of the active site [12]. The absorption spectrum of the nitrophenol moiety indicated an environment more polar than water. The pH-dependence of the absorption spectrum suggested the close proximity of a positively-charged species with a $pK_a$ about 7. The presence of a substrate or a covalently-bound inhibitor eliminated the effect of this group. It was suggested that this group was the histidine known to be at the active site, and known to be important for activity. However, on the basis of later evidence, it would be more accurate to assign this $pK_a$ of about 7 to the overall active-site 'system' (see page 164 and Fig. 4.4).

The results quoted above suggest that there are two regions at the active site of chymotrypsin, one hydrophobic and the other highly polar. It would be supposed that all substrates containing aromatic groups would arrange themselves so that these groups fitted into the hydrophobic regions. The trans-cinnamoyl group has an aromatic nucleus and this might be expected to associate with the hydrophobic region when cinnamoyl-chymotrypsin is formed. However, Bender and his group [30] report rather different results. Thus when the ultraviolet spectrum of cinnamoyl-chymotrypsin is compared with spectra of O-cinnamoyl-N-acetylserinamide in various solvents, the indication is that the cinnamoyl group in the acyl-enzyme is in a region of very high polarity (roughly equivalent to 10M lithium chloride). This unexpected result may be due to either of two reasons. The cinnamoyl group may not be at the hydrophobic binding site but in the other region of the active centre where the polarity is high. This is an unlikely situation because the presence of an aromatic nucleus in a highly-polar environment would be energetically unfavourable. Alternatively, the cinnamoyl group may reside at the non-polar site but its ultraviolet spectrum may be distorted by some other as yet unknown factor.

Direct chemical evidence for the close proximity of a histidine residue to the active serine was obtained by Westheimer and his co-workers [34], who treated chymotrypsin with p-nitrophenyl diazoacetate (X). The resulting diazoacetyl-chymotrypsin in which the diazoacetyl resided on the active serine was photolysed and the product analysed. Hydrolysis gave, among other things, carboxymethylserine, which was produced apparently by rearrangement of O-diazoacetylserine. Also identified in the hydrolysate was 1-carboxymethylhistidine (XI), which was produced by a nucleophilic attack by the imidazole species on the nearby serine derivative (XII) followed by hydrolysis of the ester bond. It is interesting that carboxymethyl-

$$NO_2{-}C_6H_4{-}OCO{-}CHN_2$$

(X)

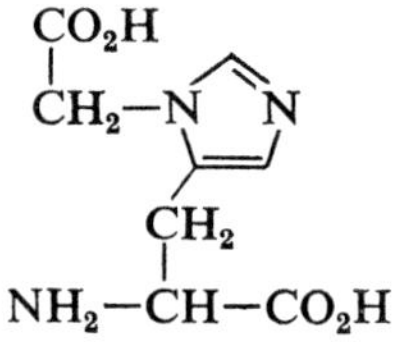

(XI)

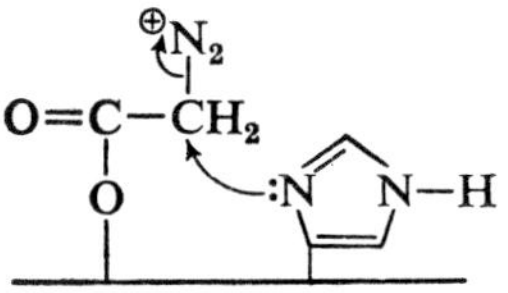

(XII)

tyrosine was also identified in the product. Westheimer did not suggest that a tyrosine residue is also in the vicinity of the active site of monomer-chymotrypsin but rather that a dimeric form of the enzyme is involved (see Fig. 4.3).

The chemical studies summarized above provide a great deal of information about the active site of chymotrypsin. Thus, the active centre contains a serine residue whose hydroxyl group is more than usually reactive and whose ester derivatives are also more than usually reactive. Near to this serine is a histidine residue which is essential for activity. Also nearby is a methionine residue, oxidation or alkylation of which appears to impede the catalytic reaction. The region encompassing the serine, histidine and methionine residues is very highly polar. Close by is another region which is non-polar, and which appears to be a major binding site for the non-polar side-chains of the substrate molecules.

*Kinetic Studies.* It is possible to distinguish in kinetic terms between three types of substrate for this enzyme. These are (a) specific substrates, derivatives of the aromatic amino-acids and similar compounds, in which an *ester* group is cleaved, (b) specific substrates which are *amides*, and (c) non-specific substrates which possess very good leaving groups, such as p-nitrophenyl acetate. Since different results were obtained using these different types of substrates there has been some doubt as to whether they are all hydrolysed by the same mechanism. However, it appears to be established now that the hydrolysis of all substrates by chymotrypsin follows the general scheme indicated in equation (4.1). According to this scheme $E$ represents the enzyme and $S$ the substrate. At first a noncovalent enzyme-substrate complex ($E—S$) is formed subject to the dissociation constant $K_s$. The next stage is formation of the acyl-enzyme ($E—S'$), with loss of the leaving group $P_1$. For example, during the hydrolysis of p-nitrophenyl acetate $E—S'$ refers to acetyl-chymotrypsin and $P_1$ to the p-nitrophenolate anion. The last step is reaction of the acyl-enzyme with some nucleophilic species, usually water, to give the product $P_2$ and regenerate the enzyme.

The best possible evidence for the involvement of an acyl-enzyme intermediate in such a reaction scheme is of course the isolation and identification of such an intermediate, and demonstration that the intermediate

$$\mathrm{E} + S \overset{K_S}{\rightleftharpoons} E—S \underset{k_{-2}}{\overset{k_{+2}}{\rightleftharpoons}} \underset{\displaystyle P_1}{\underset{+}{E—S'}} \underset{k_{-3}}{\overset{k_{+3}}{\rightleftharpoons}} E + P_2 \qquad (4.1)$$

undergoes the subsequent reactions. This has been achieved for several chymotrypsin-catalysed reactions. It has already been mentioned that it is possible to isolate acetyl-chymotrypsin from the reaction of the enzyme with p-nitrophenyl acetate. Reaction of the acetyl-derivative with water, hydroxylamine, ethanol or various amines can afford acetic acid, acethydroxamic acid, ethyl acetate or acetamides respectively. Similar results have been obtained with other acyl-derivatives [24, 35, 36].

Many other experiments support the intermediacy of an acyl-enzyme. Chymotrypsin catalyses oxygen exchange between water and the various carboxylic acids which may serve as the acyl portions of specific substrates [37]. This process is essentially the last step of the scheme (4.1) and its

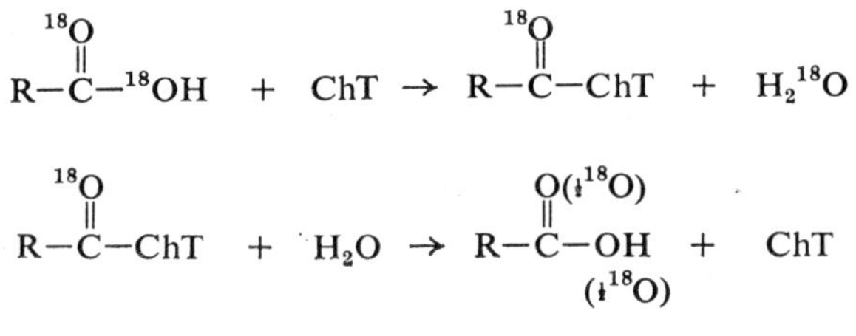

reversal. Another example is the partitioning of the acyl-enzyme between two nucleophiles. Thus, if the reaction is carried out in the presence of two nucleophilic compounds, for example water and an amine R′—$NH_2$, two products may be obtained, as shown in equation (4.2). The ratio of the two products formed will depend on $k_{+3}$ and $k_{+4}$ and on the conditions.

Therefore under the same conditions the reaction of two substrates which have the same acyl-enzyme must yield the products in the same ratios.

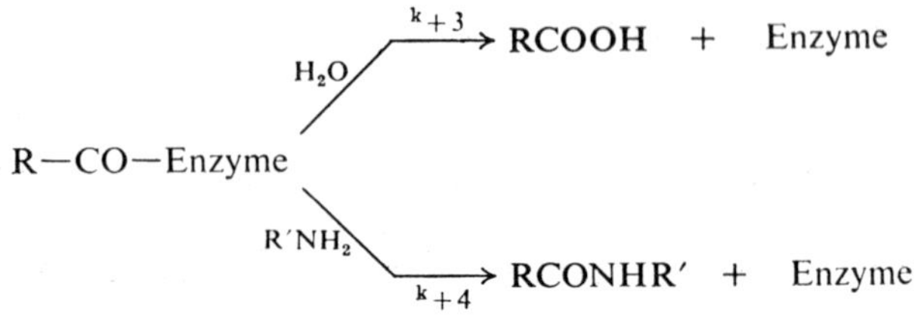


(4.2)

Epand and Wilson showed that for the reaction of chymotrypsin with various esters of hippuric acid in the presence of 0·5M hydroxylamine, the same fraction of hydroxamic acid was produced in each case [38, 39].

If the sequence shown in (4.1) is compared with the usual Michaelis–Menten scheme for an enzymic reaction in which no covalent intermediate is formed,

$$E + S \xrightarrow{K_m(\text{app})} ES \xrightarrow{k_{cat}} E + P_1 + P_2 \tag{4.3}$$

the apparent Michaelis constant which is measured experimentally is not necessarily the same as the binding constant $K_s$ and the catalytic rate constant $k_{cat}$ is not identical with either $k_{+2}$ or $k_{+3}$. It may be shown that $K_m$(app) and the real $K_s$ are related as follows:

$$K_m(\text{app}) = \{k_{+3}/(k_{+2} + k_{+3})\}K_s \tag{4.4}$$

and that the relationship between the rate constants is

$$k_{cat} = k_{+2}.k_{+3}/(k_{+2} + k_{+3}). \tag{4.5}$$

These relationships may be simplified according to the relative magnitude of $k_{+2}$ and $k_{+3}$. Thus, if $k_{+2} \gg k_{+3}$ then

$$K_m(\text{app}) = (k_{+3}/k_{+2})K_s \tag{4.6}$$

and
$$k_{cat} = k_{+3} \tag{4.7}$$

that is, $k_{+3}$ represents the rate-determining step.

If
$$k_{+3} \gg k_{+2},$$

then
$$k_{cat} = k_{+2} \tag{4.8}$$

and
$$K_m(\text{app}) = K_s. \tag{4.9}$$

In the case of a non-specific substrate possessing a good leaving group, each as p-nitrophenyl acetate, $k_{+3}$ is very much smaller than $k_{+2}$ [40, 24]. The reaction assumes an interesting form when it is followed by observing spectroscopically the release of p-nitrophenol. There is initially a rapid release of p-nitrophenol followed by a slower, zero-order process [41]. The early rapid release of the leaving group (the 'initial burst') represents the first and second stages of scheme (4.1), leading to the acyl-enzyme, which proceed until practically all of the enzyme is in the acylated form. Thereafter the slower deacylation step controls the overall reaction. In the extreme situation, under acid conditions, the deacylation step is so slow that the acyl-enzyme becomes effectively a stable species.

Bender and his co-workers [39, 42] have outlined some consequences of the scheme (4.1) for the two types of specific substrates, esters and amides. The results of inhibition studies with the D-isomers of specific substrates suggest that the leaving groups (the ester or amide moieties) contribute very little, if anything, to the binding of the substrates or inhibitors to the enzyme. The values of $K_s$ for amides and esters of the same acyl-amino-acids should therefore be similar. However, the $K_m$(app) values for these compounds are very different and this is explicable in terms of the above scheme. For specific substrates which are amides $k_{+2}$ is rate-determining and therefore equations (4.8) and (4.9) hold. By contrast, for esters of specific substrate $k_{+2}$ is greater than $k_{+3}$, hence the rate-determining step is the deacylation and equations (4.6) and (4.7) apply. Since for these esters $k_{+3}$ is rate-determining, it follows that the overall rates of hydrolysis of all esters of the same acid derivative should be the same, the same acyl-enzyme being involved. This is an important consequence of, and conversely good evidence for, the acyl-enzyme hypothesis.

It has been stated the $K_s$ should be the same for esters and amides of the same specific substrate amino-acid derivative. Because $K_s = K_m$(app) for the amide (for example N-acetyl-L-tyrosine amide) then this experimentally-determined value may be used as $K_s$ for the ester (N-acetyl-L-tyrosine ethyl ester). In this way, all the rate constants for various substrates could be evaluated. One interesting result of these analyses was that whereas for the hydrolysis of the ethyl esters of N-acetyl-L-tyrosine, phenylalanine and tryptophan the deacylation step is rate-determining, for N-acetyl-glycine ethyl ester it is the acylation step which determines the overall rate. In this respect therefore the ester of a non-specific amino-acid resembles the amide of a specific substrate. In each case the acylation step is the slower; in the former presumably because the poor binding is unfavourable, in the latter because the amide species is a poor leaving group.

In this series of papers, Bender's group also presented values for the activation parameters for some chymotrypsin-catalysed hydrolyses [43]. Of special interest are their figures for the deacylations of some acyl-chymotrypsins. These are quoted in Table 4.1.

Table 4.1. *Activation parameters for the deacylation of acyl-chymotrypsins*

| Acyl-enzyme | $k_{+3(b)}$ | $\Delta G^*_{(c)}$ | $\Delta H^*_{(c)}$ | $-T\,\Delta S^*_{(c)}$ | $\Delta S^*_{(d)}$ |
|---|---|---|---|---|---|
| Acetyl-tyrosyl- | 3540 | 14·3 | 10·3 | 4·0 | −13·4 |
| Acetyl-tryptophanyl- | 942 | 17·9 | 12·0 | 5·9 | −19·8 |
| Cinnamoyl- | 14·7 | 20·1 | 11·2 | 8·9 | −29·6 |
| Acetyl- | 1 | 20·4 | 9·7 | 10·7 | −35·9 |

From ref. [43]. (b) Rates are relative to that for acetyl-chymotrypsin, (c) kcal/mole, (d) kcal/mole/degree.

The values of $k_{+3}$ are relative to the value for acetyl-chymotrypsin. The values of $k_{+3}$ show a progressive decrease down the table corresponding to the specificity of the enzyme. However, it is clear that the enthalpies of activation ($\Delta H^*$) for all of these derivatives are very similar, and that the entropies of activation make the important contributions to the differences in the rates. Bender and his co-workers discuss these results in terms of the rigidity of the acyl-enzyme which arises from the strong binding between the acyl-group and the protein. This evidence therefore supports the view, discussed later in chapter 9, that factors related specifically to the entropy terms make a significant contribution to catalysis by enzymes.

Another important feature of chymotrypsin catalysis is the influence of pH on the various reaction constants. In general the *overall* chymotrypsin reaction appears to require (a) that a group, $pK_a$ 6·8, be in the basic form, and (b) that a group, $pK_a$ 9, be in its conjugate acid form [44]. Recent careful analysis of the pH-dependence [9, 45] has shown that the individual stages in the sequence (4.1) are influenced in the following ways. The $K_s$ factor depends on the group with $pK_a$ 9 and it has been suggested that a conformational change occurs on protonation of this group. Only the conformation which carries this protonated group is able to bind the substrate in an active enzyme–substrate complex. That this group is the N-terminal amino-group of the isoleucine-16 residue appears to be well established. Both of the subsequent stages, $k_{+2}$ and $k_{+3}$, are dependent on the basic form of the group with an apparent $pK_a$ 6·8. The original assignment of this simply as the imidazole group of a histidine residue now seems to be an oversimplification.

More information concerning the role of this prototropic system with the apparent $pK_a$ of 6·8 was obtained by studying the reaction in deuterium oxide solution. In $D_2O$ a proton-transfer process usually manifests itself in a kinetic isotope effect leading to a decrease in rate by a factor of 2–3. The deacylation process for chymotrypsin has been shown to be 2·5 times slower in $D_2O$ than in water [46, 47] and so it appears that in this step at least a proton transfer occurs, i.e. a base catalysis is involved in some way. Furthermore, in $D_2O$ the whole pH-rate curve is displaced to higher pH by about 0·6 pH units, in agreement with the above interpretation. Formally, at least, the acylation process might be expected to be very similar to that of deacylation and we assume that the same group is involved in some way in both stages.

*Substrate Binding and Specificity*. Chymotrypsin does not exhibit a narrow specificity although some aspects of substrate structure are essential for activity. The best substrates are N-acyl-amino-acid derivatives which carry cyclic groups on their side-chains (XIII) [48, 49]. These cyclic groups need not be aromatic [49]. Strong stereospecificity is observed, chymotrypsin

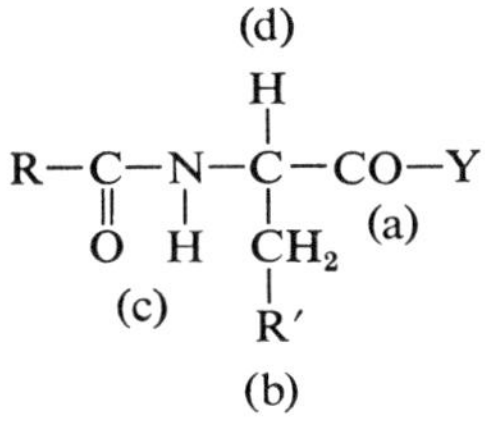


(XIII)

catalyses the hydrolysis of derivatives of L-amino-acids but hardly at all of the D-isomers. D-Isomers may be inhibitors. The nature of the acyl-group (R) usually has little effect on the reaction. Thus the substrate may be a long protein chain or it may simply be an acetyl derivative ($R = CH_3$). The situation of the amide nitrogen is important, replacement of the hydrogen by a methyl group renders the molecule ineffective as a substrate [50].

Experiments on substrate specificity and on inhibition lead to the conclusion that several binding sites exist in the enzyme molecule. These are (a) the serine residue which forms a covalent bond with the substrate, (b) the non-polar region which accommodates the hydrophobic side-chain, (c) a region which may form a hydrogen bond with the acyl-amide group of the substrate, (d) a site which corresponds to the hydrogen atom on the α-carbon (in the diagram (XIII) the corresponding regions of the substrate are indicated) [50, 51]. It has already been stated that the leaving group (Y) does not appear to contribute to the binding of the substrate to the enzyme.

Knowles and his co-workers have presented some interesting information concerning the importance of specificity in the deacylation of acyl-chymotrypsins [52]. They show that in the L-series the better the binding between the acyl-group and the enzyme the more rapid is the reaction. This is in agreement with the results of Bender, Kezdy and Gunder which are reported on page 160, which indicate that specificity factors have a great influence. However, Knowles *et al.* also show that in the D-series, that is for acyl-chymotrypsins derived from D-amino-acids, the better the binding the slower the rate of deacylation. These results suggest that for L-amino-acid derivatives, the acyl-group is held at three different points, and that this binding holds the carbonyl group in a position which is favourable for attack by the deacylating nucleophile. The stronger the binding of the side-chain the more rigidly is the carbonyl group held in this position. For a D-amino-acid derivative, if the same binding occurs then the carbonyl must be held in an unfavourable position. The stronger the binding therefore the less likely is the carbonyl group to suffer the nucleophilic attack. Such experiments emphasize the three-dimensional nature of the arrangement of the binding sites in an enzyme which shows stereospecificity.

Some exceptions to the Knowles 'better binding–better reaction' rule have been noted, especially the examples of 'wrong-way' binding by Rapp *et al.* [53].

*Three-dimensional Structure of Tosyl-Chymotrypsin.* X-Ray diffraction studies by Blow *et al.* [54] have resulted in an electron-density map at high resolution which reveals the conformation of the protein chains. This is shown in Fig. 4.2. The three peptide chains and the cystine bridges are clearly visible. The model confirms much of the information which had been obtained from chemical studies. Thus the active serine, which carries the tosyl group, is indeed very close to histidine-57, as expected. Methionine-192 is also nearby. Histidine-40 has been extimated to be at least 13 Å from histidine-57.

The positions of the chain termini (the chains are labelled A, B and C in order according to the sequence) are compatible with the loss of two small peptides from the pre-enzyme followed by relatively small conformational changes.

The three-dimensional structure does more than merely confirm the chemical results. It provides a clear explanation for the importance of the isoleucine-16 amino-group. The electron-density map makes it clear that the carboxylate anion of the side-chain of the aspartic acid residue next to the active serine (asp-194) forms an ion pair with the protonated amino-group of isoleucine-16. This is clearly shown in Fig. 4.3. This ionic interaction holds the active site in the necessary conformation. In the non-protonated form (above pH 9) or in the acetylated derivative this ion pair cannot exist and the active configuration is destroyed.

Another feature of Fig. 4.3 is the presence of a tyrosine residue (Tyr-146) from a neighbouring molecule very close to the active site. It will be recalled that Westheimer *et al.* [34] suggested the presence of such a group on the basis of studies with diazoacetyl-chymotrypsin.

However, probably the most important consequence of the elucidation of the detailed three-dimensional structure of the enzyme, is that it throws light on a hitherto unsuspected aspect of the active site, and hence may provide an important basis for our understanding of the catalytic activity of an enzyme. It has become clear that the imidazole ring of histidine-57 which has been implicated in the catalytic process as a general base, is linked by a hydrogen bond to the side-chain carboxylate of the aspartyl-102 residue buried in the interior of the protein molecule. The region

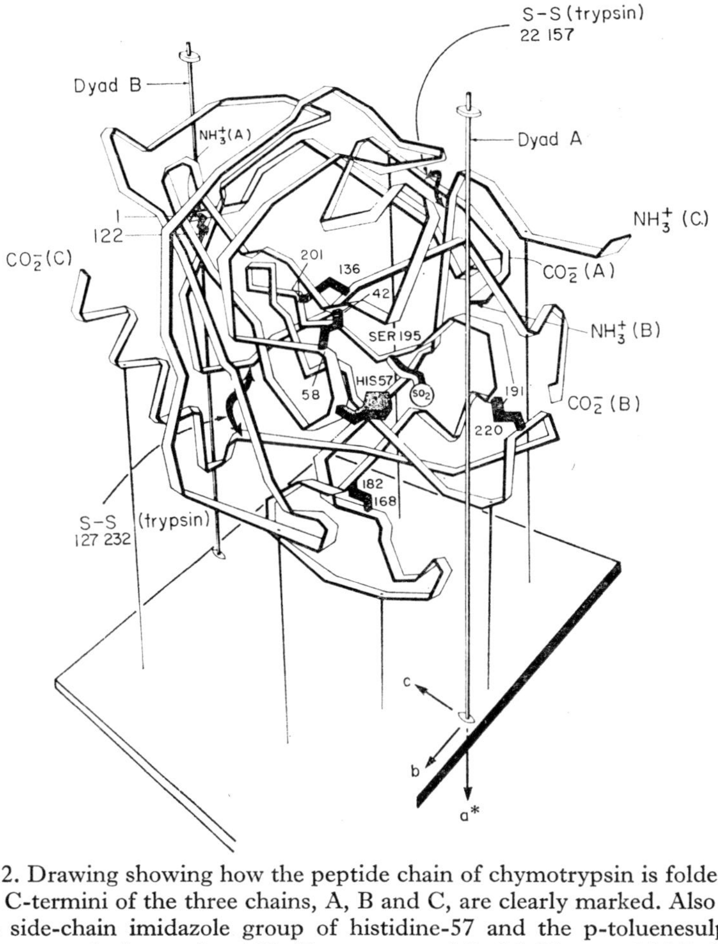


FIG. 4.2. Drawing showing how the peptide chain of chymotrypsin is folded. The N- and C-termini of the three chains, A, B and C, are clearly marked. Also shown are the side-chain imidazole group of histidine-57 and the p-toluenesulphonyl group attached to serine-195. (By courtesy of D. M. Blow and M.R.C.)

surrounding this aspartyl residue is probably highly hydrophobic and Blow and his colleagues [54a] suggest that the group with $pK_a$ 6·8, involved in the catalytic reaction, is in fact a complex system made up of the carboxylic acid, imidazole and hydroxymethyl groups of the aspartic acid, histidine and serine residues at the active site. This complex system would exist in the acidic and basic forms shown in equation (4.10).

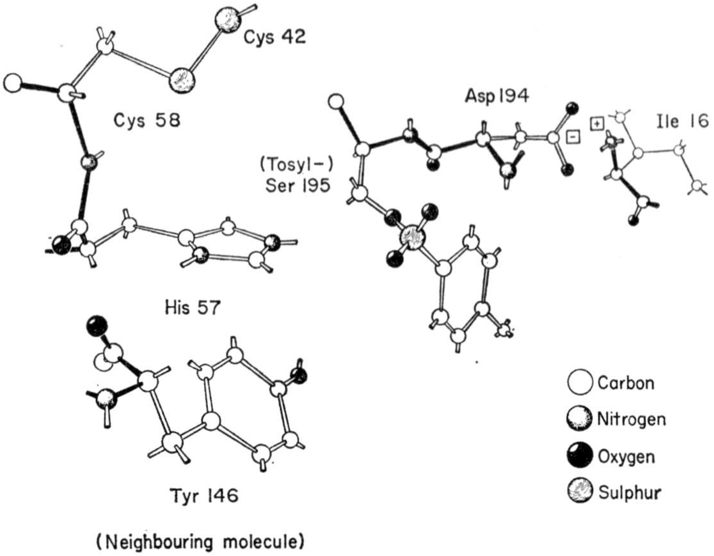


FIG. 4.3. Diagram showing how the histidine-57, serine-195, aspartate-194 and isoleucine-16 residues are arranged in the three-dimensional structure of α-chymotrypsin. The disulphide bridge between cysteines-58 and 42 is also shown, in addition to tyrosine-146 from a neighbouring molecule. (By courtesy of D. M. Blow and M.R.C.)

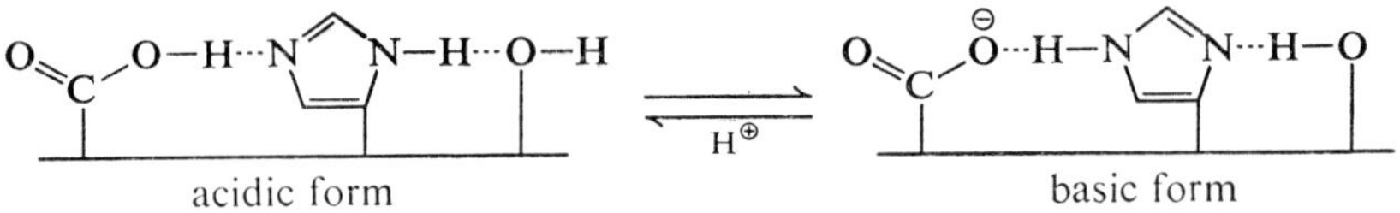


(4.10)

The basic form is envisaged as being stabilized by resonance, and at pH values in the region of 7 the serine oxygen can be considered to be carrying

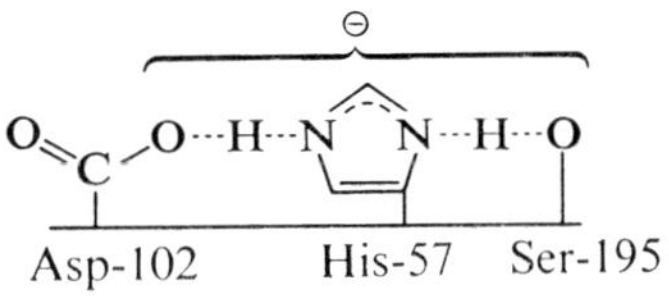


far more negative charge than if it were in an isolated serine residue, and therefore to be a more powerful nucleophile.

*The Mechanism for Chymotrypsin Catalysis.* The mechanism for the catalysis of tł e hydrolysis of an ester by chymotrypsin which is presented in Fig. 4.4 is based on the suggestions of many groups of workers [47, 41, 29b, 55] and especially on the work of Blow and his co-workers [54a]. This mechanism appears to be compatible with almost all of the evidence available.

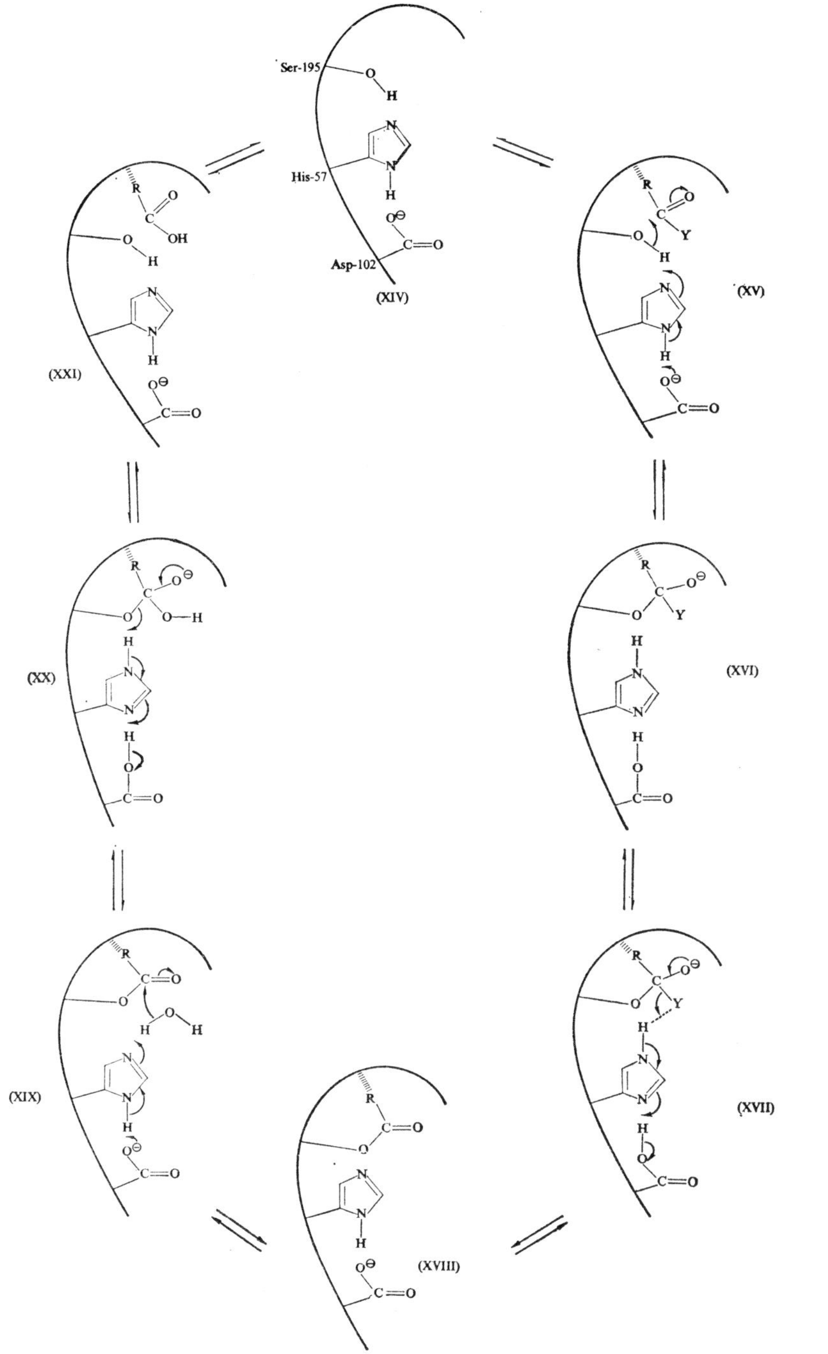


FIG. 4.4. Proposed mechanism for the chymotrypsin-catalysed hydrolysis of an ester or amide R.CO.Y

In this scheme, XIV represents the free enzyme in which the carboxylate group of asp-102, the imidazole ring of his-57 and the hydroxyl group of ser-195 are arranged in definite positions relative to each other. The substrate becomes bound (XV), and the carbonyl group is subjected to a nucleophilic attack by the oxygen of the serine hydroxyl. The great nucleophilicity of this group is attributable to the negative charge transferred from the carboxylate group in the hydrophobic interior of the enzyme, through the imidazole ring. A tetrahedral intermediate (XVI) is produced, and this process is probably the rate-determining step in the acylation stage [47]. The imidazole–carboxylic complex assists the departure of the Y-group by functioning as an acidic catalyst (XVII) and loss of HY leaves the acyl-enzyme (XVIII). The deacylation step is mechanistically similar, the nucleophilic attack by water being assisted by the same imidazole–carboxylic system (XIX). The resulting tetrahedral intermediate breaks down (XX), the departure of the leaving group (the hydroxyl of the serine) being assisted by acid catalysis. This gives the non-covalent enzyme–acid complex (XXI), which dissociates liberating the free enzyme.

The acid catalysis of the departures of the leaving groups (XVII and XX) is obligatory if the principle of microscopic reversibility is to be complied with.† It is worth noting that in both the acylation and deacylation steps the nucleophilic processes are rate-determining [47]. The nature of the active site should favour this type of process (see XV), in that in effect the negative charge on the carboxyl in the non-polar interior of the molecule is transferred to the highly-polar region of the active site.

Many other mechanisms have been suggested for the chymotrypsin catalysis. These include processes in which two imidazole residues take part [56, 47] in a concerted reaction. The amino-acid sequence of chymotrypsin (see Fig. 4.1) makes it quite clear that histidine-40 must be in the near vicinity of histidine-57 which is involved in the catalytic process. However, at the present time, a concerted reaction appears to be very unlikely. The slope of the pH/log $k_{+2}$ curve is 1·0, indicating that only one group is involved (ref. [40], page 251). In addition, the three-dimensional model of tosyl-chymotrypsin shows that the two histidine residues are too far separated to interact in this way.

Suggested mechanisms which require the direct involvement of the aspartic acid residue which is next to the active serine in the amino-acid sequence have been eliminated on the basis of oxygen isotope exchange experiments, and other evidence [57, 58, 59, 60].

† Blow *et al.* [54a] suggest a slightly different mechanism in which the nucleophilic substitution at the carbonyl takes place in one step and does not proceed through a tetrahedral intermediate. The acid catalysis of the departure of the leaving group is thus simultaneous with the nucleophilic attack. This accords with a suggested electrophilic component in chymotrypsin catalysis (see Holloway, M. R., *Ann. Rep.* 1968).

## Trypsin and other Active Serine Enzymes

*Trypsin (3.4.4.4)*. The close similarity between trypsin and chymotrypsin is indeed remarkable; the most important difference lies in the specificities of the two enzymes. Trypsin catalyses the hydrolysis of esters and amides derived from L-amino-acids which have basic side-chains, that is, from arginine or lysine [61]. The specificity is quite narrow, although as for chymotrypsin some non-specific substrates are known [62, 63].

Figure 4.5 shows the amino-acid sequences of both pre-trypsin and pre-chymotrypsin [64]. The amino-acid residues which are sequentially

| | | | | | | | | | | | | | | | | | | | | | | |
|---|---|---|---|---|---|---|---|---|---|---|---|---|---|---|---|---|---|---|---|---|---|---|
| | | | | | 1 | 2 | 3 | 4 | 5 | 6 | 7 | 8 | 9 | 10 | 11 | 12 | 13 | 14 | 15 | 16 | 17 | 18 |
| Chymotrypsinogen | | | | | Cys- | Gly- | Val- | Pro- | Ala- | Ile- | Gln- | Pro- | Val- | Leu- | Ser- | Gly- | Leu- | Ser- | Arg- | ILE- | VAL- | GLY- |
| Trypsinogen | | | | | | | | | | | | | | Val- | Asp- | Asp- | Asp- | Asp- | Lys- | ILE- | VAL- | GLY- |
| | | | | | | | | | | | | | | 1 | 2 | 3 | 4 | 5 | 6 | 7 | 8 | 9 |
| 19 | 20 | 21 | 22 | 23 | 24 | 25 | 26 | 27 | 28 | 29 | 30 | 31 | 32 | 33 | 34 | 35 | 36 | 37 | 38 | 39 | 40 | 41 |
| Asp- | Glu- | Glu- | Ala- | Val- | Pro- | Gly- | Ser- | Trp- | PRO- | Trp- | GLN- | VAL- | SER- | LEU- | Gln- | Asp- | Lys- | Thr- | GLY- | Phe- | HIS- | PHE- |
| Gly- | Tyr- | Thr- | Cys- | Gly- | Ala- | Asn- | Thr- | Val- | PRO- | Tyr- | GLN- | VAL- | SER- | LEU- | Asn- | | - | Ser- | GLY- | Tyr- | HIS- | PHE- |
| 10 | 11 | 12 | 13 | 14 | 15 | 16 | 17 | 18 | 19 | 20 | 21 | 22 | 23 | 24 | 25 | | | 26 | 27 | 28 | 29 | 30 |
| 42 | 43 | 44 | 45 | 46 | 47 | 48 | 49 | 50 | 51 | 52 | 53 | 54 | 55 | 56 | 57 | 58 | 59 | 60 | 61 | 62 | 63 | 64 |
| CYS- | GLY- | GLY- | SER- | LEU- | ILE- | ASN- | Glu- | Asn- | TRP- | VAL- | VAL- | Thr- | ALA- | ALA- | HIS- | CYS- | Gly- | Val- | Thr- | Thr- | Ser- | Asp- |
| CYS- | GLY- | GLY- | SER- | LEU- | ILE- | ASN- | Ser- | Gln- | TRP- | VAL- | VAL- | Ser- | ALA- | ALA- | HIS- | CYS- | Tyr- | Lys- | Ser- | Gly- | Ile- | Gln- |
| 31 | 32 | 33 | 34 | 35 | 36 | 37 | 38 | 39 | 40 | 41 | 42 | 43 | 44 | 45 | 46 | 47 | 48 | 49 | 50 | 51 | 52 | 53 |
| 65 | 66 | 67 | 68 | 69 | 70 | 71 | 72 | 73 | 74 | 75 | 76 | 77 | 78 | 79 | 80 | 81 | 82 | 83 | 84 | 85 | 86 | 87 |
| VAL- | Val- | Val- | Ala- | Gly- | Glu- | Phe- | Asp- | Gln- | Gly- | Ser- | Ser- | Ser- | Glu- | Lys- | Ile- | Gln- | Lys- | Leu- | Lys- | Ile- | Ala- | Lys- |
| VAL- | Arg- | Leu- | Gly- | Glu- | Asp- | Asn- | Ile- | Asn- | Val- | Val- | Glu- | Gly- | Asp- | Glu- | Gln- | Phe- | Ile- | Ser- | Ala- | Ser- | Lys- | Ser- |
| 54 | 55 | 56 | 57 | 58 | 59 | 60 | 61 | 62 | 63 | 64 | 65 | 66 | 67 | 68 | 69 | 70 | 71 | 72 | 73 | 74 | 75 | 76 |
| 88 | 89 | 90 | 91 | 92 | 93 | 94 | 95 | 96 | 97 | 98 | 99 | 100 | 101 | 102 | 103 | 104 | 105 | 106 | 107 | 108 | 109 | 110 |
| Val- | Phe- | Lys- | Asn- | SER- | Lys- | TYR- | ASN- | Ser- | Leu- | Thr- | Ile- | ASN- | ASN- | Asn- | ILE- | Thr- | LEU- | Leu- | LYS- | LEU- | Ser- | Thr- |
| Ile- | Val- | His- | Pro- | SER- | - | TYR- | ASN- | (Pro, | Leu, | Thr, | Asn) | ASN- | ASN- | Asp- | ILE- | Met- | LEU- | Ile- | LYS- | LEU- | Lys- | Ser- |
| 77 | 78 | 79 | 80 | 81 | | 82 | 83 | 84 | 85 | 86 | 87 | 88 | 89 | 90 | 91 | 92 | 93 | 94 | 95 | 96 | 97 | 98 |
| 111 | 112 | 113 | 114 | 115 | 116 | 117 | 118 | 119 | 121 | 122 | 123 | 124 | 125 | 126 | 127 | 128 | 129 | 130 | 131 | 131 | 132 | 133 |
| ALA- | ALA- | SER- | Phe- | Ser- | Gln- | Thr- | VAL- | Ser- | Ala- | Val- | Cys- | LEU- | PRO- | Ser- | Ala- | Ser- | Asp- | Asp- | Phe- | Ala- | ALA- | GLY- |
| ALA- | ALA- | SER- | Leu- | Asn- | Ser- | Arg- | VAL- | Ala- | Ser- | Ile- | Ser- | LEU- | PRO- | Thr- | Ser- | Cys- | | - | Ala- | Ser- | ALA- | GLY- |
| 99 | 100 | 101 | 102 | 103 | 104 | 105 | 106 | 107 | 108 | 109 | 110 | 111 | 112 | 113 | 114 | 115 | | | 116 | 117 | 118 | 119 |
| 134 | 135 | 136 | 137 | 138 | 139 | 140 | 141 | 142 | 143 | 144 | 145 | 146 | 147 | 148 | 149 | 150 | 151 | 152 | 153 | 154 | 155 | 156 |
| THR- | Thr- | CYS- | Val- | Thr- | Thr- | GLY- | TRP- | GLY- | Leu- | THR- | Arg- | Tyr- | Thr- | Asn- | Ala- | Asn- | Thr- | PRO- | ASP- | Arg- | LEU- | Gln- |
| THR- | Gln- | CYS- | Leu- | Ile- | Ser- | GLY- | TRP- | GLY- | Asn- | THR- | Lys- | Ser- | Ser- | Gly- | Thr- | Ser- | Tyr- | PRO- | ASP- | Val- | LEU- | Lys- |
| 120 | 121 | 122 | 123 | 124 | 125 | 126 | 127 | 128 | 129 | 130 | 131 | 132 | 133 | 134 | 135 | 136 | 137 | 138 | 139 | 140 | 141 | 142 |
| 157 | 158 | 159 | 160 | 161 | 162 | 163 | 164 | 165 | 166 | 167 | 168 | 169 | 170 | 171 | 172 | 173 | 174 | 175 | 176 | 177 | 178 | 179 |
| Gln- | Ala- | Ser- | Leu- | PRO- | Leu- | LEU- | SER- | Asn- | Thr- | Asn- | CYS- | LYS- | Lys- | Tyr- | Trp- | Gly- | Thr- | Lys- | ILE- | Lys- | Asp- | Ala- |
| Cys- | Leu- | Lys- | Ala- | PRO- | Ile- | LEU- | SER- | Asp- | Ser- | Ser- | CYS- | LYS- | Ser- | Ala- | Tyr- | Pro- | Gly- | Gln- | ILE- | Thr- | Ser- | Asn- |
| 143 | 144 | 145 | 146 | 147 | 148 | 149 | 150 | 151 | 152 | 153 | 154 | 155 | 156 | 157 | 158 | 159 | 160 | 161 | 162 | 163 | 164 | 165 |
| 180 | 181 | 182 | 183 | 184 | 185 | 186 | 187 | 188 | 189 | | | 190 | 191 | 192 | 193 | 194 | 195 | 196 | 197 | 198 | 199 | 200 |
| MET- | Ile- | CYS- | ALA- | GLY- | Ala- | Ser- | Gly- | Val- | Ser- | | - | SER- | CYS- | Met- | GLY- | ASP- | SER- | GLY- | GLY- | PRO- | Leu- | VAL- |
| MET- | Phe- | CYS- | ALA- | GLY- | Tyr- | Leu- | Glu- | Gly- | Gly- | Lys- | Asn- | SER- | CYS- | Gln- | GLY- | ASP- | SER- | GLY- | GLY- | PRO- | Val- | VAL- |
| 166 | 167 | 168 | 169 | 170 | 171 | 172 | 173 | 174 | 175 | 176 | 177 | 178 | 179 | 180 | 181 | 182 | 183 | 184 | 185 | 186 | 187 | 188 |
| 201 | 202 | 203 | 204 | 205 | 206 | 207 | 208 | 209 | 210 | 211 | 212 | 213 | 214 | 215 | 216 | 217 | 218 | 219 | 220 | 221 | 222 | 223 |
| CYS- | Lys- | Lys- | Asn- | Gly- | Ala- | Trp- | Thr- | Leu- | Val- | GLY- | ILE- | VAL- | Ser- | SER- | TRP- | GLY- | SER- | Ser- | Thr- | Cys- | Ser- | Thr- |
| CYS- | Ser- | Gly- | Lys- | Leu- | Gln- | | | | - | GLY- | ILE- | VAL- | - | SER- | TRP- | GLY- | SER- | Gly- | Cys- | Ala- | Gln- | Lys- |
| 189 | 190 | 191 | 192 | 193 | 194 | | | | | 195 | 196 | 197 | | 198 | 199 | 200 | 201 | 202 | 203 | 204 | 205 | 206 |
| 224 | 225 | 226 | 227 | 228 | 229 | 230 | 231 | 232 | 233 | 234 | 235 | 236 | 237 | 238 | 239 | 240 | 241 | 242 | 243 | 244 | 245 | 246 |
| Ser- | Thr- | PRO- | GLY- | VAL- | TYR- | Ala- | Arg- | VAL- | Thr- | Ala- | Leu- | VAL- | Asn- | TRP- | Val- | Gln- | GLN- | THR- | Leu- | ALA- | Ala- | ASN- |
| Asn- | Lys- | PRO- | GLY- | VAL- | TYR- | Thr- | Lys- | VAL- | Cys- | Asn- | Tyr- | VAL- | Ser- | TRP- | Ile- | Lys- | GLN- | THR- | Ile- | ALA- | Ser- | ASN- |
| 207 | 208 | 209 | 210 | 211 | 212 | 213 | 214 | 215 | 216 | 217 | 218 | 219 | 220 | 221 | 222 | 223 | 224 | 225 | 226 | 227 | 228 | 229 |

FIG. 4.5. The structural similarity of trypsinogen and chymotrypsinogen. Gaps in the sequence have been arbitrarily inserted to draw attention to the resulting areas of homology. Amino-acid residues that are homologous in the two proteins are shown in capital letters. (K. A. Walsh and H. Neurath, *Proc. natn. Acad. Sci., U.S.A.* **52,** 884 (1964).) (Note: Residue-102 in chymotrypsinogen and -177 in trypsinogen are now established as Asp. By courtesy of H. Neurath and the National Academy of Sciences)

common to both are shown in capital letters. The correspondence is striking, especially when one observes that many of the differences involve pairs of amino-acids which are essentially similar. For example, serine and threonine, arginine and lysine or asparagine and glutamine may be interchanged. Aromatic amino-acids often correspond as do those with other non-polar side-chains, valine, leucine and isoleucine. The disulphide bridges between corresponding half-cystine residues are also similar and the overall three-dimensional structures of the two enzymes may therefore resemble each other fairly closely. Blow emphasizes this point in Fig. 4.2 by indicating on the drawing of chymotrypsin the positions of the possible disulphide bridges of trypsin. Clearly S–S bridges in these positions are quite feasible.

If it is assumed that trypsin does have a three-dimensional structure similar to that of chymotrypsin, then the specificity of trypsin for the basic amino-acid residues is readily explained. Residue-177, which would reside in the active site in a suitable position for interaction with the substrate side-chain, has been shown to be an aspartyl residue. This provides the negatively-charged group necessary for the substrate binding.

Many similarities have been found in the mechanisms by which these two enzymes appear to function. A serine residue is involved; trypsin is inhibited by di-isopropylfluorophosphate and a phosphoserine derivative is formed [65]. The active serine is at position-183 [56, 64, 66, 67]. The variation of the reaction rate with pH shows that a basic group, $pK_a$ 6·25, is essential for activity [68] and the heat of ionization of this group was found to be 7 kcal/mole. The involvement of a histidine residue was confirmed when it was shown that trypsin is inactivated by the chloroketone derived from N-α-tosyl-lysine (TLCK). The reaction of trypsin with this reagent resulted in alkylation of a histidine residue [69] which was shown to be that at position-46 [70]. Notice that this corresponds to histidine-57 in the chymotrypsin sequence which is acylated by TPCK. TPCK has no effect on trypsin while TLCK does not inhibit chymotrypsin [69].

The overall reaction scheme (eq. 4.1) which was shown to apply to chymotrypsin catalysis also applies in the case of trypsin. Thus, an acyl-trypsin is formed as an intermediate in the reaction. Although the isolation of an acyl-trypsin has proved to be a difficult problem (for example *trans*-cinnamoyl-trypsin is extremely labile whereas the chymotrypsin derivative is isolatable [63]), it has been possible to study the hydrolysis of some non-specific substrates under conditions where the acylation and deacylation steps are distinguishable [62, 63]. Evidence that acyl-trypsin intermediates are formed during the hydrolysis of specific substrates is afforded by studies which have shown that the hydrolysis of various esters of N-benzoyl-arginine proceeded at identical maximal velocities. The methyl, ethyl, isopropyl, benzyl and cyclohexyl esters were used [63]. This situation is similar to that represented by equations (4.5) and (4.6) for chymotrypsin in which the deacylation of the common acyl-enzyme is the rate-determining step.

Some differences, however, are noticeable. For example, trypsin does not possess a methionine residue corresponding to that at position-192 in chymotrypsin and consequently no analogous alkylation or oxidation reactions can be carried out. The $K_m$ term does not appear to be influenced by a group of $pK_a$ 9·0. In the case of chymotrypsin this was the amino-group of the isoleucine-16 residue. Although isoleucine-6, corresponding to this, is liberated during the conversion of pre-trypsin into trypsin, it clearly has no effect on the binding of the substrate. This point is emphasized by the fact that although acetylation inactivates chymotrypsin it does not reduce the activity of trypsin [71, 72]. Indeed, the activity of this enzyme towards some substrates appears to be enhanced by acetylation [73].

One or more tyrosine residues appear to be involved in some way in the activity of trypsin. Thus, six out of the ten tyrosine residues of the native enzyme react with cyanuric fluoride and the product retains 75% of its activity [74]. However, when cyanuric fluoride reacts with the inactive di-isopropylphosphotrypsin, only four tyrosine residues become substituted. This suggests that the remaining two residues are very close to the active site. More evidence concerning the tyrosine residues was obtained from acetylation studies [72]. Reaction of trypsin with N-acetyl-imidazole does not normally reduce the enzymic activity. However, in the presence of urea or guanidinium hydrochloride (which tend to destroy the three-dimensional structure) some tyrosine residues which were apparently previously 'buried' become exposed and are acetylated. The product is inactive. De-acetylation with hydroxylamine reactivates the enzyme. Vallee *et al.* [72] suggest that the tyrosine residues help to hold the protein in the necessary conformation.

From the more limited information available about trypsin, it seems likely that its mechanism of action is very similar to that of chymotrypsin. Indeed the same may be said for several other peptidases which exhibit the same properties. These include chymotrypsin B, which has a specificity similar to that of chymotrypsin A, thrombin and plasmin, whose specificities resemble that of trypsin, pancreatopeptidase E (elastase; 3.4.4.7) which catalyses the hydrolysis of peptides mainly at bonds adjacent to neutral amino-acids, and subtilopeptidase A. Subtilisin BPN′ (compare subtilopeptidase A, 3.4.4.16), which is a bacterial enzyme, shows many similarities to chymotrypsin. Its X-ray crystal structure has been determined [54b] and the result shows clearly that the active site serine residue is hydrogen-bonded to a histidine residue. This histidine may itself be hydrogen-bonded to an aspartyl or asparaginyl residue. On this basis, it seems likely that the mechanism of action of subtilisin BPN′ is very similar to that of chymotrypsin.

### Carboxylic Ester Hydrolases

Many of the ester hydrolases which are listed in the Enzyme Commission group 3.1.1.–, appear to function by mechanisms which resemble those employed by the peptidases mentioned above.† They are inhibited by DFP(I) by reaction at a serine residue, their activities depend on a group (or groups) with $pK_a$ 6–7 and for some of them there is evidence for the intermediacy of acyl-enzymes. Their differences lie in their specificities and in this context they are often classified into two groups; those whose specificity is directed towards esters of choline and those which hydrolyse other esters including fatty acid esters. The distinction is usually demonstrated experimentally by the effect of physostigmine (eserine), which

$HO{\cdot}CH_2{\cdot}CH_2{\cdot}\overset{\oplus}{N}(CH_3)_3$

choline

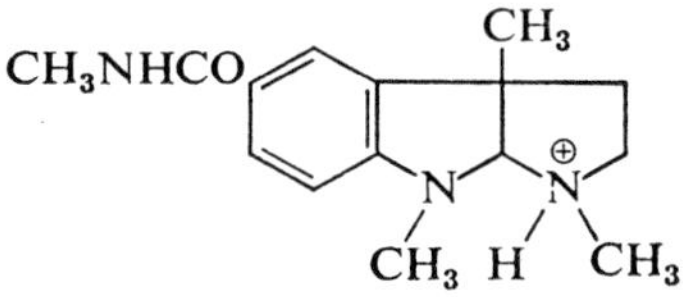

Physostigmine(eserine)

strongly inhibits cholinesterases but has no similar effect on the simple esterases.

The specificities of the esterases which are not sensitive to physostigmine overlap to a certain extent and usually depend on the number of carbon atoms in the acyl part of the ester. For example, carboxylesterase (ali-esterase; 3.1.1.1) appears to exhibit its maximum activity with esters (of *m*-hydroxybenzoic acid) derived from acids with carbon-chains containing 7–10 units. It has been suggested that various natural tissues may contain series of esterases whose specificities are for esters containing acid groups of different chain lengths [76]. On the basis of specificity studies, Malhotra and his co-worker [78] suggest that the binding site of goat carboxylesterase is a narrow hydrophobic region 16–17 Å long which binds the long chain of the fatty acid portion of the substrate. By contrast, a citrus esterase (an acetylesterase) appears to favour esters of acetic acid and there is evidence that for this enzyme the alcohol portion of a substrate makes a significant contribution to the binding [39].

The cholinesterases also show overlapping specificities. Acetylcholinesterase, whose physiological substrate is acetylcholine, is also able to catalyse

† Certainly not all of them function in this way. The acetylesterase from animal tissue listed by the Commission (C-esterase; 3.1.1.6) is not inhibited by DFP [75] and presumably acts by a different mechanism. However, Hofstee [76] refers to two other acetylesterases (plant tissue) which are inhibited by this reagent. Aryl-esterase (A-esterase; 3.1.1.2) is entirely different again, being metal-dependent [77].

the hydrolysis of propionyl- and butyryl-choline, although with not such great efficiency. Correspondingly, other enzymes in this group hydrolyse acetylcholine less readily than they do other acylcholines [80].

From pH studies of the inhibition of the cholinesterases by physostigmine and similar compounds, it is clear that the positive charge on the inhibitor is mainly responsible for its binding [81]. Therefore, the active site of a cholinesterase possesses at least two distinct regions, an anionic site which binds the positive centre of the substrate and the catalytic site which is associated with the ester link and its cleavage. In addition, for some of these enzymes, there must be a binding site for the acyl portion of the substrate which accounts for the specificities. The experimental observation that dimethylaminoethanol (XXII) binds to acetylcholinesterase more powerfully than does isoamyl alcohol (XXIII) (both are structurally similar to choline) emphasizes the coulombic nature of the attractive forces [82]. In addition the fact that the tetramethyl ammonium ion, $\overset{\oplus}{N}(CH_3)_4$, is a better inhibitor than is the unsubstituted ammonium

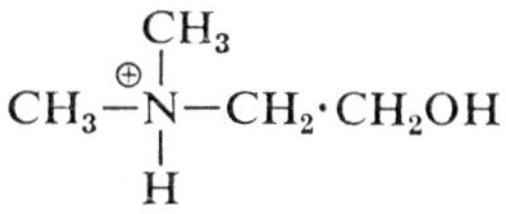

(XXII)

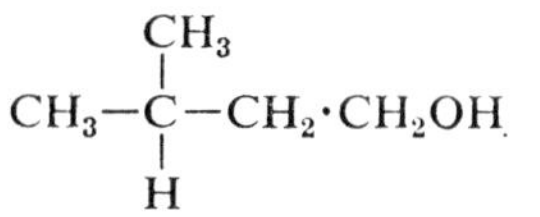

(XXIII)

ion ($\overset{\oplus}{N}H_4$) suggests that the methyl groups attached to the nitrogen atom of the substrate may also be involved in the enzyme–substrate interaction. The anionic group on the enzyme appears to have a $pK_a$ of 4–4·5 and it is almost certainly a carboxylate ion [83].

Krupka has studied the pH-dependence of the various steps of the acetylcholinesterase-catalysed hydrolysis of charged and uncharged esters (these are specific and non-specific substrates [83]). For specific substrates the deacylation step is probably rate-determining, whereas for non-specific substrates acylation determines the rate. Apparently the acylation step requires a group, $pK_a$ 5·2–5·5, in its basic form. This group is apparently 8–10 Å away from the anionic site. Deacylation depends on a group, $pK_a$ 6·3, which is estimated to be 5·5 Å from this anionic site. In addition an acidic group with $pK_a$ 9·2 is required for acylation (and possibly also for deacylation). Krupka suggests a mechanism for the action of acetylcholinesterase based on these observations, as shown in Fig. 4.6.

According to this scheme the nucleophilic attack by the hydroxyl of the active serine residue is enhanced by the base $B_2$ which is presumably the side-chain imidazole of a histidine residue. Its rather low $pK_a$ (5·5) may be the result of an unusual environment. The carboxylate anion binds the substrate's positively-charged centre. The acid group H—A ($pK_a$ 9·2)

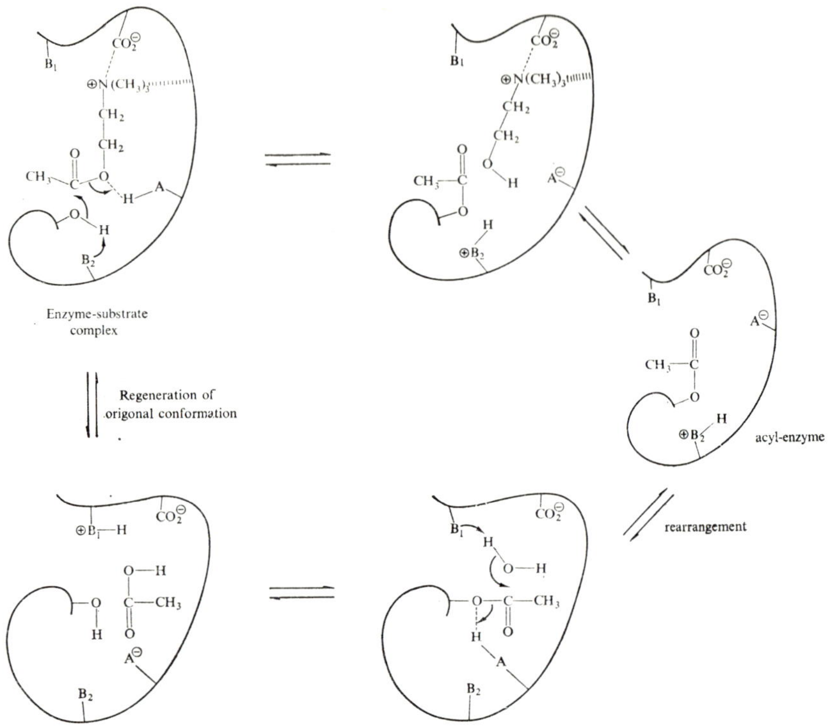


FIG. 4.6. Suggested mechanism for acetylcholinesterase-catalysed hydrolysis of acetylcholine, after Krupka [83]. (By courtesy of R. M. Krupka and the Committee of Editors, American Chemical Society Journals and reprinted by permission of the copyright owner)

assists in the departure of the leaving group. The resulting acyl-enzyme then rearranges so that the acyl-group moves towards the basic group $B_1$ (another imidazole group, $pK_a$ 6·3) which assists the nucleophilic attack of the water molecule. H—A may assist in this reaction too.

The most interesting aspect of Krupka's suggestion is that two histidine residues are involved, each assisting at a different stage. Against this mechanism, however, it must be stated that a mechanism requiring two histidine residues has been eliminated for chymotrypsin and other active serine enzymes. In view of the probable close similarity of structure and function of chymotrypsin and acetylcholinesterase, this mechanism must at present be regarded as tentative.

## Structural Relationships

The similarities between the structures of chymotrypsin and trypsin have already been mentioned. However, identity or near-identity of amino-acid sequences around the active site serine extends beyond these two enzymes. Table 4.2 shows the amino-acid sequences in the vicinity of the

Table 4.2. *Amino-acid sequences around the active serine residues of hydrolytic enzymes*

| Enzyme | Amino-acid sequences |
|---|---|
| Chymotrypsin A | Cys. Met. Gly. Asp. *Ser*. Gly. Gly. Pro. Leu. Val. |
| Chymotrypsin B | Cys. Met. Gly. Asp. *Ser*. (Gly. Gly. Pro. Leu.) Val. |
| Trypsin | Cys. Gln. Gly. Asp. *Ser*. Gly. Gly. Pro. Val. Val. |
| Pancreatopeptidase E | Cys. Gln. Gly. Asp. *Ser*. (Gly. Gly. Pro.) Leu. His. |
| Subtilopeptidase A | Asn. Gly. Thr. *Ser*. Met. Ala. Ser. Pro. His. |
| Butyrylcholinesterase | Phe. Gly. Glu. *Ser*. Ala. Gly. (Ala. Ala. Ser.) |
| Acetylcholinesterase | Glu. *Ser*. Ala. |
| Carboxylesterase | Gly. Glu. *Ser*. Ala. Gly. Gly. |

From refs. [84] and [85] and sources quoted therein.

active serine for several hydrolytic enzymes. The relationships are clear. It has been suggested that these enzymes possess a common protein 'ancestor' and that they are derived from it by genetic mutation [64, 86].

## HYDROLASES WHICH POSSESS ACTIVE CYSTEINE RESIDUES

Papain (3.4.4.10) and ficin (3.4.4.12) are the most widely studied of this group of proteolytic enzymes. Papain consists of a single polypeptide chain with 211 residues, corresponding to a molecular weight of about 23,000. Most of the details of its amino-acid sequence were elucidated by Smith and his group [87], their preliminary sequence being modified by Drenth *et al.* [88] on the basis of X-ray diffraction studies. The sequence was finally completed by Husain and Lowe [88a] and is shown in Fig. 4.7.

Early studies with papain made it clear that the thiol group of a cysteine residue was extremely important for enzymic activity. Thus, various oxidizing agents which react with the sulphhydryl group also destroy activity, while reducing agents which regenerate the sulphhydryl after oxidation cause reactivation of the enzyme [89]. Further evidence for the essential nature of the sulphhydryl group was obtained by Finkle and Smith [90] and Kimmel and Smith [91], who showed that papain is inactivated by mercurial derivatives and also that inactivation of the enzyme with iodoacetamide leads to alkylation of a cysteinyl residue.

The inhibitor TPCK, which reacts with the active serine enzymes by alkylation of the imidazole group of a histidine residue, also inhibits papain [92, 93]. In this case, however, no imidazole is affected and the evidence suggests that a cysteine residue becomes alkylated. Confirmation of this is afforded by the experimental result that the reaction of TPCK with papain depends on a group, $pK_a$ 8·3, which must be in its basic form [92]. This $pK_a$ value corresponds most closely to a sulphhydryl group.

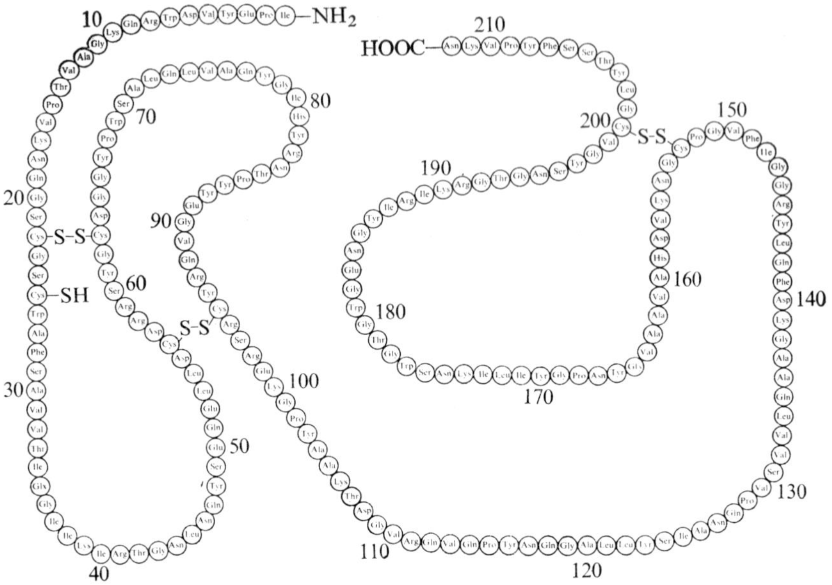


FIG. 4.7. Amino-acid sequence of the enzyme papain based on the chemical studies of Smith and his co-workers [87], and the X-ray diffraction analysis of Drenth *et al.* [88], and finally completed by Husain and Lowe [88*a*]. (By courtesy of G. Lowe and the Editorial Board of the Biochemical Journal)

It is generally accepted that papain-catalysed reactions proceed through an acyl-enzyme intermediate. Such an intermediate (thionohippuryl-papain) has been detected spectroscopically by Lowe and Williams [94], and *trans*-cinnamoyl-papain has been isolated and its deacylation studied [92]. More evidence for the acyl-enzyme hypothesis is provided by the observation that the catalytic rate constants for the hydrolysis of a series of esters of hippuric acid are all equal [95]. This indicates that deacylation of a common acyl-enzyme is rate-determining. It has also been shown that values of $k_{+3}$ (see eq. 4.1, which also holds for papain) for various esters of N-$\alpha$-benzyloxycarbonyl-L-lysine are identical [92]. It is extremely difficult to explain the various transpeptidation reactions which papain can catalyse without invoking an acyl-enzyme intermediate [96].

The nature of the acyl-enzyme derivatives also seems to be well established. That they are thio-esters (that is, a cysteine residue is involved as opposed to oxy-esters in which a serine side-chain is involved) is clear from spectroscopic data [92, 94] and from studies of the reactivities of the acyl-enzymes towards various nucleophiles [92].

The three-dimensional structure of papain obtained by Drenth and his co-workers [88] gives us a clear picture of the active site of the enzyme in addition to providing information about the amino-acid sequence. The molecule contains a deep groove running the whole width of the enzyme, as shown in Fig. 4.8. This groove contains, or consists of, the active site;

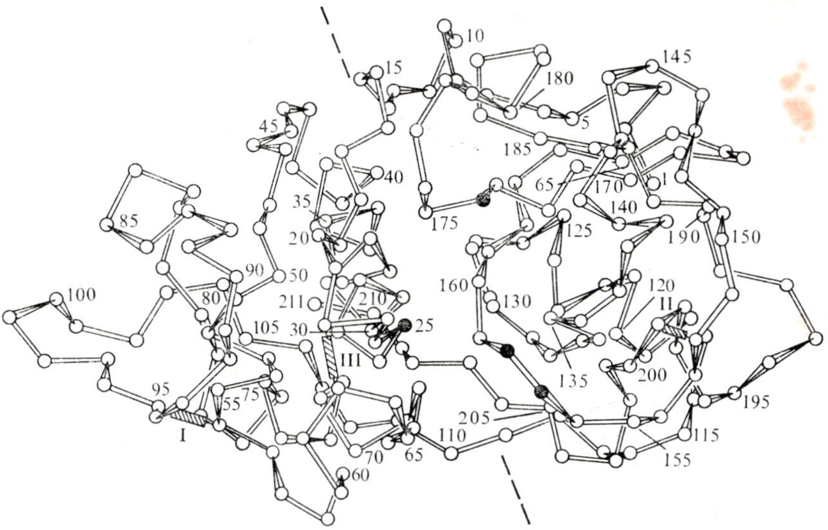


FIG. 4.8. Perspective drawing of the main chain conformation of papain. The circles represent the α-carbons of the 211 residues. (By courtesy of J. Dreuth and the Editor, *Nature*)

cysteine-25 which is clearly visible is the esssential residue for activity. Close by is histidine-158† which Husain and Lowe linked to cysteine-25 by means of 1,3-dibromoacetone [97].

When detailed study of the three-dimensional structure indicated that the histidine-158 was also hydrogen-bonded to an aspartyl residue (residue 174† which was originally identified as an asparagine), comparison with the chymotrypsin case gave rise to the speculative proposal [97a] that the active site of papain may be represented as in Fig. 4.9 and that the mechanism

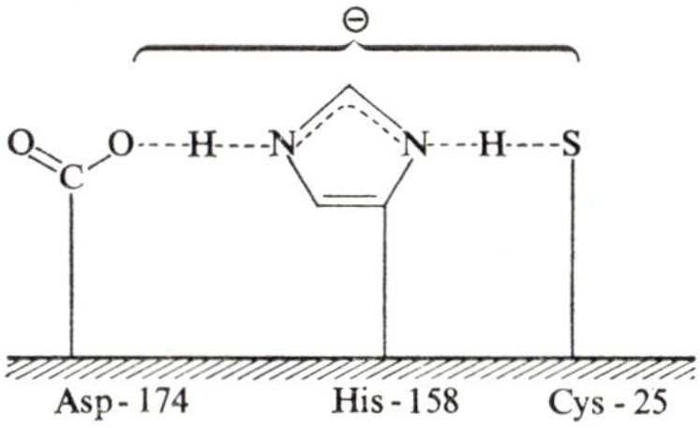


FIG. 4.9. The possible arrangement of the aspartyl, histidyl and cysteinyl residues at the active site of papain

may therefore be entirely analogous to that of chymotrypsin, the —SH of cysteine playing the role of the —OH of serine. Fig. 4.10 shows a possible mechanism for papain based on this suggestion.

† The residue numbers quoted here and in Figs 4.8, 4.9 and 4.10 are the original numbers given in ref. 88. They differ from those in Fig. 4.7.

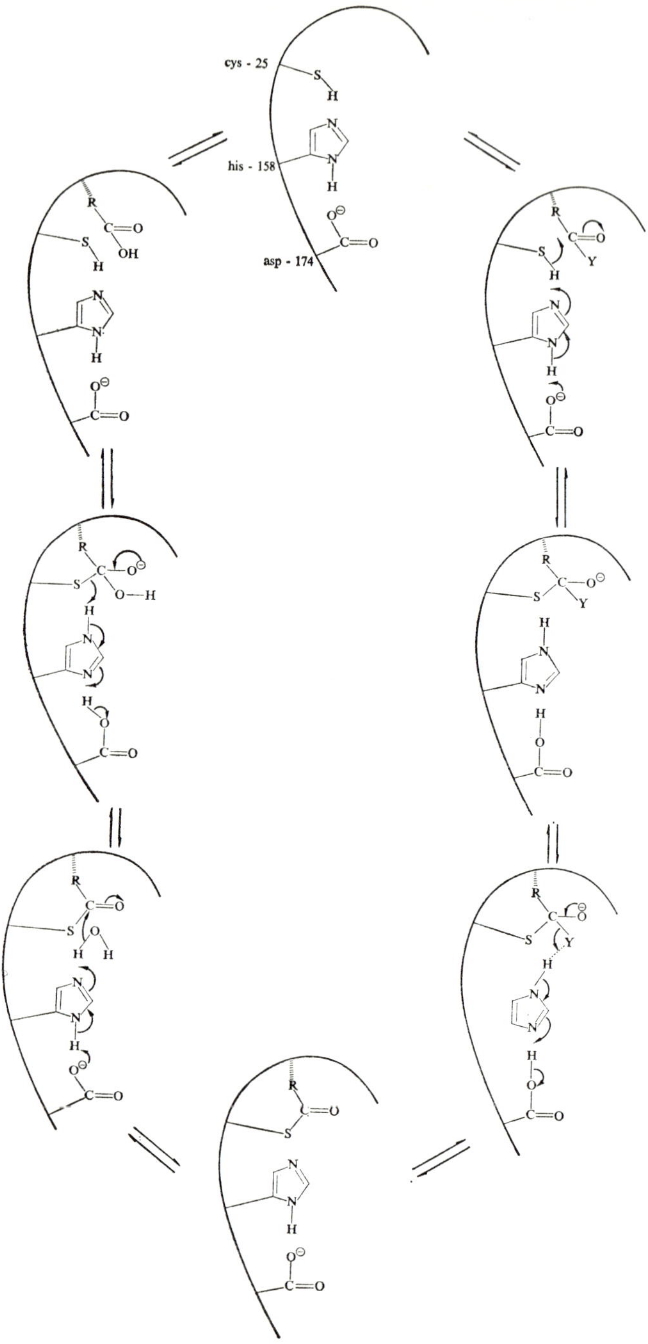


FIG. 4.10. Proposed mechanism for the papain-catalysed hydrolysis of an ester or amide substrate R.C.O.Y. based on the hypothesis that the mechanism is essentially similar to that of chymotrypsin

Both the acylation and deacylation steps are pH-dependent, an apparent $pK_a$ of about 4 being involved. The system shown in Fig. 4.9 may exhibit properties corresponding to a $pK_a$ of 4 especially in view of the complexity of the surrounding environment. An apparently similar system in chymotrypsin exhibits a $pK_a$ of about 6·5.

As in the cases of many other enzymes, relationships between substrate structure and enzyme activity have yielded much information concerning the nature and overall size of the active site of papain. As for trypsin, arginyl or lysyl peptide bonds are favoured, but the specificity is not so restricted [98]. The most sensitive substrate known is benzoyl-L-arginine amide, but many other compounds which are normally associated with other enzymes are hydrolysed in the presence of papain; these include benzyloxycarbonylglycyltyrosine, acetyltyrosine amide, leucylglycylglycine and leucine amide [96]. Some derivatives of amino-acids with acidic side-chains may also function as substrates, but in these cases pH-studies suggest that ionization of the carboxyl group decreases the susceptibility to enzymic hydrolysis [90].

Like the active serine enzymes, papain catalyses the hydrolysis not only of amides, but also of esters [90] and thiol esters [99].

For the proteolytic enzymes which have been already mentioned in this chapter, the side-chain of the amino-acid which provides the carbonyl group of the peptide bond to be broken is the major factor in determining the susceptibility of the substrate, and therefore in the binding of the substrate to the enzyme. Schechter and Berger [100] and Brubacher and Bender [101] have published data which suggest that for papain, several

$$NH_2\text{—}\underset{1}{Ala}.\underset{2}{Ala}.\underset{3}{Ala}.\underset{4}{Ala}.\underset{5}{Ala}.COOH$$

XXIV

amino-acid residues in a peptide substrate may contribute to the binding. Schechter and Berger studied the hydrolysis of various oligopeptides of alanine, some of which contained D-alanine residues in place of the L-isomer. No hydrolysis occurs at bonds involving a D-residue, or at L–L bonds

$$NH_2-P_4-P_3-P_2-P_1\underset{\uparrow}{-}P_1^1-P_2^1-P_3^1-COOH$$

Bond broken

(XXV)

in a D–L–L sequence [102]. Some examples serve to indicate the type of result which these workers obtained. If the papain-catalysed hydrolysis of penta-L-alanine (XXIV) is considered, it was found that the hydrolysis of the 2–3 peptide bond is 2·7 times slower if the alanine-5 residue is replaced by the D-isomer. Therefore replacement of a residue by its diastereoisomer at a position three residues away from the site of catalytic action of the enzyme has a strong influence on the reactivity. Similarly, replacement of L-alanine-1 by its D-isomer leads to a reduction in the rate of cleavage of the 4–5 peptide bond by a factor of 2·2. This replacement is four residues away from the bond under attack. If the structure (XXV) represents a hypothetical peptide substrate for papain then the results may be summarized in Table 4.3, which indicates the decrease in hydrolytic rate caused by a diastereoisomeric replacement at each of the residues.

Table 4.3. *Effect of diastereoisomeric replacement on the rate of hydrolysis of papain substrates*[a]

| Position of replacement (see XXV) | Approx. factor by which rate is decreased |
|---|---|
| $P_1$ | 1000 |
| $P_2$ | 100 |
| $P_3$ | 5 |
| $P_4$ | 2·5 |
| $P_1^1$ | 1000 |
| $P_2^1$ | 10 |
| $P_3^1$ | 3·5 |

[a] Taken from results quoted in ref. [100].

It is clear from these results that the effect of diastereoisomeric replacement is strongly felt over a range of seven amino-acid residues, three on the amino (right-hand) side of the hydrolysed bond and four on the carboxyl side. The effect decreases with distance from the catalytic point.

Brubacher and Bender [101] reached similar conclusions by studying the deacylation of cinnamoyl-papain in the presence of various polyglycinamides. Table 4.4 records the relative rates with which these peptides

Table 4.4. *Deacylation of* trans-*cinnamoyl-papain in the presence of polyglycinamides*[a]

| Nucleophile | Relative rate of reaction of the nucleophile with the acyl-enzyme |
|---|---|
| Glycinamide | 1·0 |
| Diglycinamide | 19 |
| Triglycinamide | 37 |
| Tetraglycinamide | 50 |

[a] Data from ref. [101].

react with the acyl-papain. If these nucleophiles are assumed to be of the same strength (their $pK_a$ values are similar) then the differences in reactivity should reflect mainly differences in binding. As the table shows, diglycinamide reacts more rapidly than does glycinamide and therefore may be assumed to bind more strongly. The tripeptide is more reactive and the tetrapeptide is a little more active still. Further increases in the length of the peptide would presumably have very little more effect. These results, which were obtained before the three-dimensional structure of the enzyme was elucidated, indicated that the active site of papain was able to accommodate and interact with a peptide containing at least seven residues, each making some contribution to the enzyme–substrate binding. The active site was suggested to be a long groove in the protein molecule, and the X-ray studies confirmed this suggestion.

Originally it was thought that the substrate specificity of papain was rather similar to that of trypsin, i.e. if $P_1$ in XXV were lysine or arginine then the peptide would be a good substrate. Recent work by Schechter and Berger [100a] using a number of synthetic peptides has shown that although the amino-acid in position $P_1$ does contribute, the next residue, in position $P_2$, makes an even greater contribution. If the residue at $P_2$ is an aromatic amino-acid the affinity between enzyme and substrate is greatly enhanced.

The evidence which has been obtained for the enzyme ficin suggests that its mode of action is probably similar to that of papain [103, 104, 105].

**Relationships between the Active Serine and Active Cysteine Enzymes**

We have already commented on the structural relationships among the active serine enzymes. Lowe [106] has pointed out that there are similar relationships between the active serine and the active cysteine enzymes. Thus, the amino-acid sequence in the vicinity of the active serine in trypsin bears close resemblances to the sequences around the active cysteine residues of papain and ficin. It should be noted that the specificities of these three enzymes are very similar although their sources differ. The sequences near the active residues are shown in Table 4.5. If the dipeptide units Ser. Cys.

Table 4.5. *Amino-acid sequences, near active residue*[a]

| | |
|---|---|
| Papain: | Lys. Asn. Glu. Gly. Ser. Cys. Gly. Ser. Cys.* |
| Ficin: | Arg. Gln. Glu. Gly. Glu. Cys. Gly. Ser. Cys.* |
| Trypsin: | Lys. Asn. Ser. Cys. Glu. Gly. Gly. Asp. Ser.* |

[a] Quoted in ref. [106]. * Active residue.

and Glu. Gly. in, for example, trypsin are interchanged the similarities become even more striking. Such 'cross-overs' are apparently commonly observed in related proteins [107]. These and other relationships raise the possibility that these two groups of proteolytic enzyme arose from common evolutionary ancestors, the unexceptional replacement of cysteine for serine or vice versa having little effect on their biological functions.

Two independent groups have studied a very interesting aspect of this interrelationship. Neet and Koshland [108] and Polger and Bender [109] have converted the active serine residue in subtilopeptidase A into a cysteine residue. The sequence of reactions is shown below: it is applicable only because this enzyme possesses no disulphide bridges. The thioacetyl-subtilopeptidase formed is analogous to acetyl-papain; deacylation occurs

$$\text{E—OH} \xrightarrow{\text{Phenylmethane sulphonyl fluoride}} \text{E—O—PMS} \xrightarrow{\text{Thioacetate}} \text{E—S—COCH}_3 \longrightarrow \text{E—SH}$$

| Subtilopeptidase | Phenylmethane-sulphonyl-subtilopeptidase | Thioacetyl-subtilopeptidase | Thio-subtilopeptidase |
|---|---|---|---|

readily. Although the product, thio-subtilopeptidase, is almost completely inactive towards normal substrates, it does possess catalytic activity towards p-nitrophenyl acetate, indeed it is about one-third as active towards this ester as is the natural subtilopeptidase. Kinetic studies make it clear that an acyl-enzyme intermediate is involved in this case also.

If, as seems likely, a carboxyl-imidazole-hydroxyl complex exists in natural subtilopeptidase A, then a carboxyl-imidazole-thiol system must exist in the thio-enzyme. Since these relationships emphasize the similarities between the active serine and active cysteine enzymes, the assignment of similar mechanisms to papain and chymotrypsin appears to be further justified.

## HYDROLASES WHICH REQUIRE METAL-IONS

The enzymes which fall into the two categories already discussed in this chapter are *endopeptidases*, that is they are able to catalyse the hydrolysis of peptide bonds in the middle of a peptide chain. This does not, however, rule out the possibility of their catalysing the hydrolysis of bonds in dipeptides or derivatives of single amino-acids. The *exopeptidases* on the other hand do not possess these properties, they are able to cleave only peptide bonds which are formed by the terminal amino-acid residue of a peptide. Thus, the carboxypeptidases cleave the C-(carboxyl)-terminal peptide bonds in a substrate while the aminopeptidases effect the hydrolysis of the peptide bond associated with the amino-terminal residue. In general, the carboxypeptidases require the presence of free carboxyl groups in their substrates, and the amino-peptidases require free amino-groups. Several enzymes, dipeptide hydrolases, require both free carboxyl and free amino-groups in their dipeptide substrates. The observation that many of these enzymes are metal-dependent gave rise to several suggestions concerning the role of the metal-ions in determining specificity and in the catalytic processes themselves (ref. [1], page 304). It is clear now, however, that such generalizations are dangerous; not all exopeptidases are metal-dependent and not all metal-dependent hydrolases are exopeptidases. In this, as in every other enzyme category, it is important to consider each enzyme individually.

### Peptidyl-L-amino-acid Hydrolase (Carboxypeptidase A.3.4.2.1)

Of all the metal-ion requiring hydrolases carboxypeptidase A has been studied in by far the greatest detail, and only now are ideas on its mechanism of action becoming established. This is due to several factors which include complicated kinetic patterns [110], uncertainty about the amino-acid residues involved in the binding of the metal-ion to the apo-enzyme, and some unusual changes in enzymic properties brought about by chemical modification, which suggested that the enzyme may function according to at least two separate and distinct mechanisms [111].

Carboxypeptidase A is a pancreatic enzyme whose biological function is that of protein digestion. Like many other hydrolytic enzymes it is produced biologically as an inactive precursor which is converted into the active form by the action of trypsin. Pre-carboxypeptidase A apparently occurs as an aggregate of three sub-units, I, II and III, I being the direct precursor of carboxypeptidase A itself [112]. Activation of I requires an endopeptidase which is very similar to chymotrypsin and which is generated from sub-unit II. No function has, as yet, been ascribed to sub-unit III.

As its names suggest, carboxypeptidase A catalyses the hydrolysis of peptides in which the amino-acid residue providing the amino-group of the peptide bond to be cleaved is the terminal residue, that is, it effects the cleavage of the C-terminal residue from a peptide or protein chain. The

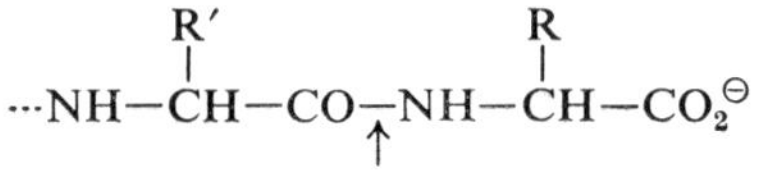

(XXVI)

absolute requirements for this enzyme are that the C-terminal residue must be of the L-configuration and that its carboxyl group must be free [110, 113]. Substrates in which the side-chain R (XXVI) is aromatic are favoured although carboxypeptidase A has a fairly broad specificity in this respect; peptides possessing almost any C-terminal residue except proline will be hydrolysed. The specificity is therefore rather similar to that of chymotrypsin with a significant difference in that the amino-acid which determines specificity in the case of chymotrypsin is to the left of the peptide bond (that is, it contributes the carbonyl group) whereas for carboxypeptidase the aromatic amino-acid contributes the amino-group. The nature of the side-chain R′ is of less importance although an acidic side-chain may decrease the rate of hydrolysis, whereas an aromatic residue may enhance it [114]. Dipeptides possessing free amino-groups are generally not hydrolysed.

Not all the features of structure (XXVI) are necessary in a substrate. The terminal residue may be an α-hydroxy-acid; benzoylglycyl-L-phenyllactic acid is a good substrate for this enzyme (115, 116]. The penultimate residue need not be a complete amino-acid but may simply be an acyl-group which possesses an electronegative substituent as in chloroacetyl-L-phenylalanine [115].

β-Phenylpropionic acid, in which the carboxylate group and the aromatic ring are separated by two carbon atoms, and which therefore bears a close steric resemblance to the specific substrates, is an effective competitive inhibitor [117].

Carboxypeptidase A consists of a single polypeptide chain of 307 residues, with molecular weight 34,600 [118, 119]. The amino-acid sequence has now been almost completely established [120, 121, 119a]. The enzyme contains one atom of zinc per molecule [122]. This zinc atom is essential for activity. It may be removed either by dialysis at low pH or by dialysis at neutral pH against buffers containing the chelating agent 1,10-phenanthroline [123]. The loss of activity from preparations treated in these ways parallels exactly the loss of zinc. The removal of the zinc atom appears to affect the overall structure of the protein very little; the optical rotations of the native and metal-free proteins are identical, as are their behaviours on sedimentation [124]. Addition of zinc ions to the apo-enzyme regenerates enzymic activity, the activity of the reconstituted enzyme being proportional to the amount of zinc added up to one atom per molecule of protein.

Certain other transition metal-ions may replace the zinc and regenerate peptidase activity in the apo-enzyme; they include Fe(II), Mn(II) Co(II) and Ni(II) [123, 125]. It is clear that these metal ions occupy the site formerly occupied by the zinc ion. Under the specific conditions of assay, the Co(II)-enzyme is more active against benzyloxycarbonylglycylphenylalanine than is the zinc-derivative [125, 126].

*The Binding of the Zinc Ion to the Apo-enzyme.* Since 1960 a considerable amount of evidence has been accumulated concerning the nature of the amino-acid residues in the apoenzyme which contribute to the binding of the zinc ion to the protein. Most of this evidence has pointed in one direction, suggesting that the thiol group from a cysteine residue functions as one ligand to the zinc ion. However, the three-dimensional structure, based on X-ray studies, reported by Lipscomb's group indicates that this is not the case [119]. Because the zinc atom is clearly essential for catalytic activity, it is worthwhile for us to review briefly the evidence which was obtained concerning its binding.

Silver(I) and p-mercuribenzoate are commonly used as thiol reagents and it is often possible to estimate the number of thiol groups per molecule of a protein by titration with these reagents, the former amperometrically and the latter spectroscopically [127]. Coombs *et al.* [120] found that native carboxypeptidase A possessed no thiol groups detectable with either silver(I) or p-mercuribenzoate. When the zinc ion had been removed, by dialysis against buffer containing 1,10-phenanthroline, titration indicated the equivalent of one thiol group per molecule. After addition of one atom of zinc per molecule to the apo-enzyme, again no thiol group could be detected. When less than one atom of zinc per molecule was added the titres were such that the sums of zinc present plus thiol detected were approximately equal to unity. The apo-enzyme also consumed 1·22 equivalents of ferricyanide, another common thiol reagent (the conditions required for this reaction were rather drastic; 4M urea, 36-molar excess of reagent, 48 hours).

More information was obtained from studies of the stability constants of various complexes of the apo-enzyme with metal-ions. The magnitudes of the stability constants of the Hg(II)-, Cd(II), Zn(II)-, Cu(II)-, Ni(II)-, Co(II)- and Mn(II)-carboxypeptidases, and the sequence in order of magnitude, correlate best with similar series of complexes involving models which possess bidentate nitrogen–sulphur ligands, such as cysteine, mercaptoethylamine and glutathione. Much poorer correlations were obtained with model systems possessing other ligands [128, 129].

Furthermore, complexometric titration showed that combination of the zinc ion with the apo-enzyme liberated two protons per molecule and that the $pK_a$ values of the two prototropic groups were 7·7 and 9·1 [130]. Stability constant measurements and complexometric titration with the pre-carboxypeptidase A indicated that the nitrogen ligand mentioned above as being involved in the binding of the zinc to the apo-enzyme was not involved in its binding to the pre-apo-enzyme [131]. A sulphur–oxygen bidentate ligand system was implicated in the latter case.

Williams [129] reported more data in support of a sulphur ligand. Cobalt(II)-carboxypeptidase A is highly coloured, having an absorption maximum at 530 nm (molar extinction coefficient, $\varepsilon = 120$). This agrees best with the properties of certain model cobalt(II)–sulphur ligand complexes in terms of both the wavelength and the intensity of the absorption. Further agreement was observed in the absorption at 285–300 nm. More spectroscopic information was obtained from studies of pyridoxal phosphate and its complex with apo-carboxypeptidase. Pyridoxal phosphate is able to form complexes with amino-thiols such as cysteine. On formation of these complexes the ultraviolet spectrum of the pyridoxal phosphate changes, important features being a new maximum at 330 nm and decreased absorption at 388 nm. The product of combination of apocarboxypeptidase A and pyridoxal phosphate showed a decrease at 388 nm and a new maximum at 325 nm [120].

The results reviewed above suggested strongly that the zinc ion was bound to the apo-enzyme through at least two (and possibly three) ligands, one of which was a sulphur atom and another was a nitrogen. The probable third ligand involved an oxygen atom. In the pre-enzyme, the nitrogen ligand apparently made no contribution and only the S—O ligands were involved.

A few results were obtained which did not support these assignments. For example, other thiol reagents, iodo- and bromo-acetate, iodoacetamide and N-ethylmaleimide, failed to react with either the native carboxypeptidase or with the apo-enzyme [120]. The amino-acid analysis of the protein showed the presence of two half-cysteine residues per molecule [118]. If one of these, as a cysteine, were involved in the metal-ion binding, why was the other not detected by the thiol reagents in the native enzyme, and why did the titrations with the apo-enzyme not indicate two cysteines per molecule? It is worth noting that when the apo-enzyme was treated with 8M urea and a detergent it failed to titrate for even the one thiol group. Two thiol groups can only be detected after reduction of the apo-enzyme [132].

Lipscomb *et al.* [119] have published a model for the three-dimensional structure of carboxypeptidase A. The extremely good resolution of their electron density maps (2·0 Å) makes their conclusions very difficult to challenge. This enzyme has been shown to be enzymically active in the crystalline state [133], and so the results of X-ray crystography are entirely relevant to our discussion of the mechanism. The structure shows that the two half-cysteine residues in fact make up a disulphide bridge which is quite far removed from the zinc atom, and that the zinc ligands are the side-chains of two histidyl and one glutamyl residues (Fig. 4.11). In the native enzyme a water molecule makes up the fourth ligand.

The evidence which suggested a sulphur ligand clearly requires reinterpretation. The lack of reaction with iodoacetamide, etc., is understandable in terms of Fig. 4.11, as is the apparently anomalous behaviour of the second cysteine residue; there is no thiol group in the molecule. The results from the titrations with silver(I) and p-mercuribenzoate fall into the same category as the evidence concerning stability constants; these are all processes which involve the formation of metal-ion–ligand complexes. Although the binding of the metal ions to the apo-enzyme involves a nitrogen–nitrogen–oxygen system, it resembles more closely models which

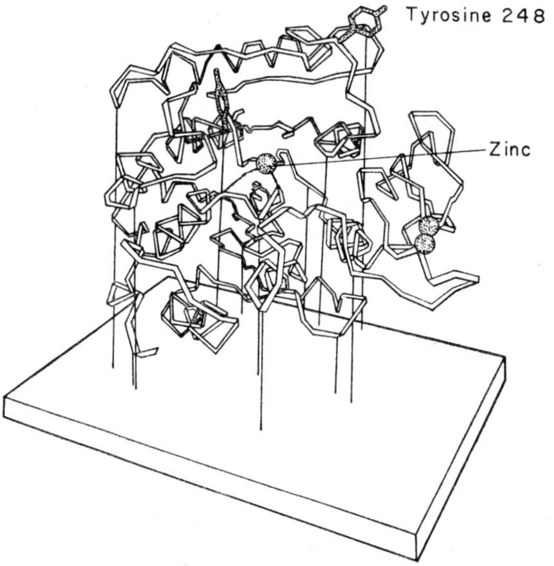


FIG. 4.11. The three-dimensional shape of the enzyme carboxypeptidase, showing the groove above and to the right of the zinc atom, into which the substrate fits. (By courtesy of W. N. Lipscombe, *New Scientist* and M.R.C.)

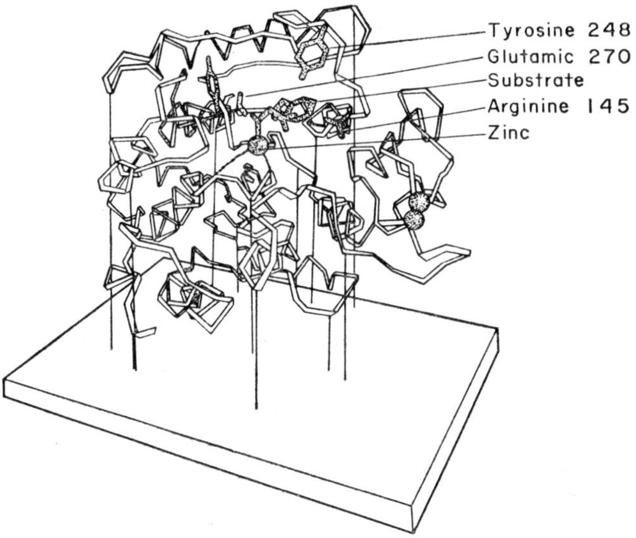


FIG. 4.12. The substrate in place in the groove of carboxypeptidase. The substrate binds to a side group of arginine and to the zinc atom. After the substrate is in position, both tyrosine and arginine move toward the substrate, in confirmation of the induced fit hypothesis of enzyme action. (By courtesy of W. N. Lipscombe, *New Scientist* and M.R.C.)

possess sulphur–nitrogen–oxygen systems. This is because the binding is apparently stronger than expected from nitrogen–nitrogen–oxygen models. Owing to the peculiar natures of the active sites of enzymes such a situation is not unlikely. Under these circumstances the reactions of silver(I) ions and p-mercuribenzoate are no longer specific for thiol groups, they merely indicate an unusual binding site. In the presence of 8M urea and a detergent this site would be expected to be destroyed (see above). The absorption of the Co(II)-enzyme might well fit in with this interpretation, although the data concerning the pyridoxal phosphate complex remain to be explained. The $pK_a$ values of the two prototropic groups shown by titration to be involved in the metal-ion binding may agree well with the X-ray structure if the higher value 9·1 may be assigned to an unusual imidazole group. If the third ligand is the carboxylate group of a glutamic acid residue it would not be detected by titration under the conditions used.

*Mechanism of Action.* From a consideration of the substrate specificity of the enzyme, we may deduce certain properties possessed by the active site. Because a free carboxylate group is required in the substrate it may be assumed that a positively-charged group exists in the protein to complement it. Similarly, some appropriate, non-polar region must be present to correspond with the side-chain of the terminal amino-acid residue. Schechter and his group [134] have suggested that the active site of carboxypeptidase A extends over 18 Å, and that there are five regions, one for each amino-acid residue of a pentapeptide substrate. This suggestion is based on results similar to those described by these workers for papain (see above).

Of the possible functions of the essential metal-ion (see chapter 2), one may be eliminated. This is the possibility that the metal is needed for the maintenance of the three-dimensional structure of the enzyme and that the lack of activity of the metal-free protein is due solely to the consequent destruction of this structure. Removal of the zinc ion causes very little change in the overall structure, indeed some potential substrates and inhibitors are able to form complexes with the apo-enzyme alone.

Figures 4.11 and 4.12 show the three-dimensional structures of carboxypeptidase A and of the glycyltyrosine–carboxypeptidase A complex. The latter affords a picture of an enzyme–substrate complex and shows clearly the portions of the enzyme which are in contact with the substrate. The carboxylate group of the substrate is associated with the guanidinium group of an arginine residue, and the active site also includes a non-polar pocket into which the aromatic side-chain of the terminal amino-acid residue fits. This pocket is quite large and is able to accommodate a range of possible side-chains, hence the fairly broad specificity of the enzyme in this respect.

The carbonyl group of the peptide bond which is to be broken replaces the water molecule as the fourth ligand to the zinc atom, the ligands being arranged more or less tetrahedrally. The zinc ion therefore makes some contribution to the binding of the substrate, but its major contribution to the catalytic process appears to be in that it polarizes the carbonyl group in preparation for a nucleophilic attack.

Association of the dipeptide with the enzyme causes some extremely interesting changes in the conformation of the protein. Thus, the side-chain of the arginyl residue which becomes involved with the carboxylate anion of the substrate moves about 2 Å in order to do so. The tyrosyl-248 side-chain which projects into the solvent in the native enzyme moves about 14 Å so that in the enzyme–substrate complex its hydroxyl group is situated very close to the peptide bond. A second tyrosine residue lies near to the substrate in the complex.

Vallee and his group [135, 136] have provided a wealth of information concerning the role of one or two tyrosine residues in the catalytic process. Acetylation of the enzyme with either acetic anhydride or acetylimidazole causes complete loss of *peptidase* activity. The extent of acetylation of two tyrosine residues parallels the activity loss and deacetylation with hydroxylamine reverses the effect. The competitive inhibitor $\beta$-phenylpropionate completely prevents the acetylation. Acid anhydrides other than acetic anhydride produce similar though less marked effects (an exception is succinic anhydride which actually causes an increase in peptidase activity, but it has been shown that this reagent does not react at the active site). Iodination also removes the enzyme's activity towards peptides; that the iodination occurs at the same side-chain which is susceptible to acetylation is shown by the observation that acetylated-carboxypeptidase may be treated with iodine and subsequently deacetylated with hydroxylamine to regenerate the peptidase activity. Nitration of a single tyrosine residue with the selective nitrating agent tetranitromethane causes a 90% loss of the enzyme's peptidase activity.

These chemical results are of course consistent with the existence of a tyrosine residue at or near the active site.

Lipscomb and his colleagues [119] have proposed a mechanism for the action of carboxypeptidase A based on the arrangement of the appropriate groups in the active site as shown by X-ray crystallography. Their suggestion is represented in Fig. 4.13 for the hydrolysis of the glycyl-tyrosine dipeptide. The carboxylate oxygen of glutamic-270 is envisaged as attacking the carbonyl of the peptide bond (XXVII), the zinc ion assisting this process (i). Breakdown of the resultant quaternary complex (ii) is assisted by acid catalysis by the hydroxyl of tyrosine-248, and the product at this stage is an acyl-enzyme involving an anhydride link. The hydrolysis of this anhydride link is viewed as an attack by water assisted by the

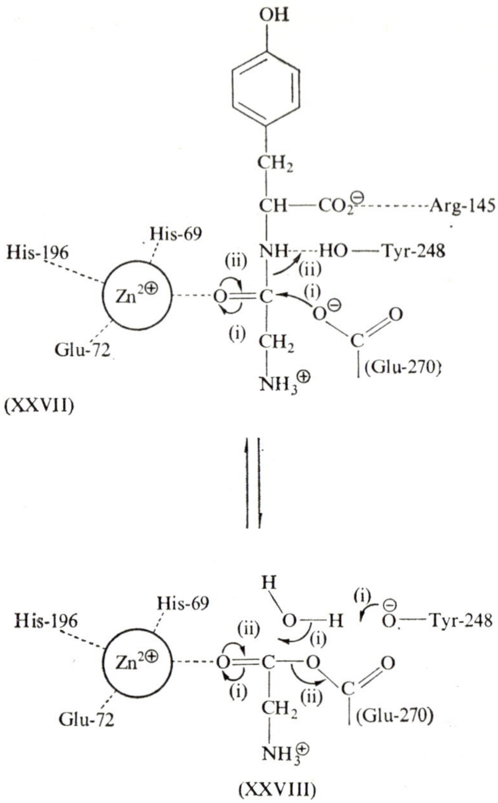


FIG. 4.13. Mechanism of action of carboxypeptidase A as suggested by Lipscomb and his co-workers [119, 119(a)]

conjugate base of tyrosine-248 (XXVIII). Lipscomb *et al.* [119] point out that when the substrate binds to the enzyme it replaces the fourth ligand to the zinc ion, the water molecule. In this way the environment around the zinc ion takes on a lower dielectric constant. This makes the polarization of the carbonyl group by the zinc ion even more effective.

Alternatively, the reaction may involve, not direct attack by the glutamic carboxyl on to the peptide carbonyl, but attack by water with assistance from the carboxyl. This alternative process would not lead to the production of an acyl-enzyme.

At present there is little evidence in favour of an acyl-carboxypeptidase. Ginodman *et al.* [137] have found that carboxypeptidase A does catalyse the exchange of $^{18}O$ in $H_2{}^{18}O$ for oxygen in the carboxyl group of N-acetylphenylalanine and this result supports an acyl-enzyme hypothesis. However, they were unable to detect exchange of tyrosine-$^{14}C$ for tyrosine in N-acetylphenylalanyltyrosine. Such exchange would be expected if the above mechanism were to apply.

For many enzymes, pH-rate profiles are able to provide information about the nature of the amino-acid residues involved in the catalytic process. This is not so in the case of carboxypeptidase A because the pH-rate profile appears to reflect mainly the metal–protein binding, which is of course essential to activity and which appears to overwhelm the effect of any other process [125].

We have already referred to the ability of this enzyme to catalyse the hydrolysis of esters, such as N-benzoylglycylphenyllactic acid. The binding between the enzyme and such a substrate is stronger than that between the enzyme and the corresponding peptide substrate. Since the catalytic constants ($k_{cat}$ in eq. 4·3) for peptide and enzyme substrates are quite similar the overall hydrolysis of an ester is more rapid than that of the corresponding peptide [138].

There is some evidence to suggest that chemical modification of a histidine residue may lead to loss of peptidase activity [120,135]. Whether this is one of the histidine residues involved in the binding of the metal-ion, or a different one, is not known. Obviously more information on this and many other points is required.

### L-Leucyl-peptide Hydrolase (Leucine Aminopeptidase, 3.4.1.1)

There is very little information available concerning the other metal-dependent hydrolases. In the case of leucine aminopeptidase this may be because there has been some difficulty in identifying the aminopeptidase obtained from various sources. The original leucine aminopeptidase was isolated from porcine kidney by Spackman *et al.* [139], and a preparation obtained from bovine eye-lens appears to be very similar to this, if not identical with it [140]. However, an aminopeptidase from porcine kidney microsomes is entirely different both in substrate specificity and in requirement for metal-ions [141, 142]. More recently Joseph and Fruton have isolated an enzyme from porcine muscle which is very similar to the kidney leucine aminopeptidase [143].

Leucine aminopeptidase is a protein with a high molecular weight (of the order of 300,000) and it almost certainly consists of a number of subunits which form an aggregate. Its specificity is such that the N-terminal peptide bond of a peptide or protein is hydrolysed. Certain derivatives of amino-acids may also function as substrates [144]. The major requirement for a substrate is that the N-terminal amino-group must be free. Substrates which possess N-terminal amino-acid residues containing aliphatic side-chains are most rapidly hydrolysed, those with aromatic side-chains less rapidly, while those with polar side-chains (arginyl, glutamyl, etc.) or no side-chains (glycine) are only very poorly hydrolysed. Peptides with N-terminal D-isomers are not affected. Schechter and Berger have shown that when peptides are hydrolysed, the presence of a D-isomer several units down the chain from the N-terminus may cause a decrease in activity [145]. It is probable therefore that the enzyme possesses a non-polar region to accommodate the side-chain of the first residue and probably also several more binding sites for the side-chains of the other peptide residues. The existence of a non-polar region is further suggested by the observation that certain long-chain aliphatic alcohols are effective competitive inhibitors [146].

The enzyme is both activated and stabilized by magnesium ions [144]. Removal of the metal-ion causes rapid, irreversible inactivation. The manganese(II) ion, though a better activator, is not so effective in stabilizing the enzyme. It is possible that the presence of the metal-ions helps to keep the enzyme in the aggregated form. Wolff and Resnick [147] found that the amino-peptidase from bovine eye-lens could be activated by Co(II), Mg(II) or Mn(II) ions but that the best activation was obtained with the combinations of Co(II) and Mg(II) or Co(II) and Mn(II). They called this effect *synergistic activation.* Because the cobalt ion could prevent the reaction of the apo-enzyme with p-hydroxymercuribenzoate, they suggested that a thiol group was involved in the cobalt binding. In view of the results obtained with carboxypeptidase such an interpretation must now be treated with caution.

The pH-binding profile for magnesium or manganese suggests that an amino-group or an imidazole may be involved [148]. It has been suggested that the amino-group of the substrate, in its unprotonated form, may also be a ligand to the metal-ion [144]. The pH-rate profile for the catalysed reaction, which appears to reflect the metal-ion binding (an optimum pH of 9–9·3) is quite compatible with this suggestion. Iodination of a tyrosine residue in eye-lens aminopeptidase causes inactivation [149] and it may be that the decrease in activity of the enzyme above pH 9 reflects the ionization of an essential tyrosine phenol group.

Although earlier workers failed to detect any transpeptidation or transamidation in the products of aminopeptidase-catalysed reactions, Hanson and Lasch [146] have reported that both lens aminopeptidase and an enzyme from rat serum are able to bring about the reactions listed below.

L-Leucine-methylcellosolve ester $\rightarrow$ L-Leucyl-L-leucine
L-Leucylamide $\rightarrow$ L-Leucyl-L-leucylamide
L-Phenylalanyl amide $\rightarrow$ L-Phenylalanyl-L-phenylalanyl amide

These results suggest the formation of an acyl-enzyme. No transpeptidation was observed for D-leucyl amide, L-valyl amide or L-tyrosyl- methylcellosolve ester.

We have already seen an example of an aminopeptidase-like hydrolysis brought about by a reagent possessing a metal-ion (chapter 1, page 41), and it is tempting to suggest a similar mode of action, with some modifications, for leucine aminopeptidase (Fig. 4.14). Thus, the amino-group and the peptide carbonyl in the substrate may become ligands to the metal-ion, which enhances the susceptibility of the carbonyl to nucleophilic attack (XXIX). If an acyl-enzyme is formed then the nucleophile (N) must

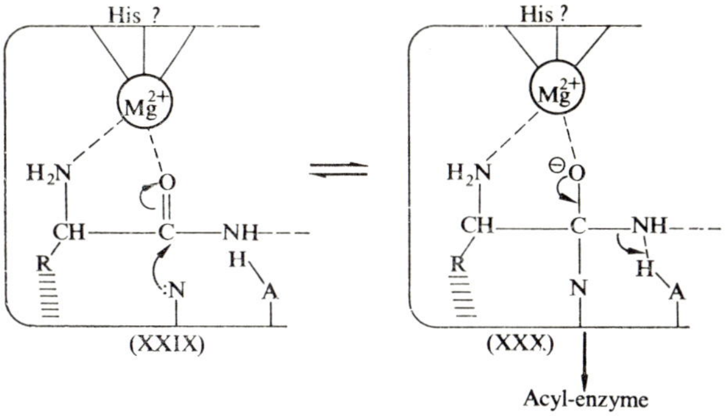


FIG. 4.14. Suggested mechanism of aminopeptidase action

come from the protein molecule. Departure of the nitrogen leaving group from the tetrahedral intermediate might be favoured by the presence of an acidic group (H—A) (XXX). This acidic group might be the tyrosine residue referred to above. This mechanism is very similar to that proposed for carboxypeptidase A.

## HYDROLASES WHICH FUNCTION AT LOW pH VALUES

### Pepsin (3.4.4.1)

Pepsin [150] is a proteolytic enzyme which functions in the acidic environment of the gastric juices. It is produced as the inactive pre-pepsin (pepsinogen) and under the acidic conditions conversion into the active form takes place spontaneously. A single-chain protein of molecular weight about 35,000, pepsin is an unusual protein in that, in keeping with its biological function, it is stable under quite strongly-acidic conditions, its isoelectric point being about pH 1. Its amino-acid composition is likewise unusual; it possesses thirty-six free carboxyl groups per molecule and only four basic residues.

In contrast to other enzymes pepsin exhibits only peptidase activity, esters or amides generally not being hydrolysed. The peptides which are hydrolysed must possess amino-acid residues of the L-configuration on both sides of the peptide bond and the hydrolysis is favoured if these two residues carry aromatic side-chains [151, 152].

It has already been stated that this enzyme functions at low pH values and indeed the pH optimum is near 2. This can vary according to the nature of the substrate [151, 152], although whether these variations are caused by ionization of groups in the substrate or by interaction of the substrate with the prototropic groups of the enzyme is not known. In an attempt to eliminate some of these effects, Clement and Snyder [153] studied the pH-dependence of the hydrolysis of N-acetyl-phenylalanyl-tyrosine methyl ester which remains neutral over the range of pH's used. The pH-rate profile showed that two prototropic groups were involved, one with $pK_a$ 1·62 in the basic and the other with $pK_a$ 3·48 in the acidic form. These $pK_a$ values suggest that two carboxyl groups take part in the catalytic process. Essentially similar results were obtained by Fruton *et al.* (151) who reported that the pH–$k_{cat}$ profile for the hydrolysis of another substrate suggested dependence on two (carboxyl) groups with $pK_a$ 3·2 and 4·5. The higher $pK_a$ values in the latter case may reflect the ionization of the substrate.

There is abundant chemical evidence to support the suggestion that at least one carboxyl group is essential for activity. Herriott *et al.* [154] showed that alkylation of carboxyl groups by sulphur mustard, $Cl\,CH_2CH_2S\,CH_2CH_2Cl$, inactivated the enzyme. Gross and Morell [155] and Erlanger *et al.* [156] independently have shown that another 'active'

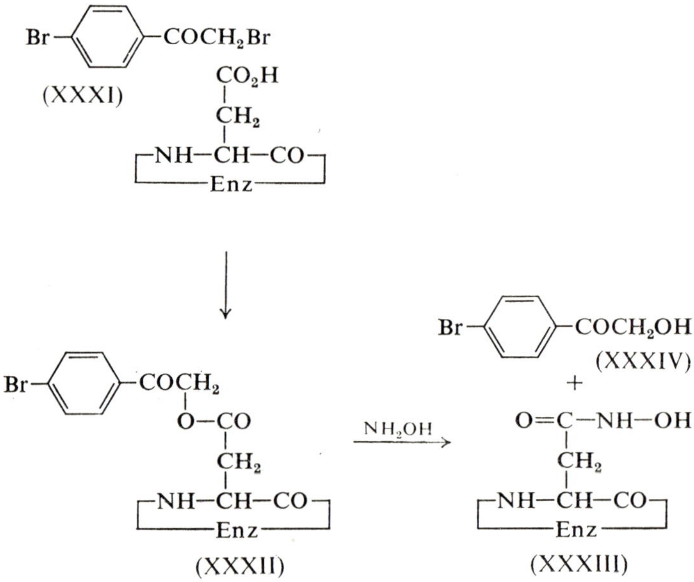


halogen compound, p-bromophenacyl bromide (XXXI) can cause inactivation of pepsin by esterification of the side-chain carboxyl group of an aspartic acid residue. The reaction sequence below shows how this was achieved. The halide (XXXI) reacted with the enzyme to give the inactive p-bromophenacyl-derivative (XXXII) in 1 : 1 stoichiometric amounts. That this reaction occurred at the active site is confirmed by the fact that it could be prevented by the presence of a substrate, and that the pH-rate profile for inactivation was similar to that for protein hydrolysis. Reaction of XXXII with hydroxylamine liberated p-bromo-α-hydroxyacetophenone and gave the hydroxamate derivative of pepsin (again 1 : 1 stoichiometry) (XXXIII) which could react with fluorodinitrobenzene to give the dinitrophenyl compound (XXXV). Lossen rearrangement [157] leads to the replacement of the original carboxyl by an amino group and after hydrolysis diaminopropionic acid (XXXVI) was identified. If a glutamic acid

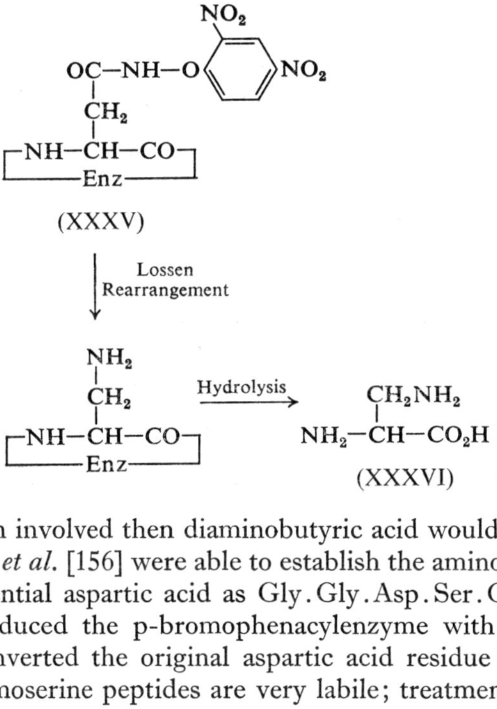


(XXXV) (XXXVI)

residue had been involved then diaminobutyric acid would have been obtained. Erlanger *et al.* [156] were able to establish the amino-acid sequence around this essential aspartic acid as Gly.Gly.Asp.Ser.Glu. Gross and Morell [155] reduced the p-bromophenacylenzyme with lithium borohydride and converted the original aspartic acid residue to homoserine (XXXVII). Homoserine peptides are very labile; treatment of XXXVII

(XXXII)

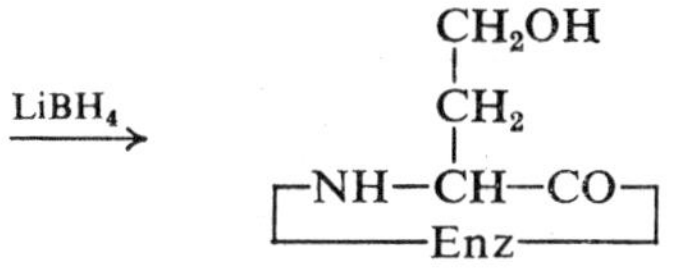


(XXXVII)

with ethanolic hydrogen chloride afforded a peptide in which the C-terminal homoserine residue could be identified.

The essential aspartic acid residue may also be esterified by the use of certain diazo-compounds, especially diazocarbonyl compounds in the presence of cupric ions [158, 159]. An example is the treatment of pepsin

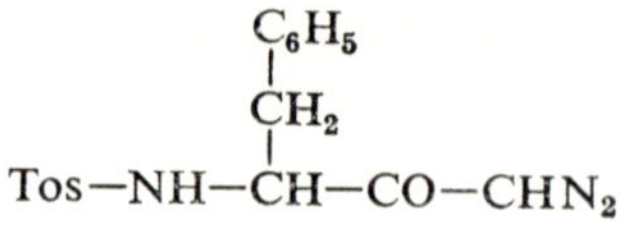

(XXXVIII)

with L-1-diazo-6-phenyl-3-toluenesulphonamidobutan-2-one (XXXVIII) by Fruton *et al.* [159]. This is the diazo equivalent of tosyl-phenylalanine chloroketone (V) which was used so effectively for chymotrypsin (see earlier). In all these cases the resultant ester derivatives react with hydroxylamine.

Apart from the carboxyl groups referred to above, the only other amino-acid side-chain so far implicated in the catalytic process is that of a tyrosine residue. Acetylation of the hydroxyl of a tyrosine, iodination and coupling with diazo-compounds have all been reported to inactivate pepsin [160].

Some of the most interesting reactions catalysed by pepsin are those involving transpeptidation, the most commonly-quoted example being the reaction with the synthetic substrate N-acetyl-L-tyrosyl-L-tyrosine. In addition to the expected products of the reaction the dipeptide L-tyrosyl-L-tyrosine was produced [161]. Now pepsin does not bring about the hydrolysis of the acetyl-amino-acid bonds and so the dipeptide could arise only from a transfer process, as shown. The acetyl-tripeptide may subsequently be cleaved by the enzyme to give tyrosine plus the starting material

$$\text{Ac.Tyr.Tyr} + \text{Ac.Tyr.Tyr} \rightarrow \text{Ac.Tyr} + \text{AcTyr.Tyr.Tyr} \qquad (4.11)$$

or acetyltyrosine plus the observed dipeptide tyrosyltyrosine.

Transfer processes of this type usually indicate a covalent intermediate in the process. In this case the transfer of the tyrosine residue in reaction (4.11) must involve an amino-enzyme intermediate (a derivative in which the amino-acid is acylated by the enzyme). The alternative possibility, an acyl-enzyme, would necessitate a nucleophilic attack by the amino-species, a process which is very unlikely at the low pH values at which pepsin functions. The intermediacy of an amino-enzyme is in direct contrast to the mechanisms of other hydrolytic enzymes mentioned in this chapter, where acyl-enzymes are involved. Support for the amino-enzyme hypothesis is provided by studies on the substrate specificity and kinetics of pepsin-catalysed hydrolytic and transfer processes. These make it clear that binding sites exist on the enzyme for the side-chain groups of the amino-acid residues on either side of the peptide bond cleaved or formed [162, 164, 165]. Binding involving both these residues is emphasized by the kinetics of hydrolysis of acetyl-L-phenylalanyl-L-tyrosine which indicate that both the products of the reaction (acetyl-L-phenylalanine and tyrosine) inhibit the enzyme [165].

Akhtar and Al-Janabi [168] have provided more evidence that an amino-enzyme derivative is produced during the reaction. They incubated-benzyloxycarbonyltyrosyltyrosine (labelled in both tyrosines, Z.Tyr*.Tyr*.OH) with pepsin and then rapidly denatured the protein. Radioactivity was incorporated which could not be removed on dialysis in 8M urea with hydroxylamine, i.e. a covalent link was involved which was not an ester link. When Z.Tyr*.Tyr.OH was used no radioactivity was incorporated, indicating that it was the C-terminal tyrosine which was incorporated.

One fragment of the substrate, therefore, becomes bound to the enzyme as an amino-enzyme derivative. However, pepsin also catalysed exchange of oxygen between $H_2{}^{18}O$ and acylated amino-acids such as benzyloxycarbonyl-L-phenylalanine (but not the D-isomer) and N-acetyl-L-phenylalanine [163]. This suggests that the other fragment of the substrate also becomes bound to the enzyme, as an acyl-enzyme. Again, Akhtar and Al-Janabi [168] provided further evidence for the existence of an acyl-enzyme intermediate.

The evidence at present available therefore indicates that (*a*) two carboxyl groups and possibly a tyrosine side-chain are involved in the catalytic process, (*b*) one of the carboxyl groups must be ionized, (*c*) both the amino-acid residues which form the substrate's peptide bond are bound to

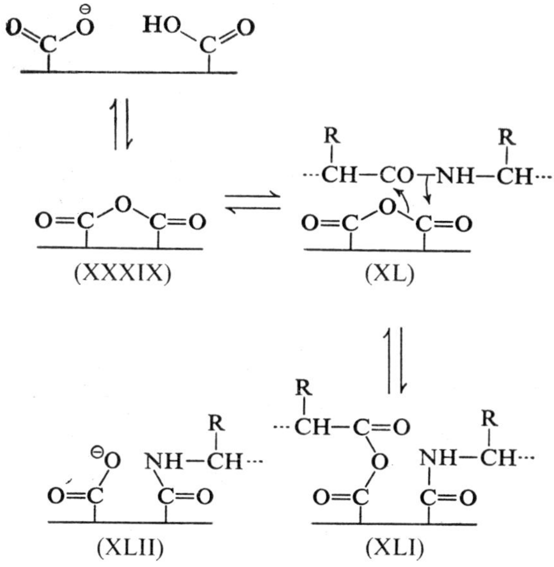


the enzyme in the Michaelis complex, (*d*) that an acyl- and amino-enzyme is formed as an intermediate in the reaction. Bender and Kezdy [166] have suggested a possible mechanism for pepsin-catalysed reactions based on most of the above considerations. The two carboxyl groups at the active site interact to form an anhydride (XXXIX) which, when the enzyme–substrate complex has been formed, undergoes a four-centre exchange with the peptide (XL). This leads to the amino- and acyl-enzyme (XLI). The acyl-enzyme portion involves a mixed anhydride and would be expected to undergo hydrolysis very rapidly giving the amino-enzyme (XLII). Hydrolysis of the amide bond in XLII, it is suggested, may be assisted by the nearby carboxyl group, in a manner similar to that described for the hydrolysis of phthalamic acid (XLIII). This results in the re-formation of the anhydride (XXXIX). Phthalic anhydride (XLIV) has been established as an intermediate in the hydrolysis of phthalamic acid [167].

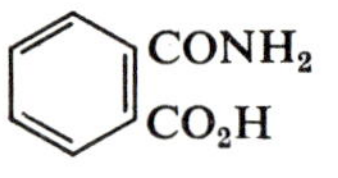

(XLIII)

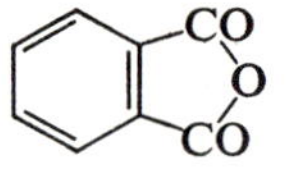

(XLIV)

This mechanism does not suggest a role for the tyrosine residue known to be essential. It is possible of course that this residue is not involved directly in the catalytic process but contributes either to enzyme–substrate binding or to maintenance of the structure of the active site. Indeed it is difficult to ascribe a role to this phenolic group at low pH which is not that of an acid catalyst, a possibility which is ruled out, at least for the rate-determining step, by the absence of a deuterium isotope kinetic effect [153].

Akhtar and Al-Janabi [168] have suggested a rather similar mechanism which also involves a 'potential' anhydride group in the enzyme.

Other proteolytic enzymes function at low pH values and it is possible that their mechanisms of action are similar to that of pepsin. They include pepsin B (3.4.4.2), rennin (3.4.4.3), gastricsin (3.4.4.–) and the cathepsins (3.4.4.23, etc.).

## REFERENCES

1. Dixon, M. and Webb, E. C., *Enzymes*, 2nd edn (Longmans, London, 1964)
2. Hartley, B. S., *Nature, Lond.* **201,** 1284 (1964); Hartley, B. S., *The Structure and Activity of Enzymes*, Eds. T. W. Goodwin, J. I. Harris and B. S. Hartley, p. 47 (Academic Press, London, 1964)
3. Desnuelle, P., *The Enzymes* Vol. 4, p. 106 (Academic Press, New York, 1960)
4. Neurath, H., Rupley, J. A. and Dreyer, W. J., *Archs biochem. Biophys.* **65,** 243 (1956)
5. Chervenka, C. H., *Biochim. biophys. Acta* **26,** 222 (1957)
6. Vaslow, F. and Docherty, D. G., *J. Am. chem. Soc.* **75,** 928 (1953)
7. Green, N. M. and Neurath, H., *The Proteins*, Eds. K. Bailey and H. Neurath, Vol. 2, part B, p. 1057 (Academic Press, New York, 1954)

8. LABOUESSE, B., OPPENHEIMER, H. L. and HESS, G. P., *Biochem. biophys. Res. Commun.* **14,** 318 (1964)
9. BENDER, M. L., GIBIAN, M. J. and WHELAN, D. J., *Proc. natn. Acad. Sci. U.S.A.* **56,** 833 (1966)
10. KNOWLES, J. R., *Biochem. J.* **95,** 180 (1965)
11. KEZDY, F. J., FEDER, J. and BENDER, M. L., *J. Am. chem. Soc.* **89,** 1009 (1967)
12. HILLE, M. B. and KOSHLAND. D. E. JR., *J. Am. chem. Soc.* **89**, 5945 (1967)
13. WEIL, L., *J. cell comp. Physiol.* **47,** Suppl. 1, 144 (1956); WEIL, L., JAMES, S. and BUCHERT, A. R., *Archs. Biochem.* **46,** 266 (1953)
14. WOOD, H. N. and BALLS, A. K., *J. biol. Chem.* **213,** 297 (1955); VISWANATHA, T. and LAWSON, W. B., *Archs biochem. Biophys.* **93,** 128 (1961)
15. KOSHLAND, D. E., JR., STRUMEYER, D. H. and RAY, W. J., JR., *Brookhaven Symp, Biol.* **15,** 101 (1962)
16. WHITAKER, J. R. and JANDORF, B. J., *J. biol. Chem.* **223,** 751 (1956)
17. SCHRAMM, V. H. J., *Z. physiol. Chem.* **348,** 232 (1967)
18. GUTFREUND, H. and STURTEVANT, J. M., *Biochem. J.* **63,** 656 (1956)
19. SCHOELLMAN, G. and SHAW, E., *Biochemistry* **2,** 252 (1963); *Biochem. biophys. Res. Commun.* **7,** 36 (1962)
20. ONG, E. B., SHAW, E. and SCHOELLMAN, G., *J. Am. chem. Soc.* **86,** 1271 (1964); *J. biol. Chem.* **240,** 694 (1965)
21. KEZDY, F. J., THOMSON, A. and BENDER, M. L., *J. Am. chem. Soc.* **89,** 1004 (1967)
22. SCHAFFER, N. K., MAY, S. C., JR. and SUMMERSON, W. H., *J. biol. Chem.* **202,** 67 (1953); JANSEN, E. J., JANG, R. and BALLS, A. K., *J. biol. Chem.* **196,** 247 (1952)
23. OOSTERBAAN, R. A., KUNST, P., VAN ROTTERDAM, J. and COHEN, J. A., *Biochim. biophys. Acta* **27,** 556 (1958)
24. BALLS, A. K. and ALDRICH, F. L., *Proc. natn. Acad. Sci. U.S.A.* **41,** 190 (1955)
25. OOSTERBAAN, R. A., VAN ADRICHEM, M. and COHEN, J. A., *Biochim. biophys. Acta* **63,** 204 (1962)
26. SCHONBAUM, G. R., ZERNER, B. and BENDER, M. L., *J. biol. Chem.* **236,** 2930 (1961); DIXON, G. H. and NEURATH, H., *J. biol. Chem.* **225**, 1049 (1957)
27. STRUMEYER, D. H., WHITE, W. N. and KOSHLAND, D. E., JR., *Proc. natn. Acad. Sci. U.S.A.* **50,** 931 (1963)
28. WEINER, H., WHITE, W. N., HOARE, D. G. and KOSHLAND, D. E., JR., *J. Am. chem. Soc.* **88,** 3851 (1966)
29. (a) DIXON, G. H., DREYER, W. J. and NEURATH, H., *J. Am. chem. Soc.* **78,** 4810 (1956); (b) ANDERSON, B. M., CORDES, E. H. and JENCKS, W. P., *J. biol. Chem.* **236,** 455 (1961)
30. BENDER, M. L., SCHONBAUM, G. R. and ZERNER, B., *J. Am. chem. Soc.* **84,** 2540 (1962)
31. WAGNER-JAUREGG, T. and HACKLEY, B. E., JR., *J. Am. chem. Soc.* **75,** 2125 (1953).
32. GUTFREUND, H. and STURTEVANT, J. M., *Proc. natn. Acad. Sci. U.S.A.* **42,** 719 (1956)

33. Kallos, J. and Avatis, K., *Biochemistry* **5,** 1797 (1966)
34. Schafer, J., Baronowsky, P., Laursen, R., Finn, F. and Westheimer, F. H., *J. biol. Chem.* **241,** 421 (1966)
35. Balls, A. K. and Wood, H. N., *J. biol. Chem.* **219,** 245 (1956)
36. Bender, M. L., Clement, G. E., Gunter, C. R. and Kezdy, F. J., *J. Am. chem. Soc.* **86,** 3697 (1964)
37. Sprinson, D. P. and Rittenberg, D., *Nature, Lond.* **167,** 484 (1951); Docherty, D. G. and Vaslow, F., *J. Am. chem. Soc.* **74,** 931 (1952); Bender, M. L. and Kemp, K. C., *J. Am. chem. Soc.* **79,** 111, 116 (1957)
38. Epand, R. M. and Wilson, I. B., *J. biol. Chem.* **238,** 1718 (1963)
39. Zerner, B. and Bender, M. L., *J. Am. chem. Soc.* **86,** 3669 (1964)
40. Bruice, T. C. and Benkovic, S. J., *Bioorganic Mechanisms* Vol. 1. (Benjamin, New York, 1966)
41. Spencer, T. and Sturtevant, J. M., *J. Am. chem. Soc.* **81,** 1874 (1959)
42. Zerner, B., Bond, R. P. M. and Bender, M. L., *J. Am. chem. Soc.* **86,** 3674 (1964)
43. Bender, M. L., Kezdy, F. J. and Gunter, C. R., *J. Am. chem. Soc.* **86,** 3714 (1964)
44. Himoe, A., Parks, P. C. and Hess, G. P., *J. biol. Chem.* **242,** 919 (1967)
45. Himoe, A. and Hess, G. P., *Biochem. biophys. Res. Commun.* **23,** 234 (1966)
46. Bender, M. L. and Hamilton, G. A., *J. Am. chem. Soc.* **84,** 2570 (1962)
47. Bender, M. L. and Kezdy, F. J., *J. Am. Chem. Soc.* **86,** 3704 (1964)
48. Bergmann, M. and Fruton, J. S., *Adv. Enzymol.* **1,** 63 (1941); Kauffman, S. and Neurath, H., *Archs. Biochem.* **21,** 437 (1949)
49. Jennings, R. R. and Niemann, C., *J. Am. chem. Soc.* **75,** 4687 (1953)
50. Bender, M. L. and Turnquest, B. W., *J. Am. chem. Soc.* **77,** 4271 (1955); Cohen, S. G. and Weinstein, S. Y., *J. Am. chem. Soc.* **86,** 5326 (1964); Hein, G. and Niemann, C., *Proc. natn. Acad. Sci. U.S.A.* **47,** 1341 (1961)
51. Hein, G. and Niemann, C., *J. Am. chem. Soc.* **84,** 4495 (1962)
52. Ingles, D. W., Knowles, J. R. and Tomlinson, T. A., *Biochem. biophys, Res. Commun.* **23,** 619 (1966)
53. Rapp, J. R., Niemann, C. and Hein, G. E., *Biochemistry* **5,** 4100 (1966)
54. Matthews, B. W., Sigler, P. B., Henderson, R. and Blow, D. M., *Nature, Lond.* **214,** 652 (1967)
54a. Blow, D. M., Birktoft, J. J. and Hartley, B. S., *Nature, Lond.* **221,** 337 (1969)
54b. Wright, C. S., Alden, R. A. and Kraut, J., *Nature, Lond.* **221,** 235 (1969)
55. Bender, M. L., *J. Am. chem. Soc.* **84,** 2582 (1962)
56. Walsh, K. A., Kauffmann, D. L., Sampath-Kumar, K. S. V. and Neurath, H., *Proc. natn. Acad. Sci. U.S.A.* **51,** 301 (1964)
57. Porter, G. R., Rydon, H. N. and Schofield, J. A., *Nature, Lond.* **182,** 927 (1958); Rydon, H. N., *Nature, Lond.* **182,** 928 (1958)
58. Bernhard, S. A., *J. cell. comp. Physiol.* **54,** Suppl. 1, 195 (1959)
59. Cohen, J. A., Oosterbaan, R. A., Jansz, H. S. and Berends, F., *Nat. meeting A.C.S.* 17C (1962)
60. Gold, A. M. and Fahrney, D. E., *Biochem. biophys. Res. Commun.* **10,** 55 (1963)

61. DESNUELLE, P., *The Enzymes,* Vol. 4, p. 119 (Academic Press, New York, 1960)
62. STEWART, J. A. and OUELLET, L., *Can. J. Chem.* **37,** 751 (1959)
63. BENDER, M. L. and KAISER, E. T., *J. Am. chem. Soc.* **84,** 2556 (1962)
64. WALSH, K. A. and NEURATH, H., *Proc. natn. Acad. Sci. U.S.A.* **52,** 884 (1964)
65. JENSEN, E. J. and BALLS, A. K., *J. biol. Chem.* **194,** 721 (1952)
66. THOMASEK, V., MIKES, O., HOLEYSOVSKY, V., KEIL, B. and SORM, F., *Biochim. biophys. Acta* **69,** 186 (1963); MIKES, O., HOLEYSOVSKY, V., THOMASEK, V. and SORM, F., *Biochem. biophys. Res. Commun.* **24,** 346 (1966)
67. TRAVIS, J. and LIENER, I. E., *J. biol. Chem.* **240,** 1967 (1965)
68. GUTFREUND, H., *Trans. Faraday Soc.* **51,** 441 (1955)
69. SHAW, E., MARES-GUIA, M. and COHEN, W., *Biochemistry* **4,** 2219 (1965)
70. PETRA, P. H., COHEN, W. and SHAW, E., *Biochem. biophys. Res. Commun.* **21,** 612 (1965)
71. NORD, F. F., BIER, M. and TERMINIELLO, L., *Archs biochem. Biophys.* **65,** 120 (1956)
72. RIORDAN, J. F., WACKER, W. E. C. and VALLEE, B. L., *Biochemistry* **4,** 1758 (1965)
73. LOCKSLEY-TRENHOLM, H., SPOMER, W. E. and WOOTTON, J. F., *J. Am. chem. Soc.* **88,** 4281 (1966)
74. HACHIMORI, Y., MATSUSHIMA, A., SUZUKI, M., INADA, Y. and SHIBATA, K., *Biochim. biophys. Acta* **124,** 395 (1966)
75. ALDRIDGE, W. N., *Biochem. J.* **53,** 110 (1953)
76. HOFSTEE, B. H. J., *The Enzymes* Vol. 4, p. 485 (Academic Press, New York, 1960)
77. MOUNTER, L. A., *J. biol. Chem.* **209,** 813 (1954); 215, 705 (1955); MOUNTER, L. A., ALEXANDER, H. L., TUCK, K. D. and DIEN, T. T. H., *J. biol. Chem.* **226,** 867 (1957)
78. MALHOTRA. O. P. and PHILLIPS, G., *Biochem. Z.* **346,** 386 (1966)
79. JANSEN, E. F., JANG, R. and MACDONNELL, L. R., *Archs Biochem.* **15,** 415 (1947); JANSEN, E. F., NUTTING, M.-D.F, and BALLS, A. K., *J. biol. Chem.* **175,** 975 (1948)
80. WHITTAKER, V. P., *Physiol. Rev.* **31,** 312 (1951); AUGUSTINNSON, K.-B., *The Enzymes* Vol. 4, p. 521 (Academic Press, New York, 1960)
81. WILSON, I. B. and BERGMAN, F., *J. biol. Chem.* **185,** 479 (1950)
82. WILSON, I. B., *J. biol. Chem.* **197,** 215 (1952)
83. KRUPKA, R. M., *Biochemistry* **5,** 1983, 1988 (1966)
84. RYLE, A. P., *Annual Reports* (*Chem. Soc.*) **63,** 614 (1966)
85. OOSTERBAAN, R. A. and COHEN, J. A., *The Structure and Activity of Enzymes,* Eds. T. W. Goodwin, J. I. Harris and B. S. Hartley (Academic Press, London, 1964)
86. HARTLEY, B. S., BROWN, J. R., KAUFFMANN, D. L. and SMILLIE, L. B., *Nature, Lond.* **207,** 1157 (1966); ROVERY, M., *Bull. Soc. Chim. biol.* **46,** 1757 (1964)
87. LIGHT, A., FRATER, R., KIMMEL, J. R. and SMITH, E. L., *Proc. natn. Acad. Sci. U.S.A.* **52,** 1276 (1964)

88. Drenth, J., Jansonius, J. N., Koekoek, R., Swen, H. M. and Wolthers, B. G., *Nature, Lond.* **218,** 929 (1968)
88a. Husain, S. S. and Lowe, G., *Biochem. J.* **114,** 286 (1969)
89. Bersin, J., *Ergebn. Enzymforsch.* **4,** 68 (1935)
90. Finkle, B. J. and Smith, E. L., *J. biol. Chem.* **230,** 669 (1958)
91. Kimmel, J. R. and Smith, E. L., *J. biol. Chem.* **207,** 515 (1954)
92. Brubacher, L. J. and Bender, M. L., *J. Am. chem. Soc.* **88,** 5871, 5880 (1966)
93. Williams, A., Doctoral Thesis, Oxford (1964), quoted in ref. [94]
94. Lowe, G. and Williams, A., *Biochem. J.* **96,** 194 (1965); Lake, A. W. and Lowe, G., *Biochem. J.* **101,** 402 (1966)
95. Lowe, G. and Williams, A., *Biochem. J.* **96,** 199 (1965)
96. Kimmel, J. R. and Smith, E. L., *Adv. Enzymol.* **19,** 267 (1957)
97. Husain, S. S. and Lowe, G., *Chem. Commun.* 310 (1968)
97a. Perutz, M., Royal Soc. Meeting, Dec. 1968
98. Bergmann, M. and Fruton, J. S., *Adv. Enzymol.* **1,** 263 (1957)
99. Johnston, R. B., *J. biol. Chem.* **221,** 1037 (1956)
100. Schechter, I. and Berger, A., *Biochem. biophys. Res. Commun.* **27,** 157 (1967)
100a. Schechter, I. and Berger, A., British Biophysical Society Meeting (London, Dec. 1968)
101. Brubacher, L. J. and Bender, M. L., *Biochem. biophys. Res. Commun.* **27,** 176 (1967)
102. Schechter, I., Abramowitz, N. and Berger, A., *Israel. J. Chem.* **3,** 101(1965)
103. Smith, E. L. and Kimmel, J. R., *The Enzymes* Vol. 4, p. 165 (Academic Press, New York, 1960)
104. Stein, M. J. and Liener, I. E., *Biochem. biophys. Res. Commun.* **26,** 376 (1967)
105. Wong, R. C. and Liener, I. E., *Biochem. biophys. Res. Commun.* **17,** 470 (1964)
106. Lowe, G. *Nature, Lond.* **212,** 1263 (1966)
107. Sorm, F. and Keil, B., *Adv. Protein Chem.* **17,** 167 (1962)
108. Neet, K. E. and Koshland, D. E., jr., *Proc. natn. Acad. Sci. U.S.A.* **56,** 1606 (1966)
109. Polger, L. and Bender, M. L., *Biochemistry* **6,** 610 (1967)
110. Neurath, H., *The Enzymes* Vol. 4, p. 11 (Academic Press, New York, 1960)
111. Vallee, B. L., Riordan, J. F. and Coleman, J. E., *Proc. natn. Acad. Sci. U.S.A.* **49,** 109 (1963)
112. Brown, J. R., Greenshields, R. N., Yamasaki, M. and Neurath, H., *Biochemistry* **2,** 867 (1963); Yamasaki, M., Brown, J. R., Cox, D. J., Greenshields, R. N., Wade, R. D. and Neurath, H., *Biochemistry* **2,** 859 (1963); Brown, J. R., Yamasaki, M. and Neurath, H., *Biochemistry* **2,** 877 (1963)
113. Neurath, H. and Schwert, G. W., *Chem. Rev.* **46,** 69 (1950)
114. Dekker, C. A., Taylor, S. P., jr. and Fruton, J. S., *J. biol. Chem.* **180,** 155 (1949); Smith, E. L., *J. biol. Chem.* **175,** 39 (1948)
115. Snoke, J. E., Schwert, G. N. and Neurath, H., *J. biol. Chem.* **175,** 7 (1948)

116. SNOKE, J. E. and NEURATH, H., *J. biol. Chem.* **181,** 789 (1949)
117. ELKINS-KAUFMAN, E. and NEURATH, H., *J. biol. Chem.* **175,** 893 (1948); **178,** 645 (1949)
118. SMITH, E. L. and STOCKELL, A., *J. biol. Chem.* **207,** 501 (1954)
119. LIPSCOMB, W. N., HARTSUCK, J. A., REEKE, G. N., QUIOCHO, F. A., BETHGE, P. H., LUDWIG, M. L., STEITZ, T. A., MUIRHEAD, H. and COPPOLA, J. C., *Brookhaven Symposia in Biology*, No. 21 (Brookhaven National Laboratory, Upton, New York, 1968); REEKE, G. N., HARTSUCK, J. A., LUDWIG, M. L., QUIOCHO, F. A., STEITZ, T. A. and LIPSCOMB, W. N., *Proc. natn. Acad. Sci. U.S.A.* **58,** 2220 (1967); LIPSCOMB, W. N., HARTSUCK, J. A., QUIOCHO, F. A. and REEKE, G. N., *Proc. natn. Acad. Sci. U.S.A.* in the press.
120. COOMBES, T. L., OMOTE, Y. and VALLEE, B. L., *Biochemistry* **3,** 653 (1964)
121. SAMPATH-KUMAR, K. S. V., CLEGG, J. B. and WALSH, K. A., *Biochemistry* **3,** 1728 (1964); BARGETZI, J.-P., THOMPSON, E. O. P., SAMPATH-KUMAR, K. S. V., WALSH, K. A. and NEURATH, H., *J. biol. Chem.* **239,** 3767 (1964); SAMPATH-KUMAR, K. S. V., WALSH, K. A., BARGETZI, J.-P. and NEURATH, H., *Aspects of Protein Structure*, Ed. G. N. Ramachandran, p. 319 (Academic Press, New York, 1963)
122. VALLEE, B. L. and NEURATH. H., *J. Am. chem. Soc.* **76,** 5006 (1954)
123. VALLEE, B. L., RUPLEY, J. A., COOMBS, T. C. and NEURATH, H., *J. Am. chem. Soc.* **80,** 4750 (1958); *J. biol. Chem.* **235,** 64 (1960)
124. RUPLEY, J. A. and NEURATH, H., *J. biol. Chem.* **235,** 609 (1960)
125. COLEMAN, J. E. and VALLEE, B. L., *J. biol. Chem.* **235,** 390 (1960)
126. NEURATH, H., RUPLEY, J. A. and VALLEE, B. L., *Amino Acids, Proteins and Cancer Biochemistry*, Ed. J. T. Edsall, p. 43 (Academic Press, New York, 1960)
127. BOYER, P. D., *The Enzymes*, Vol. 1, p. 511 (Academic Press, New York, 1959); *J. Am. chem. Soc.* **76,** 4331 (1954)
128. VALLEE, B. L., WILLIAMS, R. J. P. and COLEMAN, J. E., *Nature, Lond.* **190,** 633 (1960)
129. WILLIAMS, R. J. P., *Nature, Lond.* **188,** 322 (1960)
130. COLEMAN, J. E. and VALLEE, B. L., *J. biol. Chem.* **236,** 2244 (1961)
131. PIRAS, R. and VALLEE, B. L., *Biochemistry* **6,** 348 (1967)
132. WALSH, K. A., SAMPATH-KUMAR, K. S. V., BARGETZI, J.-P. and NEURATH, H., *Proc. natn. Acad. Sci. U.S.A.* **48,** 1443 (1962)
133. QUIOCHO, F. A. and RICHARDS, F. M., *Proc. natn. Acad. Sci. U.S.A.* **52,** 833 (1964)
134. ABRAMOWITZ, N. SCHECHTER, I. and BERGER, A., *Biochem. biophys. Res. Commun.* **29,** 862 (1967)
135. SIMPSON, R. T., RIORDAN, J. F. and VALLEE, B. L., *Biochemistry* **2,** 616 (1963); RIORDAN, J. F., SOKOLOVSKY, M. and VALLEE, B. L., *Biochemistry* **6,** 358 (1967); SOKOLOVSKY, M. and VALLEE, B. L. *Abstr.* 152*nd meeting A.C.S.* 171C (1966); COLEMAN, J. E., PULIDO, P. and VALLEE, B. L., *Biochemistry* **5,** 2019 (1966); VALLEE, B. L., *Fedn. Proc.* **23,** 8 (1964)
136. VALLEE, B. L., RIORDAN, J. F. and COLEMAN, J. E., *Proc. natn. Acad. Sci. U.S.A.* **49,** 109 (1963)

137. GINODMAN, L. M., MAL'SEV, N. I. and OREKHOVICH, V. N., *Biokhimiya* **31,** 1073 (1966)
138. NEURATH, H. and SCHWERT, G. W., *Chem. Rev.* **46,** 69 (1950)
139. SPACKMAN, D. H., SMITH, E. L. and BROWN, D. M., *J. biol. Chem.* **212,** 255 (1955)
140. HANSON, H., GLASSER, D. and KIRSCHKE, H., *Z. phys. Chem.* **340,** 107 (1965); KRETSCHMER, K. and HANSON, H., *Z. phys. Chem.* **340,** 126 (1965)
141. WACHSMUTH, E. D., FRITZE, I. and PFLEIDERER, G., *Biochemistry* **5,** 169 (1966); WACHSMUTH, E. D., *Biochem. Z.* **344,** 361 (1966)
142. PFLEIDERER, G. and CELLIERS, P. G., *Biochem. Z.* **339,** 186 (1963); HANSON, H., HUTTER, H.-J., MANNSFELDT, H.-G., KRETSCHMER, K. and SOHR, C., *Z. physiol. Chem.* **348,** 680 (1967)
143. JOSEPH, R. L. and SANDERS, W. J., *Biochem. J.* **100,** 827 (1966)
144. SMITH, E. L. and HILL, R. L., *The Enzymes*, Vol. 4, p. 37 (Academic Press, New York, 1960)
145. SCHECHTER, I. and BERGER, A., *Biochemistry* **5,** 3371 (1966)
146. HANSON, H. and LASCH, J., *Z. physiol. Chem.* **348,** 1525 (1967)
147. WOLFF, J. B. and RESNICK, R. A., *Biochim. biophys. Acta* **73,** 588 (1963)
148. SMITH, E. L. and SPACKMAN, D. H., *J. biol. Chem.* **212,** 271 (1955)
149. WACHSMUTH, E. D., *Biochem. Z.* **346,** 446 (1967)
150. BOVEY, F. F. and YANARI, S. S., *The Enzymes* Vol. 4, p. 63 (Academic Press, New York, 1960)
151. FRUTON, J. S. and BERGMANN, M., *J. biol. Chem.* **127,** 627 (1939)
152. BAKER, L. E., *J. biol. Chem.* **193,** 809 (1951)
153. CLEMENT, G. E. and SNYDER, S. L., *J. Am. chem. Soc.* **88,** 5338 (1966)
154. HERRIOTT, R. M., ANSON, M. L. and NORTHROP, J. H., *J. gen. Physiol.* **30,** 185 (1946)
155. GROSS, E. and MORELL, J. L., *J. biol. Chem.* **241,** 3638 (1966)
156. ERLANGER, B. F., VRATSANOS, S. M., WASSERMANN, N. and COOPER. A. G., *J. biol. Chem.* **240,** PC3447 (1965); *Biochem. biophys. Res. Commun.* **23,** 243 (1966)
157. GOULD, E. S., *Mechanism and Structure in Organic Chemistry* p. 623 (Holt, Rinehart and Winston, London, 1959)
158. DELPIERRE, G. R. and FRUTON, J. S., *Proc. natn. Acad. Sci. U.S.A.* **54,** 1161 (1965); RAJAGOPALAN, T. G., STEIN, W. H. and MOORE, S., *J. biol. Chem.* **241,** 4295 (1966); KOSLOV, L. V., GINODMAN, L. M. and OREKHOVICH, V. N., *Biokhimiya* **32,** 1011 (1967); HAMILTOM, G. A., SPONA, J. and CROWELL, L. D., *Biochim. biophys. Res. Commun.* **26,** 193 (1967)
159. DELPIERRE, G. R. and FRUTON, J. S., *Proc. natn. Acad. Sci. U.S.A.* **56,** 1817 (1966)
160. PHILPOTT, J. ST. L. and SMALL, P. A., *Biochem. J.* **32,** 542 (1938); HERRIOTT, R. M., *J. gen. Physiol.* **19,** 283 (1935); HERRIOTT, R. M., *J. gen. Physiol.* **31,** 19 (1947); LI, C. H., *J. Am. chem. Soc.* **67,** 1065 (1945)
161. NEUMANN, H., BERGER, A. and KATCHALSKI, E., *Biochem. J.* **73,** 33 (1959)
162. FRUTON, J. S., FUGII, S. and KNAPPENBERGER, M. H., *Proc. natn. Acad. Sci. U.S.A.* **47,** 759 (1961); MAL'SEV, N. I., GINODMAN, L. M., OREKHOVICH, V. N., VALUERA, T. A. and AKIMORA, L. N., *Biokhimiya* **31,** 983 (1966)

163. SHARON, N., GRISARO, V. and NEUMANN, H., *Archs biochem. Biophys.* **97,** 219 (1962); KOSLOV, L. V., GINODMAN, L. M. and OREKHOVICH, V. N., *Doklady, Akad. Nauk S.S.R.* **172,** 1207 (1967)
164. JACKSON, W. T., SCHLAMOWITZ, M. and SHAW, A., *Biochemistry* **5,** 4105 (1966)
165. JACKSON, W. T., SCHLAMOWITZ, M. and SHAW, A., *Biochemistry* **4,** 1537 (1965)
166. BENDER, M. L. and KEZDY, F. J., *Ann. rev. Biochem.* **34,** 49 (1965)
167. BENDER, M. L., CHOW, Y.-L. and CHLOUPE, F., *J. Am. chem. Soc.* **80,** 5380 (1958)
168. AKHTAR, M. and AL-JANABI, J. M., *Chem. Commun.* 859, 1002 (1969)

# CHAPTER 5

# HYDROLYSIS II

This chapter will be concerned with the possible mechanisms by which compounds other than carboxylic acid derivatives may be hydrolysed in the presence of enzymes. Such compounds as phosphate esters, polyphosphates, glycosides and various purine derivatives may undergo hydrolysis by mechanisms which at first sight at least may be quite different from those encountered in the case of carboxylic acid derivatives, although certain parallels may be drawn, especially where enzymes are involved.

## HYDROLYSIS OF PHOSPHATE ESTERS

The non-enzymic hydrolysis of phosphate esters [1] is made complicated by several factors, not the least of which is the fact that the esters may possess ionizable hydroxyl groups. Even where these are absent, as in the case of phosphotriesters, more than one reaction mechanism is applicable. In alkaline solution, hydrolysis of trimethyl phosphate occurs with P—O bond fission, as shown by experiments with isotopic substitutions [2], and the reaction has been depicted as a direct nucleophilic attack by hydroxide ion on to the central phosphorus atom.

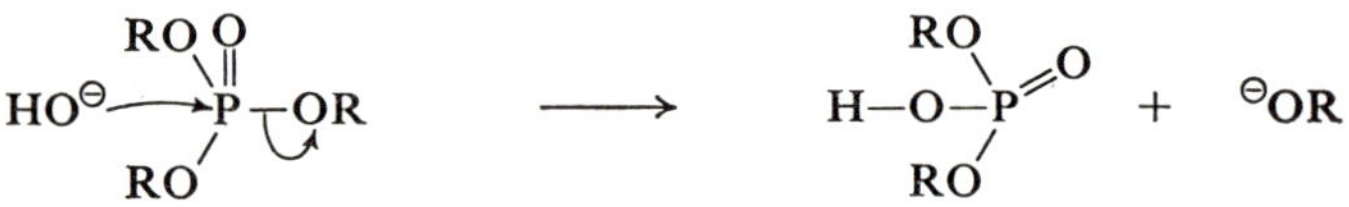

This mechanism differs from that of the similar alkaline hydrolysis of carboxyl esters in that the pentacovalent adduct (I) analogous to (II) does *not* appear to be formed as an intermediate [2].

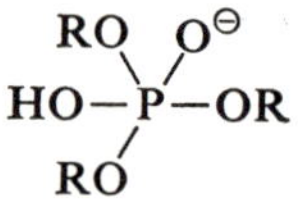

(I)

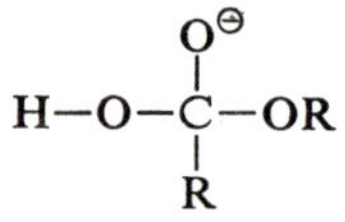

(II)

In neutral solution trimethyl phosphate is hydrolysed by a process which gives rise to C—O bond fission [2]. A nucleophilic attack occurs at the methyl group in this case.

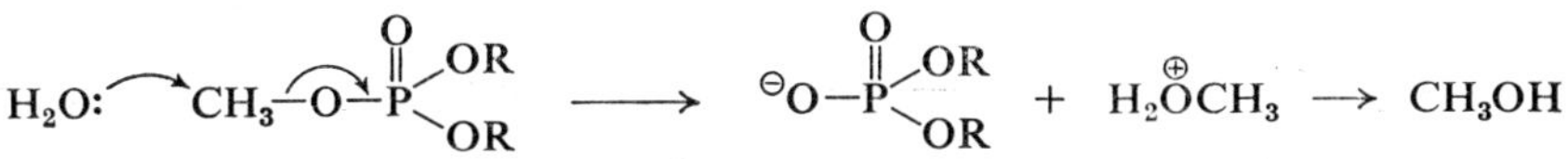

This type of mechanism, in which the alkyl group is attacked, is more important in phosphate ester hydrolysis than is the analogous process in the hydrolysis of carboxyl esters, although in the latter case such a mechanism is not unknown.

The presence of the hydroxyl group in a phospho*di*ester is a complicating factor. The P=O double bond in phosphodiesters, phosphomonoesters and phosphoric acid itself renders the hydroxyl groups very acidic. This is because the anion obtained on ionization is stabilized by resonance (III) similar to that which stabilizes the carboxylate anion. In the case of

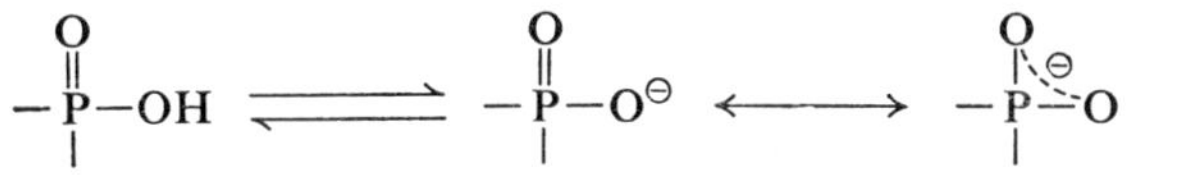 (III)

the phosphoric acid molecule the ionization of the first hydroxyl group occurs very readily, the $pK_a$ being very low (below 2) (equation 5.1). The second hydroxyl group is less easily ionized since it is more difficult to remove a proton from a species already possessing a negative charge (equation 5.2). The third ionization is correspondingly more difficult (equation 5.3) [3].

 (5.1)

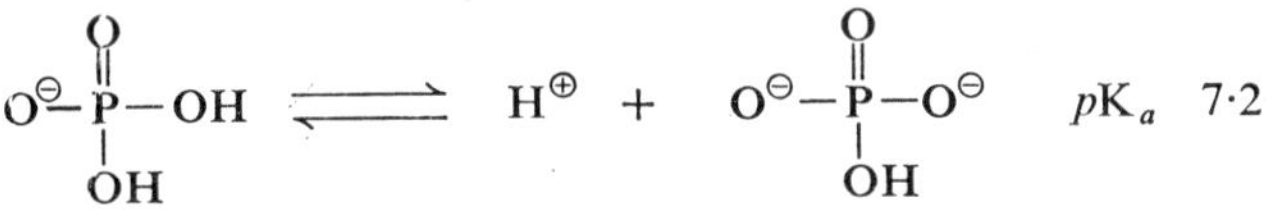 (5.2)

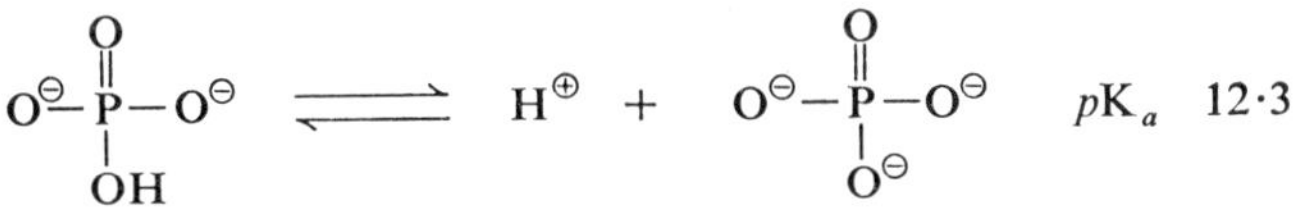 (5.3)

The $pK_a$ for the hydroxyl group of most phosphodiesters is not much different from that for the first ionization of phosphoric acid. This means that for all pH values above 2 a phosphodiester will exist predominantly in the anionic state.

Among the mechanisms by which dimethyl phosphate may be hydrolysed are the following which both give rise to P—O bond cleavage [4]:

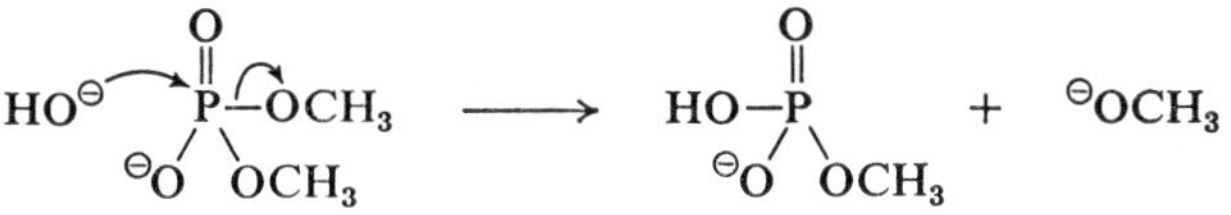 (5.4)

(5.5)

In equation (5.4) is seen the hydrolysis of the anionic species brought about by hydroxide ion, while equation (5.5) shows a formulation of the hydrolysis of the neutral (un-ionized) species, the latter being relatively insignificant except at very low pH values.

Hydrolysis of diesters also occurs with C—O bond fission by processes which must involve attack on the alkyl-group carbon atoms.

In connection with phosphodiesters, it is important to note that certain cyclic derivatives undergo hydrolysis at abnormally high rates. For example, ethylene phosphate (IV) is hydrolysed about $10^7$ times faster than dimethyl phosphate in both alkaline and acidic media [5]. In general, five-

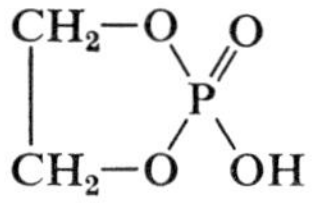

(IV)

membered ring phosphates like IV show this enhanced reactivity while six- or seven-membered ring compounds do not [6]. The hydrolysis of ethylene phosphate (IV) proceeds exclusively with P—O bond fission in both acid and alkali [7] and this indicates direct nucleophilic attack by water or hydroxide ion on to the phosphorus atom. The presence of the ring structure therefore not only increases the rate of the reaction but also controls to a certain extent the mechanism by which the reaction proceeds.

Two factors have been suggested to account for the large enhancement of the reactivities of cyclic phosphodiesters. In the first place the five-membered ring system is strained to a considerable extent [8] and hydrolysis releases this strain. This strain may be due to the angle requirements of a phosphorus atom in a five-membered ring and to the eclipsing of the hydrogen atoms or unpaired electrons on the ester oxygen atoms [8]. In addition the partial double-bond character of a P—O bond which can stabilize the open-chain phosphoester is diminished in a cyclic ester.

It will be seen later (chapter 6) that a cyclic phosphodiester is produced as an intermediate in the ribonuclease-catalysed hydrolysis of ribonucleic acid, and it is interesting to note that despite the very high reactivity of such a compound in water, its hydrolysis is further catalysed by the enzyme. However, the above comments do tend to justify the formulation of that reaction as involving a nucleophilic attack by a water molecule on phosphorus (see page 270).

In the case of the monoesters of phosphoric acid, the ionization of the hydroxyl groups is again a major factor. At pH values above neutrality the dianionic form predominates and hydrolysis is slow unless the alkyl (or aryl) group lost is such that a powerful leaving group is formed. In such a case displacement by hydroxide ion can occur, giving rise to P—O bond cleavage [9]. Most striking, however, in esters such as methyl phosphate, is the hydrolysis of the monoanionic form which appears to be the major process at pH 4. At this pH the hydrolysis of this monoanion is very much more rapid than that of the dimethyl phosphate anion [4] and this suggested that the process requires the presence of an un-ionized hydroxyl rather than an alkoxyl group. Several groups of workers have proposed that the hydrolysis of methyl phosphate anion proceeds through the formation of the unstable metaphosphate ion (V) which reacts rapidly with water to give inorganic phosphate† [10]. One possible route for the formation of the metaphosphate ion is via the hydrogen-bonded complex (VI) between the phosphate ester and water.

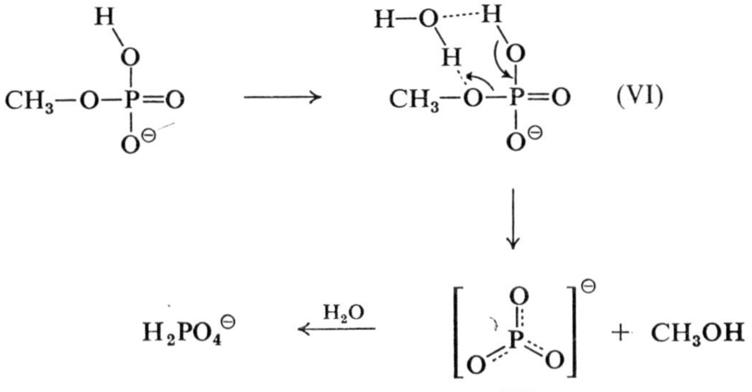


Any pathway involving a metaphosphate intermediate gives rise to P—O bond cleavage.

Other processes by which monomethyl phosphate may be hydrolysed in aqueous solution include attack by neutral water molecules on the methyl group of either the non-ionized state or, in very strongly acidic media, the protonated form (VII). These two processes lead, of course, to C—O bond cleavage.

(VII)

P—O bond fission can also be detected at high $H^{\oplus}$ concentrations, due to attack by water on to the phosphorus atom of the protonated species (VII).

In summary, three mechanisms appear to be available for the non-enzymic hydrolysis of phosphate esters. The first, nucleophilic attack on to the central phosphorus atom, gives rise to P—O bond fission while the second, in which displacement takes place on the alkyl carbon, leads to C—O bond cleavage. The third, which is particularly observed in the case

† Often represented by $P_i$.

of monoester monoanions gives rise to a metaphosphate ion intermediate. Enzyme-catalysed hydrolyses of phosphate esters appear to proceed almost invariably with P—O bond cleavage by mechanisms similar to the first of these.

Alternative-pathway or nucleophilic catalysis, which was discussed in chapter 1 mainly in connection with hydrolysis of carboxylic acid derivatives, has also been demonstrated in the non-enzymic hydrolysis of phosphate esters and other phosphate derivatives. Of special interest is the case where the attacking nucleophile is provided by a hydroxyl group in the same molecule [11]. Thus a phosphodiester such as VIII can react rapidly in acid or base to expel the group RO giving a cyclic ester (IX). This will

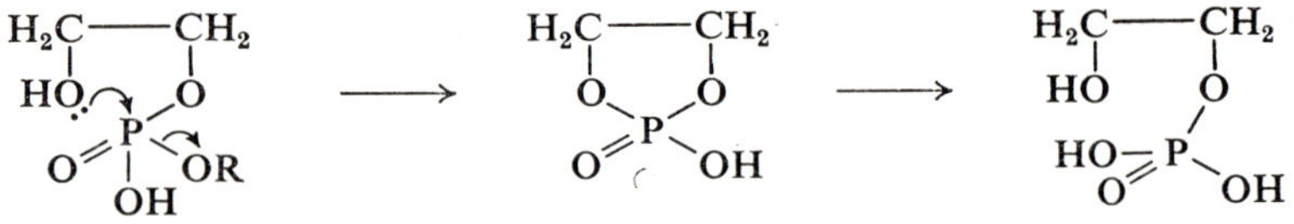

(VIII) (IX) (X)

undergo rapid hydrolysis to X as described earlier and the net result is an intramolecular catalysis of the hydrolysis of a diester to a monoester. This type of process is important in the alkaline hydrolysis of ribonucleic acid in which 3′,5′-cyclic phosphate intermediates are obtained as intermediates. Further hydrolysis yields 3′-mononucleotides (with some 2′-mononucleotides) (compare Fig. 6.10, for the ribonuclease-catalysed hydrolysis of RNA).

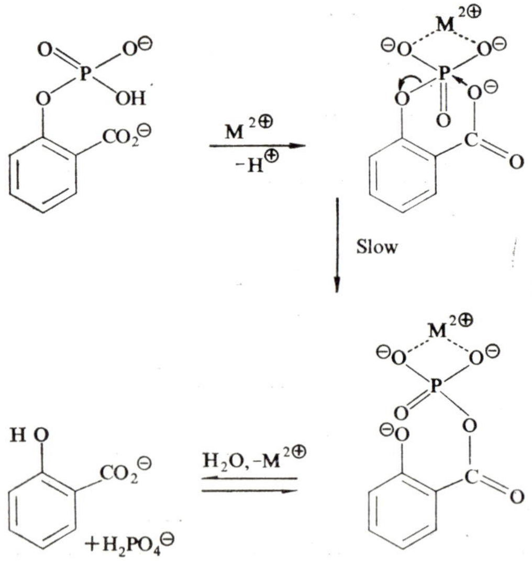

FIG. 5.1. Proposed mechanism for the metal-ion catalysed hydrolysis of salicyl phosphate. (Ref. [12])

A nucleophilic attack on the phosphorus atom of a phosphate ester species will be less favoured if that species is anionic, due to the repulsion of the negative charge. Conversely, protonation tends to enhance the possibility of such an attack. Formation of a complex between an ionized phosphate group and a metal-ion would also be expected to increase the rate of such a nucleophilic reaction. The hydrolysis of salicyl phosphate, for example, is strongly subject to metal-ion catalysis [12] and in a tentative mechanism (Fig. 5.1) the phosphate dianion is depicted as functioning as a bidentate ligand, the metal-ion enhancing the susceptibility of the phosphorus to the nucleophilic attack. Notice that the process also involves an intramolecular nucleophilic catalysis.

**Enzyme-catalysed Hydrolysis of Phosphate Esters**

Of the phosphate ester hydrolases, that which has been studied most is *orthophosphoric monoester hydrolase* (*alkaline phosphatase*, 3.1.3.1). This is a zinc-containing enzyme [13] with a molecular weight (*E. coli*) of about 86,000 [14], which catalyses the hydrolysis of a wide range of phosphate monoesters. Its designation as *alkaline* phosphatase refers to the fact that its optimum pH is near 8 in contrast with another type of phosphomonoesterase (*acid phosphatase*, 3.1.3.2) which functions optimally at lower pH values.

It is firmly established that for alkaline phosphatases from various sources (mammalian and bacterial) the hydrolysis takes place by a two-step process involving the intermediate formation of an enzyme-phosphate derivative.

$$\text{Enzyme} + \text{HPO}_3\text{OR}^{\ominus} \xrightleftharpoons{\text{Phosphorylation}} \text{Enzyme-phosphate} + \text{ROH} \quad (5.6)$$

$$\text{Enzyme-phosphate} + \text{H}_2\text{O} \xrightleftharpoons{\text{Hydrolysis}} \text{Enzyme} + P_i \quad (5.7)$$

This type of reaction pathway was indicated when it was shown that in addition to phosphatase activity the enzyme possessed the ability to function as a transferase [15, 16]. Various alcohols can act as acceptors for the transfer of phosphate from the phospho-enzyme intermediate. Morton [19] observed competition between water and alcohol as acceptors for the phosphate group indicating that water and acceptor alcohol are both bound to the same site on the enzyme–phosphate complex.

Hydrolysis in the presence of alkaline phosphatase takes place with P—O bond fission in the substrate [17], and, in agreement with the above mechanism, the enzyme catalyses the incorporation of $^{18}$O from $H_2{}^{18}O$ into phosphate [18] as required by equation (5.7) and its reverse.

Other studies which have confirmed indirectly the involvement of a phosphoryl-enzyme intermediate have included kinetic studies in which it has been shown that, under conditions where the hydrolysis step (5.7) is rate-determining, alkaline phosphatase catalyses the hydrolysis of a wide range of phosphate substrates at similar rates [16, 20, 21].

The rate of the dephosphorylation step decreases in acidic solution [21, 22] and this has allowed the isolation of the phosphoryl-enzyme at low pH values [23, 24]. The phosphorylation of the protein can be achieved in the presence either of phosphate ester substrates or of inorganic phosphate. The incorporation reactions of phosphate esters and inorganic phosphate are mutually competitive [24] showing that phosphate ions and esters are bound to the enzyme at the same site. Phosphorylations in the presence of substrates are, however, more rapid than those in the presence of $P_i$.

Hydrolysis of the phosphoryl-enzymes (mammalian and bacterial) has yielded O-phosphorylserine [23, 24], and the amino-acid sequence around this 'active site' serine has been shown to be Thr.Asp.Ser.Ala.Ala [25]. This sequence is rather similar to that at the active site of chymotrypsin and the other hydrolytic enzymes shown to be very closely related to it. At present, however, it appears unlikely that alkaline phosphatase can be included in that group of hydrolases. Its molecular weight is very much higher, and the involvement of a histidyl residue in the catalytic process in a manner similar to that found in chymotrypsin is by no means established. Indeed, photo-oxidation studies by Vallee and his co-workers [13] suggest that no histidine is implicated in the catalytic reaction, although such a residue may contribute to the binding of the metal-ion. The requirement for a metal-ion is another major point of difference between the phosphatases and the active serine peptidohydrolases.

Schwartz [22] has suggested a plausible mechanism for the action of alkaline phosphatase, based on the available information. This is essentially as shown in Fig. 5.2. This mechanism implies, in the enzyme–substrate complex (XI), a nucleophilic attack by the serine hydroxyl, with loss of the

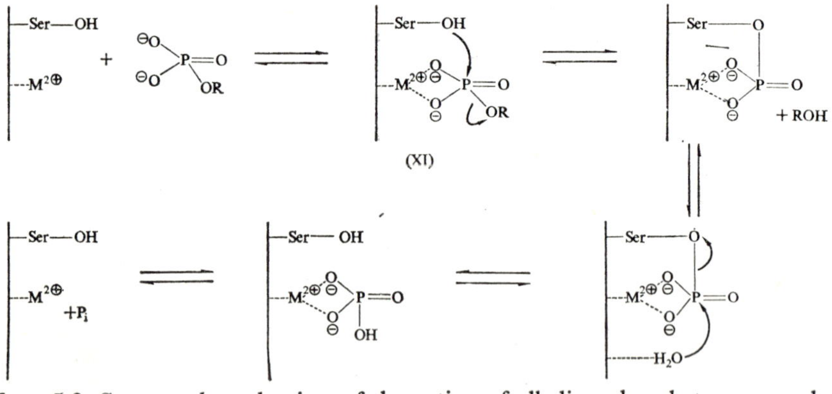


FIG. 5.2. Suggested mechanism of the action of alkaline phosphatase on a phosphate $RO.PO_3^{2\ominus}$. (Ref. [22]) (By courtesy of J. H. Schwartz and the National Academy of Sciences)

alkoxyl (OR) group. From our discussions of other enzyme-catalysed reactions we may assume that in the active site of the enzyme there exists some factor which enhances the nucleophilicity of the serine hydroxyl group and another which, probably by protonation, assists the departure of the leaving group. As in the case of other metallo-enzymes it is difficult to draw conclusions about these factors from the pH-rate profiles. The metal-ion is envisaged as providing a binding site for the dianionic phosphate species, and the formation of the metal-ion–phosphate complex will favour the nucleophilic attack on to the phosphorus atom.

It is probable that alkaline phosphatases from various sources function by similar mechanisms. The enzyme isolated from human placenta has been shown to catalyse phosphate transfer reactions, and to be dependent on the presence of metal-ions [26].

A phosphoryl-enzyme intermediate has been suggested for the reaction of a phosphodiesterase, *phospholipase D* (*phosphatidylcholine phosphatidohydrolase*, 3.1.4.4). In the presence of this enzyme the choline-phosphate ester bond of a phosphatidyl choline (XII) is cleaved [27, 28]. Yang *et al.* [27] have reported that phosphatidyl-ethanol, -ethanolamine and -glycerol

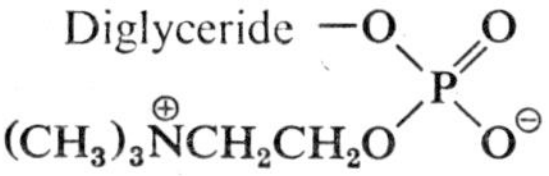


Phosphatidyl choline
(XII)

may be obtained from phosphatidyl-choline in the presence of the enzyme (i.e. transferase activity). In addition the enzyme was able to catalyse exchange between choline and phosphatidyl-choline. Because phospholipase D was inhibited by PCMB, these workers suggested that the enzyme possessed an active cysteine residue rather than an active serine, and that therefore there might be a phosphothioester link in the phosphoryl enzyme intermediate.

## HYDROLYSIS OF POLYPHOSPHATES

Reactions in which polyphosphates are hydrolysed non-enzymically do not appear to differ fundamentally from those by which phosphates are hydrolysed [1]. In acidic solution the reaction involves attack by water on to the conjugate acid of the pyrophosphate species and a possibly similar attack by water on the neutral pyrophosphate molecule.

The hydrolysis of the polyphosphate monoanion may proceed via a metaphosphate intermediate similar to that envisaged for the phosphate monoester monoanion (page 207).

An interesting example of a base-catalysed cleavage of the pyrophosphate molecule is afforded by the reaction of tetrabenzyl pyrophosphate (XIII) with propanol in the presence of lutidine or collidine [29]. The formation of propyl dibenzyl phosphate in this reaction shows that the substituted

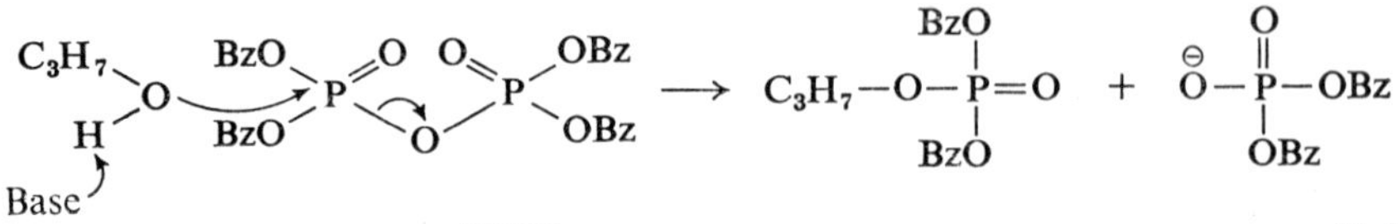

(XIII) (5.8)

pyrophosphate species is susceptible to nucleophilic attack as shown (equation 5.8). Notice that this mechanism is very similar to that proposed for the phosphotransferases.

The hydrolysis of certain polyphosphates, such as various nucleoside triphosphates, may be catalysed by bivalent metal-ions. This is very important because almost all enzymes which catalyse reactions of nucleoside triphosphates show an absolute requirement for metal-ions. On the basis of their studies, Lowenstein and co-workers [30] have assumed that a 1 : 1 ATP–metal complex is formed in which the metal-ion is bound to oxygen atoms of the $\beta$- and $\gamma$-phosphate groups. Cohn and Hughes [31] confirmed by n.m.r. measurements that the ions of Zn, Ca and Mg bind predominantly to the $\beta$- and $\gamma$-phosphate groups of ATP (XIV).

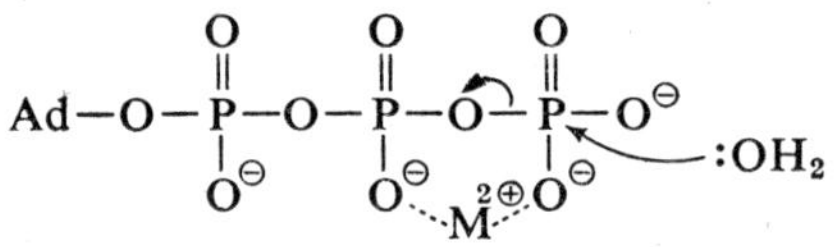

(XIV)

The observation that, in the Cu(II)-catalysed hydrolysis of ATP, $^{18}O$ from $H_2{}^{18}O$ is found in the terminal phosphate group (that is in the $P_i$ liberated) lends support to the suggested mechanism for the reaction, as shown.

Selwyn [31a] has suggested that the lanthanide-catalysed hydrolysis of ATP serves as a model for mitochondrial ATPase. For the non-enzymic reaction he proposes that the attacking nucleophile is a hydroxide ion bound to the metal-ion such that the ligands are in an octahedral configuration (5.9).

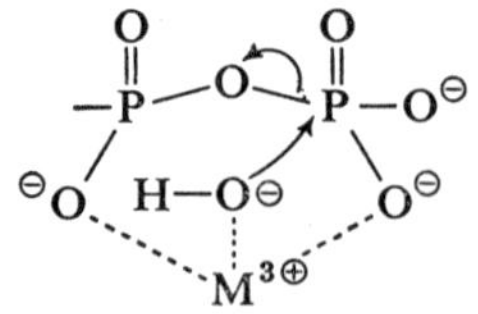

(5.9)

**ATP Phosphohydrolase (ATPase, 3.6.1.3)**

The ability of various proteins to bring about the hydrolysis of ATP to ADP and $P_i$ is of great importance in biochemistry because such a hydrolysis causes the release of a large amount of energy. It is significant that ATPase activity is found especially at two sites where energy conversion is involved in direct biochemical processes. These are in muscle (myosin), where muscle contraction is associated with work, and at membranes where ATPase activity is closely linked to the 'cation pump'—the active transport of cations across membranes.

One form of ATPase (membranes and other sources) was shown [32, 33, 34], to be dependent on both $Na^{\oplus}$ and $K^{\oplus}$ ions. This effect is in addition to the dependence on magnesium ions which is normally found with ATP reactions. Skou [32] suggested that the reaction might involve a phosphoryl-enzyme intermediate because he was able to demonstrate catalysis of ATP–ADP exchange. Later, other workers showed that labelled phosphate ($^{32}P$) from the terminal residue of ATP could be incorporated into the protein in preparations of (Na and K)-dependent ATPase [35, 36]. Sodium ions increased the amount of incorporation of phosphate into the protein and subsequent addition of potassium ions caused rapid loss of the phosphate bound.

On the basis of these results the earliest formulations of the reaction were as in equation (5.10). Thus the formation of the phosphoenzyme was

$$\mathrm{E + ATP} \xrightarrow{Na^{\oplus}} \underset{\mathrm{ADP}}{\overset{\mathrm{E{-}P}}{+}} \xrightarrow{K^{\oplus}} \mathrm{E + P_i} \qquad (5.10)$$

viewed as being brought about by the sodium ions, dephosphorylation being induced by the potassium ions. However, various studies soon made it clear that the scheme shown in (5.10) is an oversimplification, in the *in vitro* situation at least [37, 38, 39]. Certain non-parallelisms were noted between the phosphorylation and ATP hydrolysis processes. In addition, the enzymic reaction had different properties over different temperature ranges [37, 38]; over the range 0–25° an energy of activation of 29·5 kcal/mole was observed, while above 25° the value was 18·4 kcal/mole. In the absence of potassium ions ($Mg^{2\oplus}$ and $Na^{\oplus}$ only) the activation energy was apparently constant over the whole range of temperatures.

All the results obtained appear to be consistent with two modes of action for the (Na and K)-dependent ATPase. At low temperatures the major pathway is dependent on sodium ions alone and involves a direct hydrolysis without a phosphoenzyme complex. At higher temperatures another process becomes important; this requires both sodium and potassium ions and involves a phosphoenzyme intermediate. Thus, equation

(5.11) gives the best description of the mode of action of this enzyme [37]. It must not be supposed that these two mechanisms are mutually exclusive.

$$\mathrm{E + ATP} \begin{array}{l} \xrightarrow{\mathrm{Na}^{\oplus}} \mathrm{E + ADP} + P_i \text{ (low temperature)} \\ \xrightarrow{\mathrm{Na}^{\oplus}} \mathrm{E{-}P + ADP} \xrightarrow{\mathrm{K}^{\oplus}} \mathrm{E} + P_i \text{ (higher-temperature)} \end{array} \qquad (5.11)$$

The protein containing bound phosphate which may be isolated from the system stimulated by sodium ions alone is stable in acidic media [40] but loses the phosphate in alkali, by a process which follows the first-order kinetics [37]. In addition, treatment with hydroxylamine causes it to lose phosphate, and these results lead to the conclusion that the phosphate is bound to the enzyme by an acyl-phosphate link [41], that is, a mixed anhydride with the side-chain carboxyl group of an aspartyl or glutamyl residue (the α-carboxyl group of a terminal amino-acid residue has not yet been eliminated). Figure 5.3 shows a partial mechanism of action for

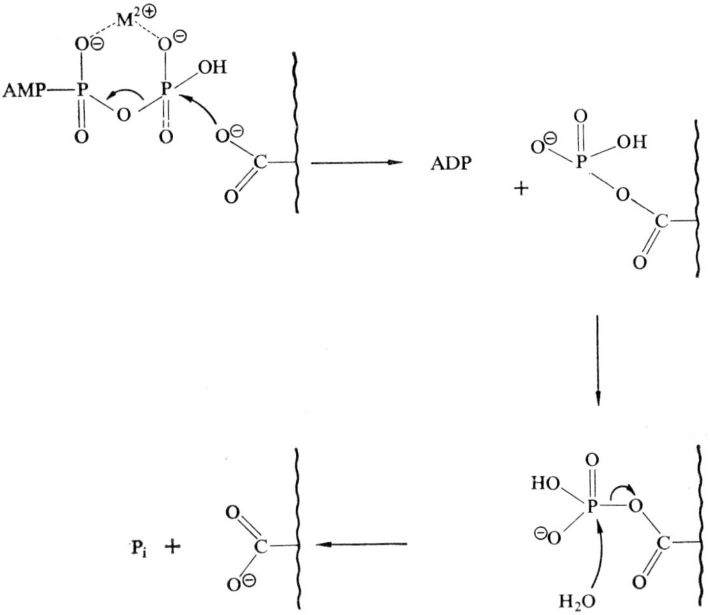


FIG. 5.3. Suggested partial mechanism for ATPase

ATPase based on this conclusion. At present it is extremely difficult to formulate roles for the sodium and potassium ions involved in the reaction of the (Na and K)-dependent ATPase.

Myosin, a protein isolated from muscle, also has ATPase activity. *In vitro*, its optimum activity is found when activated by calcium ions at pH 9. However, it has been assumed that under physiological conditions the enzyme is activated by magnesium ions at pH 7 [42]. It has been shown that one thiol group per $2 \times 10^5$ g of myosin reacts with the thiol-reagent N-ethylmaleimide, one lysine residue per $2 \times 10^5$ g reacts with trinitrobenzenesulphonate, and that one pyrophosphate molecule combines with approximately $2 \times 10^5$ g of myosin in a reversible complex [44]. Therefore the chemical molecular weight is $2 \times 10^5$ or some multiple of this. Since Tonomura *et al.* [43] isolated two different substituted lysine peptides from myosin treated with trinitrobenzenesulphonate, then the molecular weight has been assumed to be $4 \times 10^5$.

Koshland and co-workers [42, 45] showed that myosin caused more exchange of $^{18}O$ from water into inorganic phosphate than would normally be expected and deduced the existence of an exchanging intermediate in the ATPase reaction. This was suggested to be a phospho-myosin. Strong evidence for the formation of a phospho-myosin intermediate came from studies of the pre-steady state situation for the hydrolysis of ATP by myosin [43]. Tonomura and co-workers observed an initial rapid extra release of phosphate from the ATP (the terminal phosphate). This took place at a 1 : 1 ratio of myosin to ATP (based on a molecular weight of $4 \times 10^5$ for myosin) and appeared to be due to an initial irreversible reaction between ATP and myosin. Tonomura also studied the process by a rapid reaction technique [46] and proposed the mechanistic scheme as shown in equation (5.11) which indicates the way in which hydrogen ions are produced and absorbed in the various steps. E—P is envisaged as

$$\begin{array}{ccccc}
E + S + H^{\oplus} & \overset{k_{+1}}{\rightleftharpoons} & E_1S + H^{\oplus} & \xrightarrow{k_{+2}} & E + ADP + P_i + 2H^{\oplus} \\
 & & {\scriptstyle k_{+3}}\downarrow & & \uparrow{\scriptstyle k_{+5}} \\
 & & E_2SH & \xrightarrow{k_{+4}} & E-P + ADP + H^{\oplus}
\end{array} \qquad (5.12)$$

being a phospho-myosin. The rate of formation of this phospho-myosin is faster than that of the ATPase reaction in the steady state and the extra liberation of the stoichiometric amount of $P_i$ takes place by the decomposition of the phospho-myosin on addition of the trichloracetic acid which is used during the analysis of the system.

Under conditions where the rapid initial extra-liberation of $P_i$ would be observed (that is, when the phospho-myosin was present) nitrothiophenol could combine with the protein [47]. The nitrothiophenyl-myosin (NTP-myosin) produced had absorption properties which were similar to those expected if the NTP-group were attached to a carbonyl group as in a thio-ester [48] and in addition was stable at acidic or neutral pH values, unstable in alkali. An NTP-peptide could be obtained by enzymic digestion of NTP-myosin. The NTP-group was liberated from such a peptide by treatment with hydroxylamine. These results strongly suggest that the NTP-group was attached to the protein by a link involving the side-chain carboxyl of a glutamyl or aspartyl residue. Hence, the phosphate group may be bound to the myosin as an acylphosphate (Fig. 5.4).

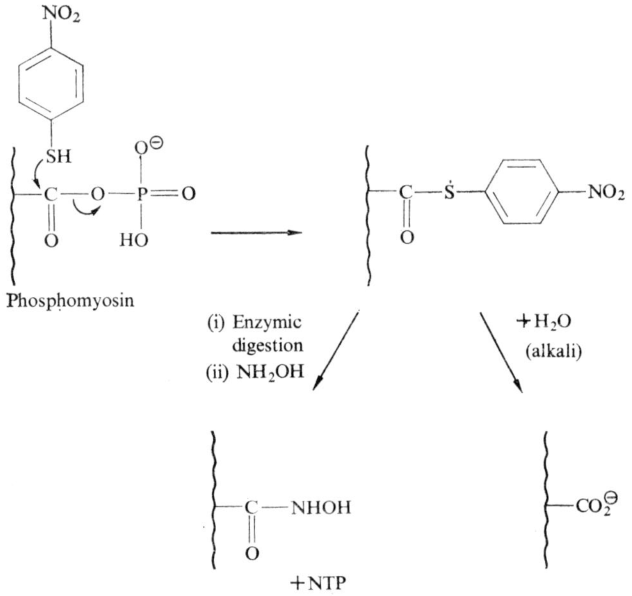


FIG. 5.4. Reactions of phosphoryl-myosin with nitrothiophenol

It is interesting to note that Koshland and co-workers [49], who observed the high degree of exchange between $H_2{}^{18}O$ and the terminal phosphate residue of ATP in the case of Mg(II)-activated myosin ATPase, reported that this did not occur when the myosin was activated by Ca-(II)-ions. Therefore, the two metal-ions are able to influence the reaction in different ways. It appears that Ca(II)-activation favours the direct hydrolysis mechanism, whereas the Mg(II)-ions lead to hydrolysis via the exchangeable intermediate.

We see therefore that myosin-catalysed hydrolysis of ATP bears a striking similarity to that catalysed by the (Na and K)-dependent ATPase. Thus, they both appear to involve two types of mechanism, one direct and the other involving a phosphorylated protein intermediate. In addition an acyl-phosphate intermediate appears to be involved in both cases. Both

ATPase activities appear to show complex temperature dependences [38, 45] which may reflect conformational changes. Therefore the partial mechanism in Fig. 5.3 probably applies to both ATPases. In this figure, the hydrolysis of the enzyme-phosphate is depicted as a nucleophilic attack by water on the phosphorus atom.

## HYDROLYSIS OF GLYCOSIDES

### Non-enzymic Hydrolysis of Glycosides

Glycosides are essentially acetals and, like the hydrolysis of acetals, non-enzymic glycoside hydrolysis is subject to specific acid catalysis, that is, to catalysis by hydrogen ions. As indicated in equation (5.13) the hydrolysis yields an alcohol and a sugar unit as a hemiacetal.

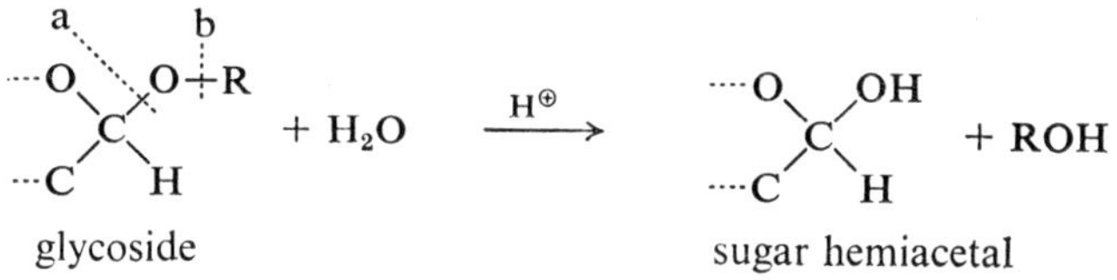


(5.13)

Either of two positions of bond cleavage (a or b) could give rise to the products shown and it is clearly of importance in a study of the reaction mechanism to establish which actually occurs. This has, in fact, been achieved by Burton *et al.* [50], who used water enriched with $^{18}O$ and showed that, in all but one very exceptional case, glycoside hydrolysis occurs by hexose–oxygen bond cleavage (position a). This establishes the $C_1$ atom of the glycoside as the centre of substitution. Yeast α-glucosidase and almond β-glucosidase also catalyse hydrolytic reactions in which hexose–oxygen fission is found to occur [51].

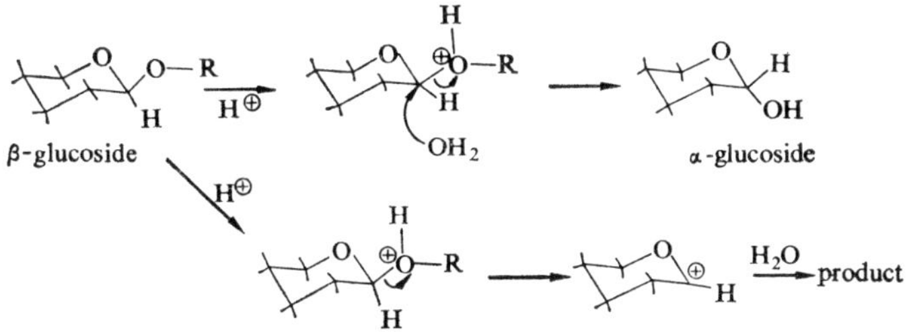


FIG. 5.5. Two mechanisms for the acid-catalysed hydrolysis of a β-glucoside. In the upper mechanism, a direct nucleophilic attack by water on the conjugate acid of the glycoside leads to inversion at C-1. In the lower mechanism, the conjugate acid decomposes by a unimolecular process giving a carbonium ion which subsequently reacts with water. The former is designated A2, while the latter is designated A1

Two types of mechanism are possible for an acid-catalysed substitution and these are represented in Fig. 5.5 for the hydrolysis of a $\beta$-glucoside. The first, which is designated A2 (compare $S_N2$), involves a direct nucleophilic attack by water on to the conjugate acid of the glucoside and is bimolecular. Such a process occurs with inversion of the configuration at the $C_1$ carbon. In the second mechanism (A1) the conjugate acid of the glucoside decomposes by a unimolecular reaction giving a carbonium ion which reacts in a subsequent step with a water molecule. In some cases such a reaction will lead to racemization at $C_1$ since the water molecule may approach from either side; however, in the case of many glycosides it is very difficult to predict the steric course of the A1 reaction.

In general the results of kinetic studies have led most workers to conclude that the A1 type of mechanism operates in the non-enzymic hydrolysis of glycosides [52]. The closed-ring carbonium ion shown in Fig. 5.5 is the most likely type of intermediate (an alternative, open-ring form was eliminated by oxygen-isotope-effect measurements [53]). Such a carbonium ion would be stabilized by a donation of electrons from the oxygen

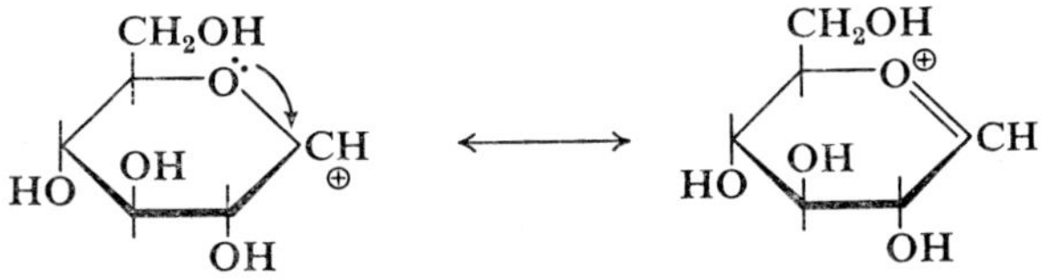


(XV)

atom of the ring so that the molecule would have much of the character of the structure of XV. The compound XV would not exist in the 'chair' form usually taken up by saturated six-membered rings (XVI); the double-bond character of the $C_1$—O bond would force the molecule into a half-chair configuration (XVII).

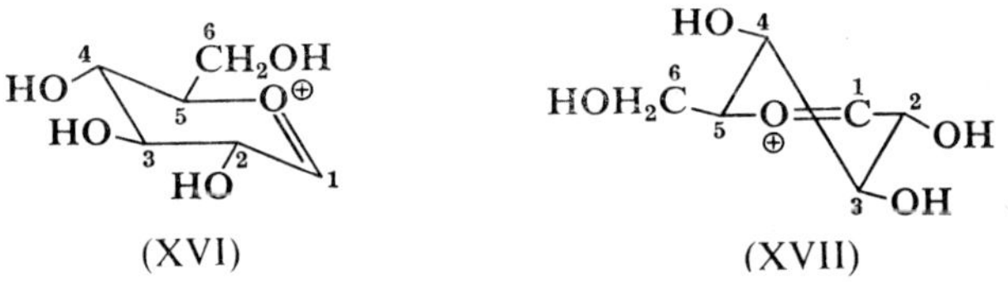


(XVI) (XVII)

**N-Acetylmuramide Glycanohydrolase (Muramidase, Lysozyme, 3.2.1.17)**

Among the enzymes, lysozyme holds a special place; this is because it was the first enzyme whose three-dimensional structure was elucidated. This work was carried out before very many chemical studies had been made, and the groups in the enzyme which are important for binding and for the catalytic process were determined from the structure rather than from chemical information. This is in contrast with the other enzymes where extensive chemical information had been obtained before the X-ray crystallography had been completed.

Lysozyme brings about the lysis of certain bacterial cells; its natural substrate appears to be a type of polysaccharide which occurs in the bacterial cell wall, from which it releases N-acetyl-amino sugars [54]. The enzyme also degrades chitin which is made up of N-acetyl-glucosamine units [55]. From lysozyme digests of bacterial cell walls, Salton and Ghuysen [56] isolated a tetrasaccharide with the structure shown (XVIII) [57]. This compound was degraded in the presence of lysozyme by hydrolysis at the point shown, giving two molecules of a disaccharide derived from N-acetylglucosamine and N-acetylmuramic acid. Bacterial cell wall

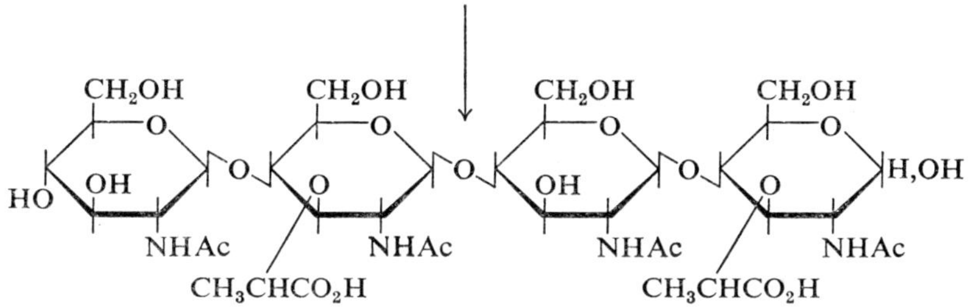


(XVIII)

polysaccharide appears to be a polymer containing alternate units of N-acetylglucosamine and N-acetylmuramic acid, joined by $\beta$-(1–4) linkages. The similarity between this and chitin which is the $\beta$-(1–4)-linked polymer of N-acetylglucosamine is clear.

Lysozyme is a single-chain protein containing 129 amino-acid residues. The amino-acid sequence was determined by Jollès and co-workers [58] and by Canfield [59], and is shown in Fig. 5.6.

The four intramolecular disulphide bridges are essential for the enzymic activity of the molecule. If they are broken by reduction or oxidation then catalytic activity is lost. Partially-reduced lysozyme can regain its activity on oxidation with oxygen [59]. It seemed likely that at least some of the tryptophan residues were involved in enzymic activity since destruction of these residues caused inactivation [60]. Hen egg-white lysozyme contains only one histidyl residue. In many other enzymes the side-chain imidazole group of at least one histidyl residue has been implicated in the catalytic process, and some workers were interested in the possibility that this single histidyl residue might be involved in the catalytic reactions of lysozyme. However, Jollès, Spotorno and Jollès [61] found that a lysozyme from duck egg-white contained *no* histidyl residue and was nevertheless fully active. Therefore no role is assigned to a histidyl residue in the mechanism of action.

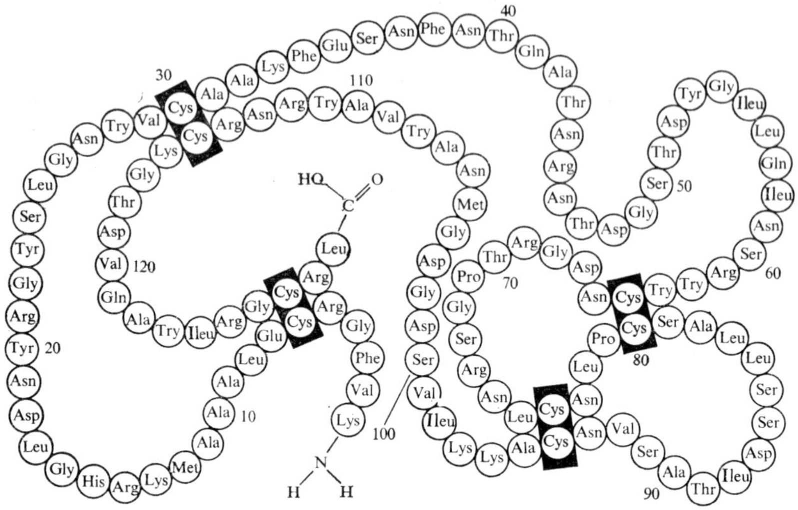


FIG. 5.6. Amino-acid sequence of the enzyme lysozyme as determined by Jollés *et al.* [58] and by Canfield [59]. (By courtesy of R. E. Canfield and the American Society of Biological Chemists)

X-ray crystallographic studies with lysozyme and lysozyme–inhibitor complexes have been very fruitful [62]. Since lysozyme has been shown to retain its catalytic activity in the crystalline state [63], it is perfectly justifiable to apply conclusions drawn from the X-ray structure to catalytic processes in solution. In general, the molecule has polar groups on its surface, in contact with solvent, while the non-polar groups predominate in the interior. The enzyme is divided into two by a large cleft (compare papain) and some hydrophobic residues lie on the surface of this cleft (tryptophans-62, -63 and -108, isoleucine-98 and valine-109).

N-Acetylglucosamine inhibits lysozyme, but glucosamine does not [64], an observation which suggests that the acetyl group in a substrate or inhibitor may be important for binding. The trimer of N-acetylglucosamine, chitotriose is hydrolysed by lysozyme but only extremely slowly [64, 65]. Larger polymers of N-acetylglucosamine are hydrolysed more rapidly, the rate increasing with increasing substrate chain length up to the hexamer [65, 66].

X-ray studies [62] of the lysozyme–N-acetylglucosamine and the lysozyme–chitotriose complexes showed that the sugar residues were bound within the cleft of the molecule, and that there were up to six positions where the sugar residues could bind. Chitotriose formed a stable complex in which the three sugar residues covered three apparent binding sites. From these and other results it was concluded that during the hydrolysis of a polysaccharide substrate, six sugar units were accommodated in the cleft of the enzymes with binding at specific points. The chitotriose was hydrolysed only very slowly because the major mode of binding of this trisaccharide to the enzyme was not productive. The non-productive

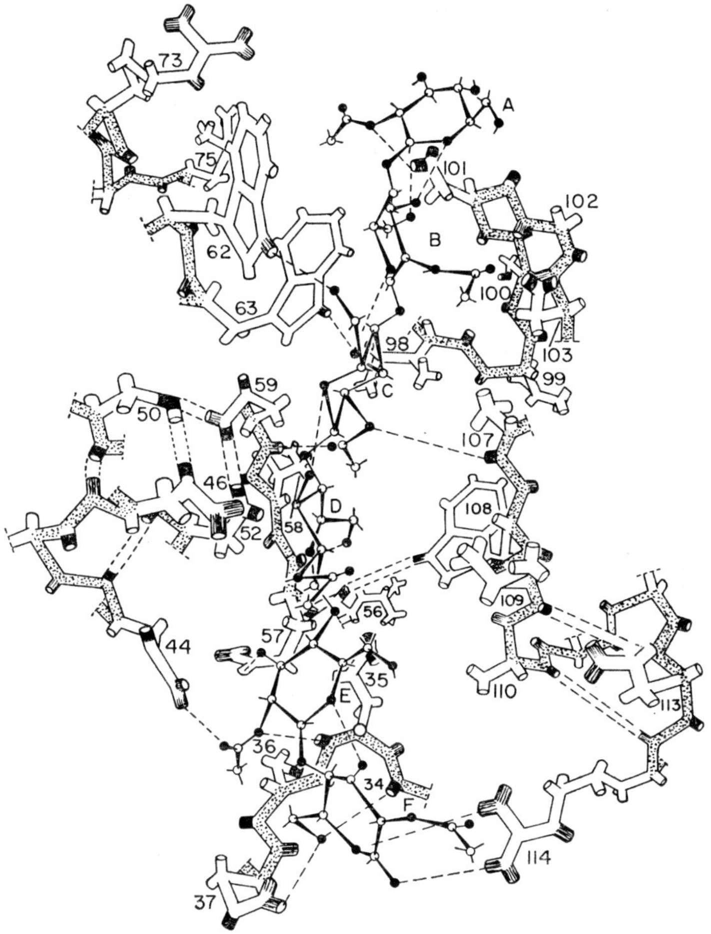


FIG. 5.7. Diagram showing part of the three-dimensional structure of lysozyme and its substrate. The six rings of the hexasaccharide bound to the enzyme are labelled A, B, C, D, E and F, respectively. Note the hydrogen bonds involved in the binding of the substrate to the enzyme, especially those with the side-chain groups of tryptophan-62 and tryptophan-63. Note, too, that ring D is not in the normal chair-form and that close to this ring are the side-chains of aspartic acid-52 and glutamic acid-35. (Ref. [62]) (By courtesy of D. C. Phillips and the Brookhaven National Laboratory)

complex was viewed as representing a portion of the enzyme–substrate complex. In Fig. 5.7 is shown a representation of the lysozyme–substrate complex. Six sugar residues are present in the cleft. The first three residues

A, B and C represent the position of the trisaccharide in the non-productive enzyme–trisaccharide complex. By constructing a model, Blake *et al.* [62] were able to show how the three other sugar units D, E and F could bind to the enzyme. Suitable binding could only be achieved if the ring of residue D were assumed to be in the half-chair form similar to that shown in the carbonium ion (XVII). With D in this state the model could be constructed so that hydrogen bonds and non-polar interactions in acceptable forms became apparent.

It was also concluded that the glycosidic bond between residues D and E was the one cleaved by the action of the enzyme. This was in agreement with results obtained by Rupley [66] who showed that the hexamer of N-acetylglucosamine was cleaved into a tetramer plus a dimer, while the pentamer was cleaved in two ways giving either tetramer plus monomer or trimer plus dimer. This is in complete agreement with the hypothesis of a binding site accommodating six residues in which the link between the fourth and fifth residues is broken. A hexamer can bind best in one way on such a site whereas for a pentamer two modes of binding are equally favoured.

In this case, as in the case of other hydrolytic reactions of glycosides, the reaction proceeds with hexose–oxygen fission [66] and therefore the $C_1$ carbon of residue D is the atom which is the centre of substitution. From the model, it is seen that the side-chain carboxyl group of the Asp-52 residue is about 3 Å from this $C_1$ atom. In addition the side-chain of Glu-35 is at about the same distance from the linking oxygen atom between residues D and E.

The side-chain of Glu-35 apparently lies in a non-polar region and it has been suggested [52] that this carboxyl group remains un-ionized even at relatively high pH values and therefore is able to function as a general-acid catalyst for the cleavage of the glycosidic linkage. The side-chain of Asp-52, however, is in a polar region and is probably ionized at most pH values.

Lowe [67], who studied the lysozyme-catalysed hydrolysis of several chitoboise $\beta$-phenyl glucosides, showed that the catalytic reaction possessed the characteristics of nucleophilic or basic catalysis, as well as acid catalysis.

The suggested mechanism of action of lysozyme [52, 62] may be summarized as follows (Fig. 5.8). On formation of the enzyme–substrate complex, ring D is forced into the half-chair configuration (XIX). Thus the ring is already part of the way along the reaction path towards a carbonium ion. The oxygen atom of the D–E linkage is protonated by the carboxyl group of Glu-35 and the $C_1$—O bond is cleaved heterolytically. This process is assisted by the carboxylate group of the Asp-52 residue which also serves to stabilize the resulting carbonium ion (XIX, XX). The hydrolysis is complete when a water molecule attacks the carbonium ion

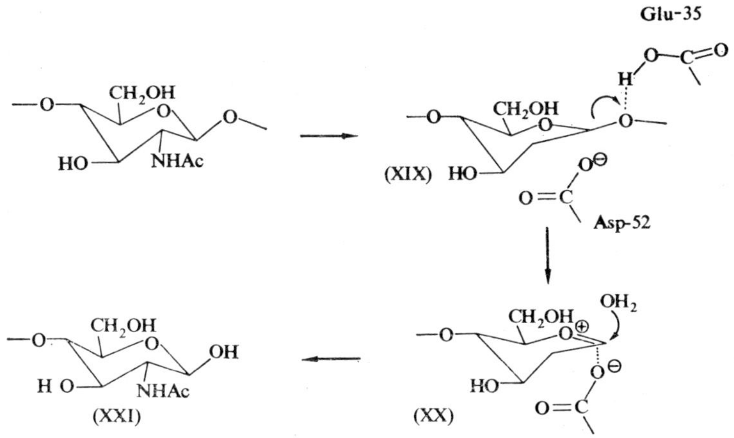


FIG. 5.8. Suggested mechanism of action of lysozyme [52, 62]

from the same side as that from which the E–F disaccharide residue left; therefore the reaction proceeds with retention of configuration (XXI). It is suggested that the assistance, as described, by the Asp-52 carboxylate group is sufficient to explain the nucleophilic or basic character of the reaction as required by Lowe's results [67].

Vernon [52] has pointed out that this mechanism helps to explain the *catalysis* by the enzyme, three factors being important. These are the general-acid catalysis by the Glu-35 carboxyl, the distortion of the ring of residue D which then favours the production of the carbonium ion and thirdly, the stabilization of carbonium ion by the nearby carboxylate anion of Asp-52.

It is of interest that transferase activity has been reported for lysozyme [68]. For many enzymes, transferase activity has been interpreted as indicating a *covalent* intermediate. In the case of lysozyme a covalent intermediate might be suggested in which the carboxylate group of Asp-52 actually forms a covalent bond with the $C_1$-carbon of residue D. However, the model apparently shows this to be extremely unlikely, although the interaction between the carboxylate ion and the carbonium ion may possess some covalent character [52]. In any case it is clearly not essential that the intermediate be covalently bound; the main requirement is that its lifetime is sufficiently long for the leaving residue to diffuse away from the enzyme and for the acceptor to become bound. In most other cases this probably does mean the involvement of a covalent intermediate.

**Other Glycosidases**

The polysaccharides whose hydrolysis is brought about by lysozyme, chitin and the bacterial cell wall polysaccharide, are structural in function. The amylases, on the other hand, hydrolyse polysaccharides which are produced in living systems for the purpose of storage. These polysaccharides, such as amylose and glycogen, consist of glucose units joined by $\alpha$-1,4-glucosidic linkages. $\alpha$-Amylase [69] is an *endo*-glycosidase, attacking the polysaccharide chains randomly, giving rise to the production of oligosaccharides. The configuration at the $C_1$-carbon of the new reducing ends produced is $\alpha$, and thus $\alpha$-amylase produces retention of configuration. By contrast, $\beta$-amylase (70) is an *exo*-glycosidase, cleaving successive maltose units from the non-reducing end of the polysaccharide chain. The maltose produced has the $\beta$-configuration at the reducing end and the reaction of this enzyme proceeds with inversion.

$\alpha$-Amylases from various sources all appear to be metallo-enzymes requiring one atom of $Ca^{2\oplus}$ per molecule of enzyme [70, 71]. Removal of the calcium ion causes complete loss of activity [70, 72] and addition of calcium to the metal-free preparations restores the activity. It is likely that the calcium ions help to hold the proteins in the active configuration.

Amylases are also influenced by the presence of anions such as chloride ion. These ions do not interact with the enzyme at the active site, since the formation of the enzyme–substrate complex is unaffected by their presence [73]. They probably bring about conformational changes; physical properties such as solubility and electrophoretic mobility are altered [74].

As in the case of almost all glycoside hydrolyses, studies with $^{18}O$-labelled water have shown that $C_1$—O bond cleavage takes place when $\alpha$-amylase acts on starch or glycogen [75, 76]. In addition, most workers have concluded that in almost all cases the functional groups involved in the catalytic reaction are a carboxyl group ($pK$ 4–5) and an imidazolium group ($pK$ approx. 8) [77, 78]. Mayer and Larner [76] proposed that, after the enzyme–substrate complex had been formed, the glycosidic bridge oxygen became protonated (by the imidazolium) and that a carbonium ion was then formed. A water molecule then attacked the carbonium ion, giving the $\alpha$-product due to the stereochemical arrangement of the enzyme–carbonium ion complex. This mechanism is therefore similar to that proposed for lysozyme. Fischer and Stein [69] suggested that the retention of configuration might be explained by a two-step reaction in which the carboxylate anion known to be involved makes a direct nucleophilic displacement on to the $C_1$-carbon, assisted by general-acid catalysis by the imidazolium group. This latter mechanism, of course, implies a covalently bound glycosyl-enzyme intermediate. Thoma [77] has calculated that the involvement of functional groups in this way, even when they act in a concerted manner, would not be sufficient to explain the high efficiency of $\alpha$-amylase. He suggests that, as in lysozyme, a distortion of the substrate takes place on binding and that this effect makes a large contribution to the catalytic efficiency.

β-Amylase reacts in such a way that inversion takes place at the central $C_1$-carbon and therefore has been considered to act by a mechanism fundamentally different from that of α-amylase [79]. However, Thoma points out that the amylases could function by essentially similar mechanisms, the only difference being in the direction of approach of the water molecule in the latter stages of the process (XXII). Small changes in the three-

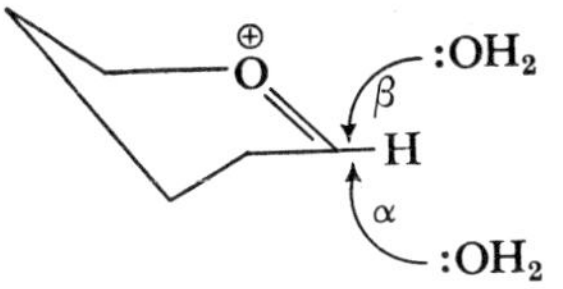


(XXII)

dimensional structure of the enzyme–carbonium ion complex would allow either inversion or retention of configuration. It is certainly tempting to speculate whether all glucosyl-enzyme intermediates (glycosyl-transferases as well as glucanohydrolases) might not exist in the carbonium ion form. However, it may become very difficult to distinguish between a possible covalent link between glycosyl-residue and enzyme and an ionic link with some covalent character.

Against Thoma's suggestion for the mechanism of action of β-amylase must be placed the observations that the plant β-amylases are completely inactivated by thiol-group reagents, which may mean that a thiol group could have a role in the catalytic process [70, 80].

The involvement of at least a carboxyl group in the catalytic reaction of almond β-glucosidase is indicated by the results of Hofstee [81], which showed an essential ionizable group with $pK_a$ approx. 4. Furthermore the presence of a carboxyl group at the active site was demonstrated by the

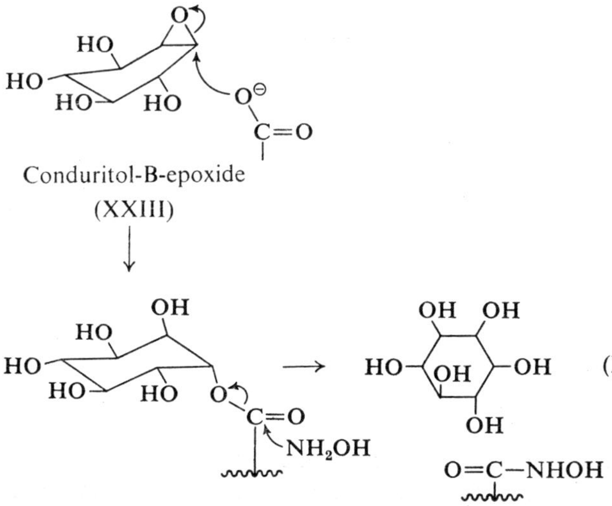


(XXIV)

reaction of the enzyme with conduritol-B-epoxide (XXIII), which is structurally similar to the natural substrate of the enzyme [82]. The reaction produces total inactivation, one mole of inhibitor being bound per mole of enzyme. (+)-Inositol (XXIV) was released when the inactivated protein was treated with hydroxylamine. These reactions are shown below. Conduritol-B-epoxide is bound to the active site of the enzyme and the epoxide ring opened by a nucleophilic attack by a carboxylate ion. The product is an ester which is susceptible to cleavage by hydroxylamine.

## PURINE AMINOHYDROLASES

These enzymes catalyse the deamination of amino-purines and their derivatives. The reaction involves the hydrolysis of the cyclic amidine system and an amino-group is replaced by a hydroxyl group (the product, an enol, usually takes up the keto-form at normal pH values), ammonia being liberated. Thus, adenine aminohydrolase (adenine deaminase, 3.5.4.2) catalyses the production of hypoxanthine (XXVI) from adenine (XXV), and guanine aminohydrolase (3.5.4.3) converts guanine (XXVII) into xanthine (XXVIII).

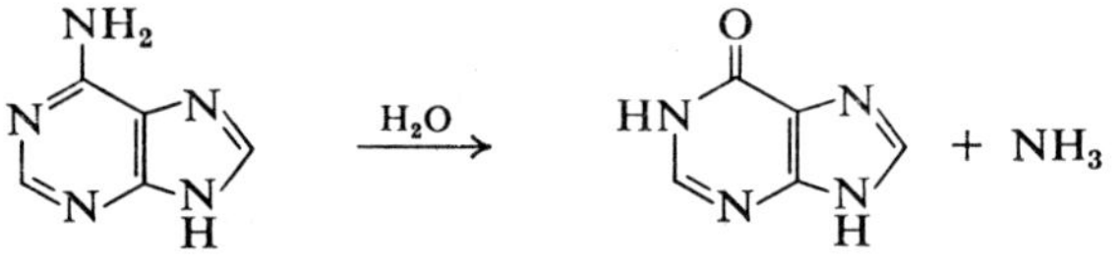

(XXV) (XXVI)

(XXVII) (XXVIII)

Enzymes have been reported which are specific for adenine, adenosine, AMP, ADP or ATP [83]. In particular, a bacterial adenosine aminohydrolase (3.5.4.4) has been studied in some detail. This enzyme exhibits a fairly broad specificity towards compounds related to adenosine, including adenine and adenosine nucleotides [84]. Its activity towards adenylic acid oligonucleotides decreases as the size of the molecule is increased. The $K_m$ for adenosine varies with pH and depends on two prototropic groups, with $pK_a$ values at 3·3 and 8·9 respectively. The group with $pK_a$ at 3·3 belongs to the substrate and it is probable that the group with $pK_a$ 8·9 is part of the active site of the enzyme [84].

The action of the enzyme is not limited to deamination, indeed it catalyses the hydrolysis of compounds with various groups on the 6-position in place of the amino-group. Thus, 6-hydrazino-, 6-chloro-, 6-methylamino- and 6-methoxy-adenosine are all substrates [85]. In addition, these compounds are hydrolysed at limiting velocities which are very similar to that of adenine itself. This is good evidence for the existence of a common intermediate in the reaction of all of these compounds, probably a purinyl-enzyme.

It is probable, therefore, that the hydrolysis takes place by a mechanism such as that shown in Fig. 5.9. A nucleophilic group in the active site of

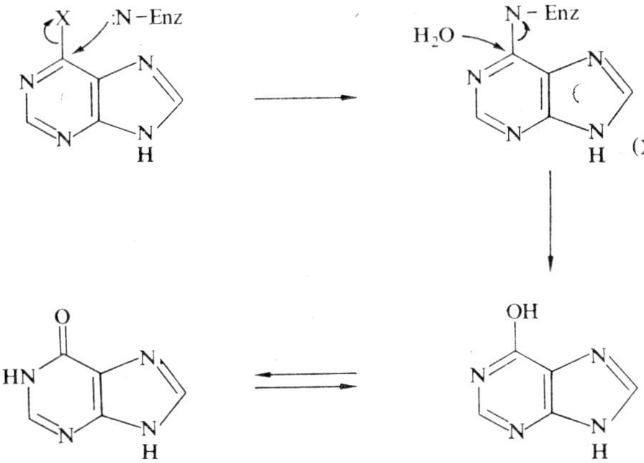


FIG. 5.9. Probable mechanism of action of the purine aminohydrolases

the enzyme displaces X from the C-6 position giving the purinyl-enzyme (XXIX) which is itself attacked by water giving the enol form of the product. The nature of the nucleophilic group at the active site of the enzyme is not known. No indication is given by the pH-dependence of the $V_{max}$, since this is constant over a wide pH range. It is possible that the same group of $pK_a$ 8·9 which is involved in enzyme–substrate binding is also involved in this catalytic process too [84]. This group may be a thiol group; the enzyme is inactivated by mercurials. However, this is doubtful because the enzyme is resistant to oxidations and alkylations [84]. We have already seen (carboxypeptidase) that mercurials may not be specific for thiol groups.

An adenosine aminohydrolase from calf intestinal mucosa [86], a bacterial adenine aminohydrolase [87] and a bacterial ATP deaminase [88] have all been shown to catalyse the general hydrolysis of 6-substituted derivatives. In addition, guanine deaminase hydrolysed various groups from the 2-position of 6-hydroxypurine derivatives [89]. The reaction scheme shown in Fig. 5.9 may therefore be quite general.

## REFERENCES


1. Bruice, T. C. and Benkovic, S. J., *Biorganic Mechanisms* Vol. 2, p. 1 (Benjamin, New York, 1966)
2. Bunton, C. A., Llewellyn, D. R., Vernon, C. A. and Welch, V. A., *J. chem. Soc.* 1636 (1961)
3. Mahler, H. R. and Cordes, E. H., *Biological Chemistry* p. 192 (Harper and Row, New York, 1966)
4. Bunton, C. A., Mhala, M. M., Oldham, K. G. and Vernon, C. A., *J. chem. Soc.* 3293 (1960)
5. Westheimer, F. H., Chem. Soc. (London), Special Publications, **8,** 1 (1957); Lecoq, J., *Compt. rend.* **242,** 1902 (1956)
6. Khorana, H. G., Tener, G. M., Wright, R. S. and Moffatt, J. G., *J. Am. chem. Soc.* **79,** 430 (1957); Cherbuliez, E., Probst, H. and Rabinowitz, J., *Helv. chim. Acta* **42,** 1377 (1959)
7. Hoake, P. C. and Westheimer, F. H., *J. Am. chem. Soc.* **83,** 1102 (1961)
8. Kaiser, E. T., Panar, M. and Westheimer, F. H., *J. Am. chem. Soc.* **85,** 602 (1963)
9. Desjobert, A., *Bull. Soc. chim. Fr.* **683** (1963); see also ref. [1].
10. Bunton, C. A., Llewellyn, D. R., Oldham, K. G. and Vernon, C. A., *J. chem. Soc.* 3574 (1958); Butcher, W. W. and Westheimer, F. H., *J. Am. chem. Soc.* **77,** 2420 (1955); Jencks, W. P., *Brookhaven Symposia in Biology* **15,** 134 (Brookhaven National Laboratory, Upton, New York, 1962); Kumamoto, J. and Westheimer, F. H., *J. Am. chem. Soc.* **77,** 2515 (1955)
11. Brown, D. M., Magrath, D. I., Neilson, A. H. and Todd, A. R., *Nature, Lond.* **177,** 1124 (1956)
12. Hofstetter, R., Murakami, Y., Mont, G. and Martell, A. E., *J. Am. chem. Soc.* **84,** 3041 (1962)
13. Tait, G. H. and Vallee, B. L., *Proc. natn. Acad. Sci. U.S.A.* **56,** 1247 (1966); Plocke, D. J., Levinthal, C. and Vallee, B. L., *Biochemistry* **1,** 373 (1962)
14. Fife, W. K., *Biochem. biophys. Res. Commun.* **28,** 309 (1967)
15. Morton, R. K., *Diss. Faraday Soc.* **20,** 149 (1955)
16. Wilson, I. B., Dayan, J. and Cys, K., *J. biol. Chem.* **239,** 4182 (1964)
17. Cohn, M., *J. biol. Chem.* **180,** 771 (1949)
18. Stein, S. S. and Koshland, D. E., jr., *Archs biochem. Biophys.* **39,** 229 (1952)
19. Morton, R. K., *Biochem. J.* **70,** 139, 15 (1958); Garen, A. and Levinthal, C., *Biochim. biophys. Acta* **38,** 470 (1960)
20. Heppel, L. A., Harkness, D. R. and Hilmoe, R. J., *J. biol. Chem.* **237,** 841 (1962)
21. Fernley, H. N. and Walker, P. G., *Nature, Lond.* **212,** 1435 (1960)
22. Schwartz, J. H., *Proc. natn. Acad. Sci. U.S.A.* **49,** 871 (1963)
23. Engstrom, L. and Agren, G., *Acta chem. scand.* **12,** 357 (1958); Engstrom, L., *Biochim. biophys. Acta* **52,** 36, 49 (1961); *Biochim. biophys. Acta* **56,** 606 (1962)
24. Schwartz, J. H. and Lipmann, F., *Proc. natn. Acad. Sci. U.S.A.* **47,** 1996 (1961)
25. Schwartz, J. H., Crestfield, A. M. and Lipmann, F., *Proc. natn. Acad. Sci. U.S.A.* **49,** 772 (1963)

26. Georgatsos, J. G., *Archs biochem. Biophys.* **121,** 619 (1967)
27. Yang, S. F., Freer, S. and Benson, A. A., *J. biol. Chem.* **242,** 477 (1967)
28. Dawson, R. M. C. and Hemington, N., *Biochem. J.* **102,** 76 (1967)
29. Dudek, G. O. and Westheimer, F. H., *J. Am. chem. Soc.* **81,** 2641 (1959)
30. Tetas, M. and Lowenstein, J. M., *Biochemistry* **2,** 350 (1963)
31. Cohn, M. and Hughes, T. R., *J. biol. Chem.* **235,** 3250 (1960); *J. biol. Chem.* **237,** 176 (1962)
31a. Selwyn, M. J., *Nature, Lond.* **219,** 490 (1968)
32. Skou, J. C., *Biochim. biophys. Acta* **23,** 394 (1957)
33. Skou, J. C., *Phys. Rev.* **45,** 596 (1965)
34. Judah, J. D. and Ahmed, K., *Biol. Rev. Cambridge Phil. Soc.* **39,** 160 (1964)
35. Ahmed, K. and Judah, J. D., *Biochim. biophys. Acta* **104,** 112 (1965)
36. Post, R. L., Sen, A. K. and Rosenthal, A. S., *J. biol. Chem.* **240,** 1437 (1965)
37. Nagano, K., Mizuno, N., Fujita, M., Tashima, Y., Nakao, T. and Nakao, M., *Biochim. biophys. Acta* **143,** 239 (1967)
38. Kanazawa, T., Saito, M. and Tonomura, Y., *J. Biochem. (Tokyo)* **61,** 555 (1967)
39. Tokiwa, T. and Tonomura, Y., *J. Biochem. (Tokyo)* **57,** 616 (1965)
40. Nagano, K., Kanazawa, T., Mizuno, N., Tashima, Y., Nakao, T. and Nakao, M., *Biochem. biophys. Res. Commun.* **19,** 759 (1965); Bader, H., Post, R. L. and Jean, D. H., *Biochim. biophys. Acta* **143,** 229 (1967)
41. Chignell, C. F. and Titus, E., *Biochim. biophys. Acta* **159,** 343 (1968); Hokin, L. E., Sastry, P. S., Galsworthy, P. R. and Yoda, A., *Proc. natn. Acad. Sci. U.S.A.* **54,** 177 (1965); Bader, H., Sen, A. K. and Post, R. L., *Biochim. biophys. Acta.* **118,** 106, 116 (1966)
42. Levy, H. M. and Koshland, D. E., jr., *J. biol. Chem.* **234,** 1102 (1959)
43. Kanazawa, T. and Tonomura, Y., *J. Biochem. (Tokyo)* **57,** 604 (1965)
44. Sekine, T., Barnett, C. M. and Keilley, W. W., *J. biol. Chem.* **237,** 2769 (1962); Kubo, S., Tokura, S. and Tonomura, Y., *J. biol. Chem.* **235,** 2835 (1960); Tonomura, Y. and Morita, F., *J. Biochem. (Tokyo)* **46,** 1367 (1959)
45. Levy, H. M., Sharon, N., Lindemann, E. and Koshland, D. E., jr., *J. biol. Chem.* **235,** 2628 (1960)
46. Tokiwa, T. and Tonomura, Y., *J. Biochem. (Tokyo)* **57,** 616 (1965)
47. Imamura, K., Kanazawa, T., Tada, M. and Tonomura, Y., *J. Biochem. (Tokyo)* **57,** 627 (1965)
48. Kitagawa, S., Chiang, K.-K. and Tonomura, Y., *Biochim. biophys. Acta* **82,** 83 (1964)
49. Levy, H. M. and Koshland, D. E., jr., *J. Am. chem. Soc.* **80,** 3164 (1968); Koshland, D. E., jr., Budenstein, Z. and Kowalsky, A., *J. biol. Chem.* **211,** 279 (1954)
50. Bunton, C. A., Lewis, T. A., Llewellyn, D. R. and Vernon, C. A., *J. chem. Soc.* 4419 (1955)
51. Bunton, C. A., Lewis, T. A., Llewellyn, D. R., Tristram, H. and Vernon, C. A., *Nature, Lond.* **174,** 560 (1954)
52. Vernon, C. A., *Proc. R. Soc.* **167B,** 389 (1967)
53. Banks, B. E. C., Meinwald, Y., Rhind-Tutt, A. J., Sheft, I. and Vernon, C. A., *J. chem. Soc.* 3240 (1961)

54. MEYER, K., PALMER, J. W., THOMPSON, R. and KHORAZO, D., *J. biol. Chem.* **113,** 303 (1936); MEYER, K. and HAHNEL, E., *J. biol. Chem.* **163,** 723 (1946); EPSTEIN, L. A. and CHAIN, E., *Br. J. exp. Path.* **21,** 339 (1940)
55. BERGER, L. R. and WEISER, R. S., *Biochim. biophys. Acta* **26,** 517 (1957)
56. SALTON, M. R. J. and GHUYSEN, J. M., *Biochim. biophys. Acta* **36,** 552 (1959); **45,** 355 (1960)
57. JEANLOZ, R. W., SHARON, N. and FLOWERS, H. M., *Biochem. biophys. Res. Commun.* **13,** 20 (1963)
58. JOLLÉS, P., *Proc. R. Soc.* **167B,** 350 (1967)
59. CANFIELD, R. E., *J. biol. Chem.* **238,** 2698 (1963)
60. HAYAISHI, K., IMOTO, T. and FUNSTEN, M., *J. Biochem.* (*Tokyo*) **54,** 381 (1963); HARTDEGEN, F. J. and RUPLEY, J. A., *Biochim. biophys. Acta,* **92,** 625 (1964)
61. JOLLÈS, J., SPOTORNO, G. and JOLLÉS, P., *Nature, Lond.* **208,** 1204 (1965)
62. BLAKE, C. C. F., MAIR, G. A., NORTH, A. C. T., PHILLIPS, D. C. and SARMA, V. R., *Proc. R. Soc.* **167B,** 365 (1967); BLAKE, C. C. F., KOENIG, D. F., MAIR, G. A., NORTH, A. C. T., PHILLIPS, D. C. and SARMA, V. R., *Nature, Lond.,* **206,** 757 (1965); BLAKE, C. C. F., JOHNSON, L. N., MAIR, G. A., NORTH, A. C. T., PHILLIPS, D. C. and SARMA, V. R., *Nature, Lond.* **206,** 378 (1965)
63. BUTLER, L. G. and RUPLEY, J. A., *J. biol. Chem.* **242,** 1077 (1967)
64. WENZEL, M., LEUK, H.-P. and SCHUTTE, E., *Z. Physiol. Chem.* **327,** 13 (1964)
65. RUPLEY, J. A., *Biochim. biophys. Acta* **83,** 245 (1964)
66. RUPLEY, J. A., *Proc. R. Soc.* **167B,** 416 (1967)
67. LOWE, G. SHEPPARD, G., SINNOTT, M. L. and WILLIAMS, A., *Biochem. J.* **104,** 893 (1967)
68. POLLOCK, J. J., CHIPMAN, D. M. and SHARON, N., *Archs biochem. Biophys.* **120,** 235 (1967); POLLOCK, J. J., CHIPMAN, D. M. and SHARON, N., *Biochem. biophys. Res. Commun.* **28,** 779 (1967); KRAVCHENKO, N. A., *Proc. R. Soc.* **167B,** 429 (1967).
69. FISCHER, E. H. and STEIN, E. A., *The Enzymes* Vol. 4, p. 313 (Academic Press, New York, 1960)
70. FRENCH, D., *The Enzymes* Vol. 4, p. 345 (Academic Press, New York, 1960)
71. VALLEE, B. L., STEIN, E. A., SUMMERWELL, W. N. and FISCHER, E. H., *J. biol. Chem.* **234,** 2901 (1959)
72. FISCHER, E. H. and HASELBACH, *Helv. chim. Acta* **34,** 325 (1951)
73. MYRBACK, K., *Z. Physiol. Chem.* **159,** 1 (1926)
74. MUUS, J., *Compte. rend. Trav. Lab. Carlsberg, ser. chim.* **28,** 317 (1953) quoted in ref. [69]
75. HALPERN, M. and LEIBOWITZ, J., *Biochim. biophys. Acta* **36,** 29 (1959)
76. MAYER, F. C. and LARNER, J., *Biochim. biophys. Acta* **29,** 465 (1958)
77. THOMA, J. A., *J. Theoret. Biol.* **19,** 297 (1968)
78. WAKIM, J., Ph.D. thesis, Indiana University (1965), quoted in ref. [77]; DAHL, C., M.S. thesis, Iowa State University, quoted in ref. [77]; THOMA, J. A. and KOSHLAND, D. E., JR., *J. molec. Biol.* **2,** 69 (1960); GREENWOOD, C. T., MACGREGOR, A. W. and MILNE, E. A., *Archs biochem. biophys.* **459,** 466 (1965); *Die Starke* **7,** 219 (1965)

79. KOSHLAND, D. E., JR., *Mechanism of Enzyme Action* p. 608, Eds. W. D. McElroy and B. Glass (Johns Hopkins Press, Baltimore, 1954)
80. WEILL, C. E. and CALDWELL, M. L., *J. Am. chem. Soc.* **67,** 212, 214 (1945); ENGLARD, S., SORAF, S. and SINGER, T. P., *J. biol. Chem.* **189,** 217 (1951)
81. HOFSTEE, B. H. J., *J. Am. chem. Soc.* **80,** 3966 (1958); *Archs biochem. Biophys.* **59,** 398 (1955)
82. LEGLER, G., *Z. Physiol. Chem.* **345,** 197 (1966); **349,** 767 (1968); *Biochim. biophys. Acta* **151,** 728 (1968)
83. DIXON, M. and WEBB, E. C., *Enzymes* 2nd edn (Longmans, London, 1964)
84. WOLFENDEN, R., SHARPLESS, T. K. and ALLAN, R., *J. biol. Chem.* **242,** 977 (1967)
85. WOLFENDEN, R., *J. Am. chem. Soc.* **88,** 3157 (1966)
86. CHASSY, B. M. and SUHADOLNIK, R. J., *J. biol. Chem.* **242,** 3655 (1967)
87. HARTENSTEIN, R. C. and FRIDOVICH, I., *J. biol. Chem.* **242,** 740 (1967)
88. CHUNG, S. and AIDA, K., *J. Biochem.* (*Tokyo*) **61,** 1 (1967)
89. ROY, J. E. and ROY, K. L., *Can. J. Biochem.* **45,** 1263 (1967)

# CHAPTER 6

# GROUP TRANSFER REACTIONS

## METHYL GROUP TRANSFER

On page 5 of chapter 1 we represented one of the simplest types of reaction encountered in organic chemistry. This was the reaction of chloride ion with methyl bromide. As a result of a nucleophilic substitution the methyl group is transferred from a bromine to a chlorine atom. The methyl bromide in this reaction functions as a methyl group donor and the chloride ion as an acceptor. Nucleophilic substitution is in fact the most common mechanism for group transfer especially where it is a simple alkyl group which is transferred. A good methyl group donor is therefore a molecule which is highly susceptible to nucleophilic attack, that is, a compound in which the methyl group carries a good leaving group. This is the case for the methyl bromide mentioned above, and also for the other methylating agents well known in organic chemistry such as methyl iodide, dimethyl sulphate and protonated diazomethane.

One of the most important methyl group donors encountered in enzyme chemistry also falls into this category. This is S-adenosyl methionine (I)

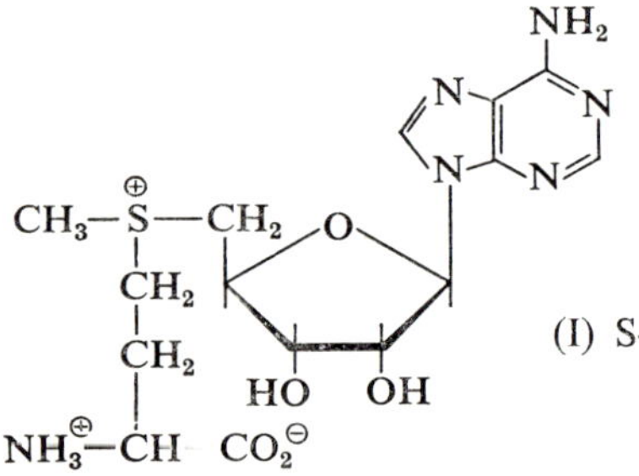


(I) S-Adenosine methionine (SAM)

in which the positively-charged sulphonium species constitutes a good leaving group in the event of a nucleophilic attack on the methyl. Such a reaction would yield S-adenosyl homocysteine. Since this is the usual product of the enzyme-catalysed reactions in which S-adenosyl methionine functions as a methyl donor it seems certain that these are in fact nucleophilic substitutions.

Methyl group transfer is an important stage in the synthesis of many biological metabolites. For example, the final step in the synthesis of methionine involves the transfer of a methyl group to homocysteine. At least one of the enzymes which catalyse this reaction is extremely interesting in that a derivative of vitamin $B_{12}$ is involved. The structure of vitamin $B_{12}$ is shown in Fig. 6.1 [1]. Important features of this molecule are the cobalt ion which accommodates four nitrogen ligands from the highly-substituted reduced porphyrin-like corrin ring, and the dimethyl benzimidazole group which provides the fifth ligand and which is also attached to the corrin ring via a ribose phosphate bridge. The sixth ligand to the metal-ion as shown is a cyanide ion. The entire structure except for the

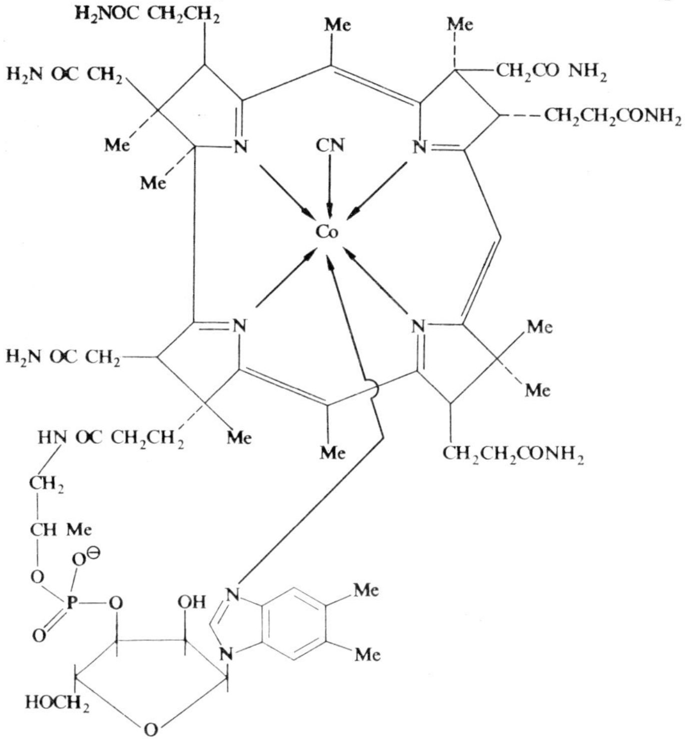


FIG. 6.1. Vitamin $B_{12}$, cyanocobalamin. (By courtesy of S. Moore and the American Society of Biological Chemists)

cyanide ion is known as cobalamin and the vitamin $B_{12}$ shown in Fig. 6.1 is referred to as cyano-cobalamin. The usual isolation procedure gives the vitamin in the cyano-cobalamin form but it is possible to replace the cyanide ion by other ligands such as water or the hydroxide ion giving aquo- or hydroxo-cobalamin respectively.

The derivatives of vitamin $B_{12}$ which function in group transfer reactions may be obtained by reduction of aquo- or cyano-cobalamin, followed by reaction with an appropriate alkylating agent. The reduction stage may be brought about by such reagents as sodium borohydride, zinc and acetic acid or reduced ferredoxin [2], or enzymically [3]. This reduction converts the cyano-cobalamin into a form ($B_{12S}$) in which the cobalt species can react with methyl iodide or some other methyl donor to give methyl-cobalamin.

The unusual feature of methyl-cobalamin is that it possesses a cobalt–carbon bond and in addition it is apparently produced by a reaction in which the metal-ion functions as a nucleophile. If the bond between the cobalt and carbon atoms is assumed to be of the normal covalent type then we may formulate the reaction as follows (equation 6.1). The cobalt ion

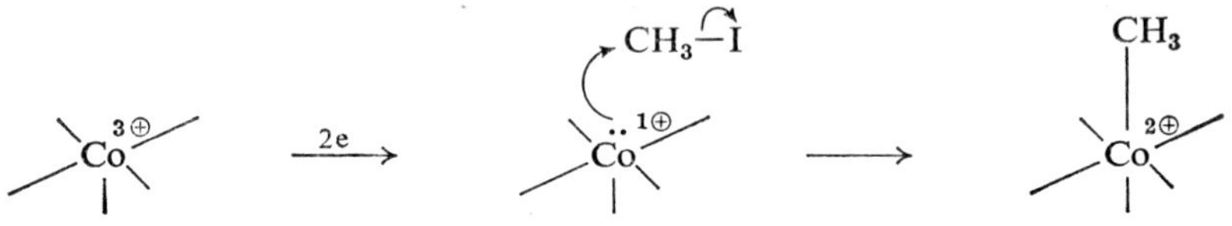

(6.1)

acquires during the reduction two electrons which enable it to participate in the nucleophilic process.

Reaction of reduced cobalamin with 5′-tosyl†-2′,3′-isopropylidene adenosine gives a 5′-deoxyadenosyl-cobalamin derivative (II) [4]. 5′-Deoxyadenosyl-cobalamin is often known as the $B_{12}$ coenzyme. It is produced

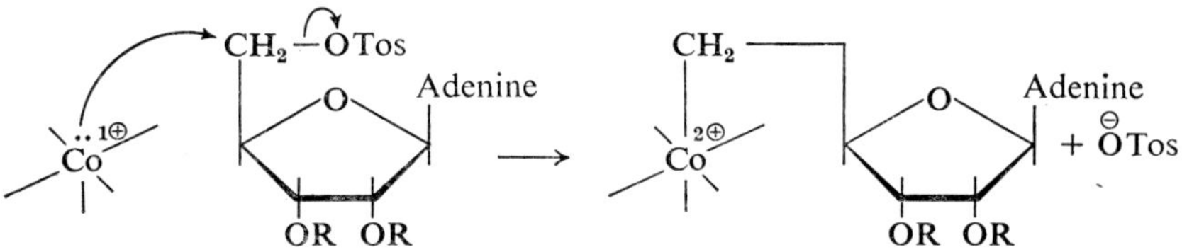

(II)

enzymically by systems which apparently first reduce cobalamin and then transfer the deoxyadenosyl group from ATP [3].

The similarity between this univalent cobalt system and other nucleophiles encountered in organic chemistry is emphasized by the reaction of $B_{12S}$ with acid chlorides and anhydrides; acyl-cobalamins are produced [5].

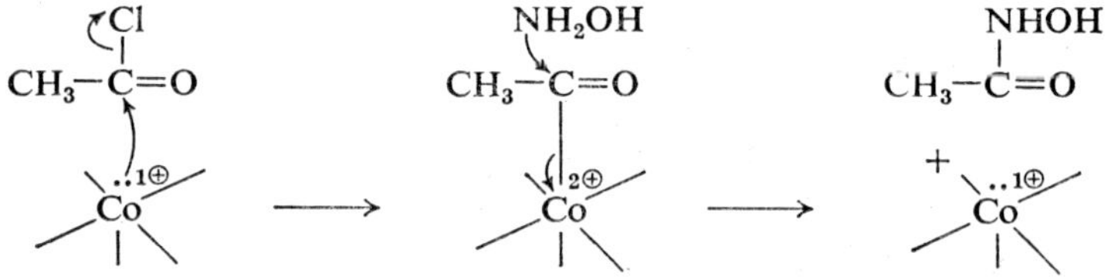

These acyl-cobalamins, like other acyl derivatives, are themselves susceptible to nucleophilic reagents such as hydroxylamine. Thus, the reaction of acetyl-cobalamin with hydroxylamine yields acethydroxamic acid and regenerates the reduced cobalt species.

† -p-toluenesulphonyl: a very good leaving group in nucleophilic substitutions.

Therefore the cobalt system functions reversibly; it can act both as a nucleophile and as a leaving group. On this basis we can see that it should function as an effective agent for group transfer reactions.

Another interesting and important aspect of the chemistry of alkyl cobalamins is the effect of light. These derivatives are stable only in the dark and strong light brings about a homolytic bond cleavage (equation 6.2). As a result an alkyl radical is produced and the cobalamin is converted into a form ($B_{12R}$) which has an oxidation level one electron higher than that of the fully-reduced state $B_{12S}$. This form $B_{12R}$ does not function

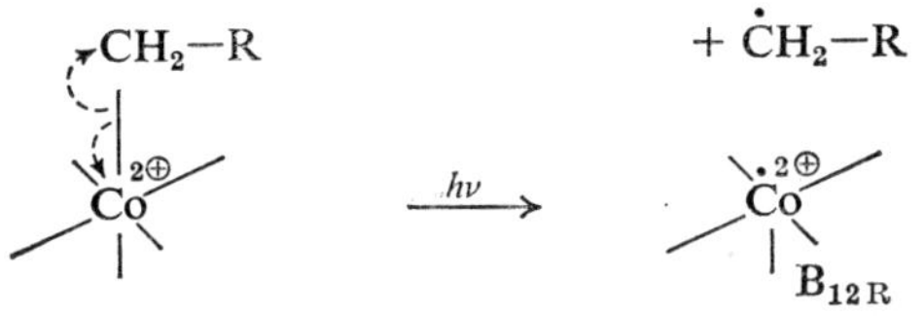


(6.2)

as a nucleophile. In the presence of oxygen it may be converted into aquocobalamin, while certain reducing agents can convert it into the $B_{12S}$ form [2, 6, 7, 8]. The fate of the alkyl radical depends on the nature of the group R (for example, see reference [8].)

The last stage in the synthesis of methionine in many living systems is the transfer of a methyl group from $N^5$-methyltetrahydrofolate (III) to homocysteine. This reaction is catalysed by the cobamide-containing enzyme $N^5$-methyl-THF: homocysteine methyltransferase (2.1.1.–). The overall reaction appears to require the presence of small (catalytic) quantities of S-adenosylmethionine (SAM) [9, 6]. The observation that the

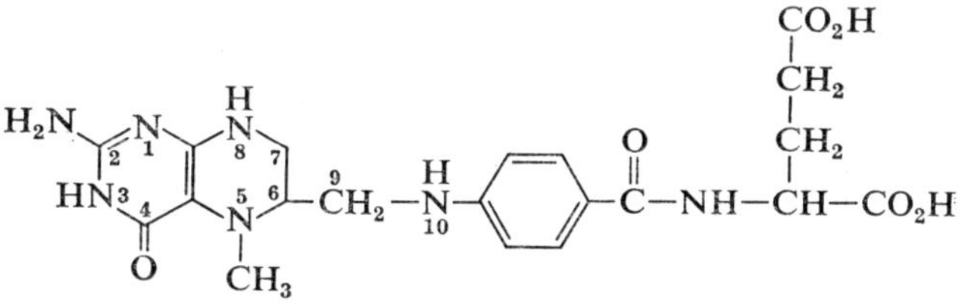


(III) $N^5$-methyltetrahydrofolic acid ($N^5$-methyl-THF)

enzyme would catalyse the transfer of the methyl group from methylcobalamin to homocysteine led to the suggestion that a methyl-cobalamin might be an intermediate in the reaction. This methyl transfer was not dependent on SAM [6, 9]. It was shown that free methyl-cobalamin was not an intermediate [10]. When $N^5$-methyl-THF labelled with $^{14}C$ in the methyl group was used as the methyl donor in the overall reaction, a labelled protein was isolated which contained at least 0·3 nmole $^{14}C$ per nmole of $B_{12}$ [11]. The labelled methyl group in this protein could be transferred stoichiometrically to homocysteine. The presence of SAM was not required for this second methyl transfer stage. It was concluded therefore that the reaction involved at least two stages; first transfer of the methyl group from the methyl donor, $N^5$-methyl-THF, to the cobalamin portion of the enzyme, and secondly transfer from the methyl-cobalamin-enzyme to the methyl acceptor, homocysteine. The first stage requires SAM as a co-factor while the second does not.

More evidence for the involvement of a methyl-cobalamin intermediate was obtained by propylation of the enzyme using propyl iodide [11]. This caused inactivation and the effect could be reversed by exposing the preparation to light (see also reference [7]). $N^5$-Methyl-THF in the presence of SAM could prevent the inhibition by propyl iodide presumably by ensuring that all the cobalamin is in the methyl form. The inhibition is therefore attributed to the formation of a propyl-cobalamin-enzyme species.

The role of the SAM in the reaction has been the subject of some speculation. In large quantities it can function as the methyl donor in the overall reaction. In this case the mechanism can be assumed to involve transfer of methyl from SAM to the cobalamin-enzyme by a nucleophilic attack as previously indicated, followed by nucleophilic attack by the sulphur of homocysteine on the methyl-cobalamin.

SAM + cobalamin-enzyme → S-Adenosylhomocysteine + methyl-cobalamin-enzyme

Homocysteine + methyl-cobalamin-enzyme → Methionine + cobalamin-enzyme.

If $N^5$-methyl-THF is used as the methyl donor then some SAM must be present. Weissbach *et al.* [9] state that the catalytic activity of the SAM is related to its ability to methylate the enzyme. Indeed a chemical methylating agent can replace the co-factor. It may be therefore that the SAM functions as an intermediate methyl acceptor donor between the $N^5$-THF and the cobalamin. It is possible that two specific methyl transfer reactions are brought about by the one enzyme. Alternatively the possibility may exist that two distinct but closely-related enzymes are involved.

The involvement of methyl-cobalamin has been implicated in enzymic reactions which result in the formation of methane from methanol by cell-free extracts of *Methanosarcina barkeri*. The system can produce methyl-cobalamin from aquo-cobalamin and methanol, and can also convert methyl-cobalamin into methane. Both these reactions are inhibited by propyl iodide, the inhibition being reversed by light. Stadtman and Blaylock therefore suggest that the conversion of methanol into methane by this organism proceeds through a methyl-cobalamin intermediate [12].

$$CH_3OH + \text{cobalamin} \rightarrow CH_3 - \text{cobalamin} \xrightarrow{\text{reduction}} CH_4 + \text{cobalamin}$$

## CARBOXYL TRANSFER—BIOTIN ENZYMES [16]

The carboxyltransferases (2.1.3.–) catalyse the transfer of a carboxyl group from a donor to a suitable acceptor. An essential component of methylmalonyl-CoA: pyruvate carboxyltransferase is vitamin H, d-biotin (IV). Another group of enzymes, the carbon dioxide ligases (carboxylases, 6.4.1.–), also possess biotin as a functional agent. In these two groups of enzymes the biotin appears to be involved in the reaction in the same way and for this reason we shall consider them together.

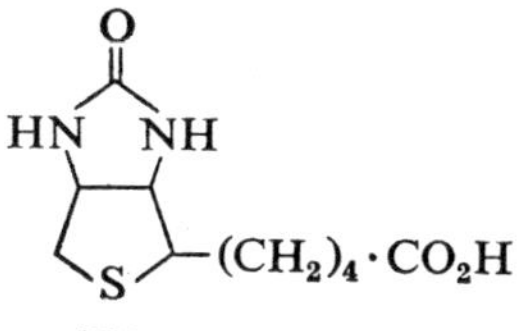


(IV)

Methylmalonyl-CoA carboxyltransferase (2.1.3.1) catalyses the transfer of $CO_2$ from the 2-position of methylmalonyl-CoA, which is a thioester (see later, page 241) to the $\beta$-carbon of pyruvate, thus giving oxalacetate (equation 6.3).

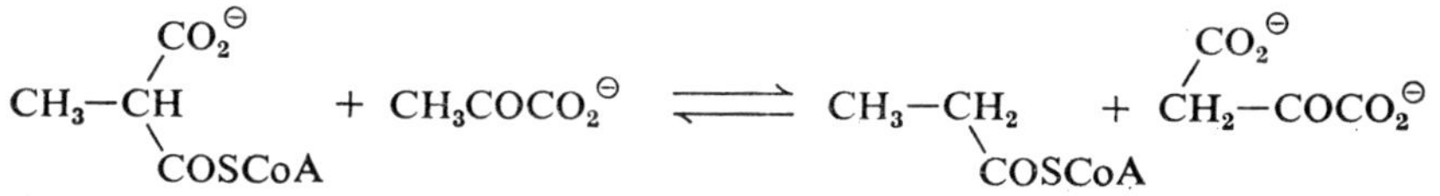


methylmalonyl-CoA pyruvate propionyl-CoA oxalacetate

(6.3)

The carboxylases, by contrast, are placed in the sixth category in the enzyme classification because they are true ligases in that a synthetic reaction is performed with simultaneous cleavage of ATP to ADP and inorganic phosphate ($P_i$). For example, propionyl-CoA carboxylase (6.4.1.3) catalyses the carboxylation of propionyl-CoA according to the equation (6.4). The carboxyl group in the product is derived from bicarbonate.

$$\text{ATP} + \text{HCO}_3^{\ominus} + \text{propionyl-CoA} \xrightleftharpoons{\text{Mg}^{2\oplus}} \text{ADP} + P + \text{methylmalonyl-CoA} \qquad (6.4)$$

The above equation represents the overall reaction, which is dependent on the presence of magnesium ions. The overall reaction (6.4) has been shown to proceed in two stages, as follows.

$$\text{Step 1: ATP} + \text{HCO}_3^{\ominus} + \text{enzyme} \xrightleftharpoons{\text{Mg}^{2\oplus}} \text{ADP} + P_i + \text{CO}_2\text{-enzyme} \qquad (6.5)$$

$$\text{Step 2: CO}_2\text{-enzyme} + \text{propionyl-CoA} \rightleftharpoons \text{methylmalonyl-CoA} + \text{enzyme} \qquad (6.6)$$

Thus, the first stage involves production of a carboxylated enzyme. It is this reaction which requires the presence of the magnesium ions. In the second stage the carboxyl group is transferred from the protein to the acceptor.

A great deal of evidence has been obtained in favour of this two-step mechanism for the reaction. This evidence includes various isotope-exchange experiments. For example, in the absence of propionyl-CoA, the first step (6.5) still takes place. Therefore, an exchange of labelled inorganic phosphate ($^{32}P_i$) with ATP may be demonstrated [13]. Similarly the enzyme catalyses exchange of [$^{32}$P] ADP or [$^{14}$C] ADP with ATP. These reactions are dependent on the presence of magnesium ions. In the absence of magnesium only step 2 (6.6) can take place, and it has been demonstrated that a $^{14}$C label in propionyl-CoA becomes incorporated into methylmalonyl-CoA [14, 24].

More direct evidence for the reaction scheme was obtained when Kaziro and Ochoa [15] obtained a radioactively-labelled enzyme on incubating propionyl-CoA carboxylase with ATP, $Mg^{2\oplus}$ and [$^{14}$C]$HCO_3^-$. The protein becomes labelled through the forward direction of step 1. In addition labelled protein could be obtained when the enzyme was incubated with methylmalonyl-CoA (labelled in the free carboxyl group) in the absence of magnesium ions (step 2, reverse direction) [16]. The incorporation of $CO_2$ into the enzyme corresponded to one molecule of $CO_2$ per residue of biotin in the protein. The labelled, bound $CO_2$ of these preparations could be transferred to a suitable acceptor by the normal process of the enzymic reaction. In acidic media the carboxyl-enzyme is unstable, decomposing to give the free enzyme and $^{14}$C-labelled carbon dioxide.

Degradative studies showed that the $CO_2$ becomes bound to the protein through the biotin residue [17, 18]. The biotin residue is itself covalently attached to the protein by means of an amide link between the carboxyl

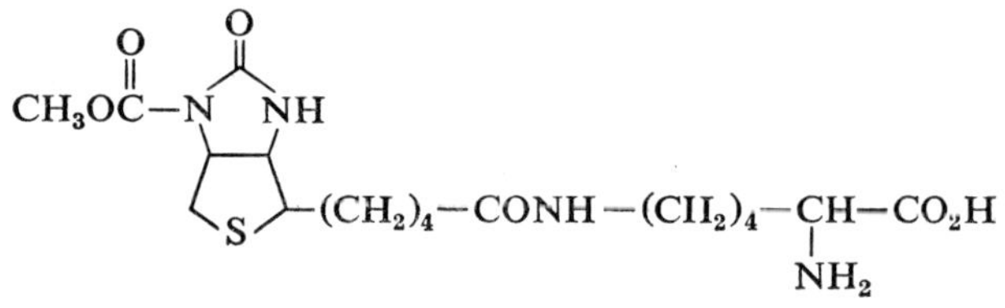


(V; 1′-N-Carbomethoxybiocytin)

group and the ε-amino-group of one lysine residue in the protein. The $^{14}CO_2$-biotin–enzyme complex was treated with diazomethane, which stabilizes the $CO_2$ bound. After digestion of the protein with pronase 1′-N-carbomethoxybiocytin (V) was identified, which possessed the radioactive label in the carbomethoxy group. 1′-N-Carbomethoxybiocytin could be converted into 1′-N-carbomethoxybiotin by the use of the enzyme biotin-amide amidohydrolase (biotinidase, 3.5.1.12), a hydrolytic enzyme which catalyses the cleavage of biotin amides.

Similar results were obtained with other carboxylases [16, 19] and with methylmalonyl-CoA carboxyltransferase [20]. In the latter case, the carboxyltransferase process therefore involves transfer of the carboxyl group from the donor to the enzyme and hence to the acceptor. Essentially, two steps similar to equation (6.6) take place.

Although the pathways of the overall reactions have been established by the methods described above, less is known about the mechanisms of the various stages. However, some relevant comments can be made. In all cases studied so far the molecule which accepts the carboxyl group from the $CO_2$-biotin–enzyme complex possesses a so-called active methylene group. This is a —$CH_2$-group adjacent to a carbonyl group derived from either an ester or a ketone. The presence of the carbonyl group increases the acidity of the methylene and hence in the presence of a base a proton may be removed and a resonance-stabilized carbanion formed. Such carbanions function effectively as nucleophiles.

$$-CH_2-\overset{\overset{O}{\|}}{C}- + \text{Base} \rightleftharpoons -\overset{\ominus}{C}H-\overset{\overset{O}{\|}}{C}- + H-\overset{\oplus}{\text{Base}}$$

The acceptance of a carboxyl group in the enzymic reaction is probably therefore a nucleophilic process.

The nitrogen atoms in biotin also have nucleophilic properties. For example, the reaction of biotin with methyl chloroformate gives 1′-N-carbomethoxybiotin (VI) as the major product [21]. Carboxylation of biotin in the enzyme is therefore almost certainly a reaction in which the biotin functions as a nucleophile [22].

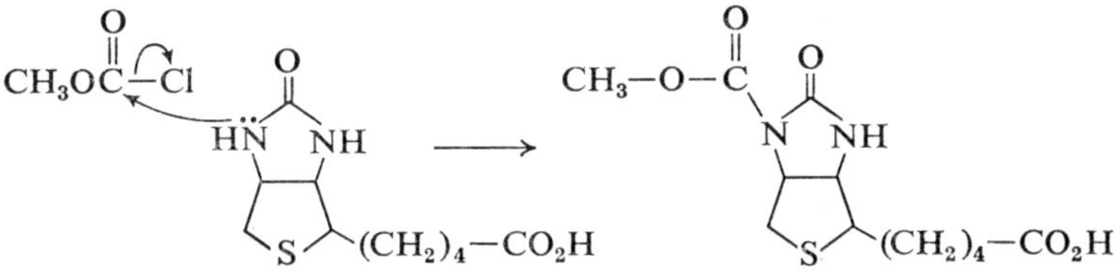


(VI)

In addition to the magnesium-ion-dependence of the first stage of the pyruvate-CoA carboxylase (6.4.1.1) reaction (equation 6.5), Utter and co-workers [23] have shown that the enzyme requires manganese ions. These manganese ions are tightly bound (non-dialysable), in contrast to the magnesium ions which are dialysable, and are required for the second step of the reaction; that is, the transfer of the carboxyl group from the biotin-enzyme to the final acceptor (pyruvate). They suggest that in this reaction the $Mn^{2\oplus}$ ion is bound to an oxygen of the N-carboxyl group, thus making the carboxyl carbon more susceptible to nucleophilic attack by the pyruvate species. Caplow [22] has suggested that in addition to the oxygen from the carboxyl group, the carbonyl group from the ureido-portion of the N-carboxybiotin is also bound to the metal-ion. The reaction is envisaged as passing through a tetrahedral intermediate, and by complexing with the ureido-carbonyl of the biotin species the metal-ion is considered to accelerate the breakdown of the tetrahedral structure.

This possible mechanism for the transfer of the carboxyl from the carboxybiotin intermediate is shown below. It can apply to the second stages of both the carboxylase and the transcarboxylase reactions. $X^{\ominus}$ represents in a general way the nucleophilic acceptor species which is produced from the active methylene compound, probably by base catalysis

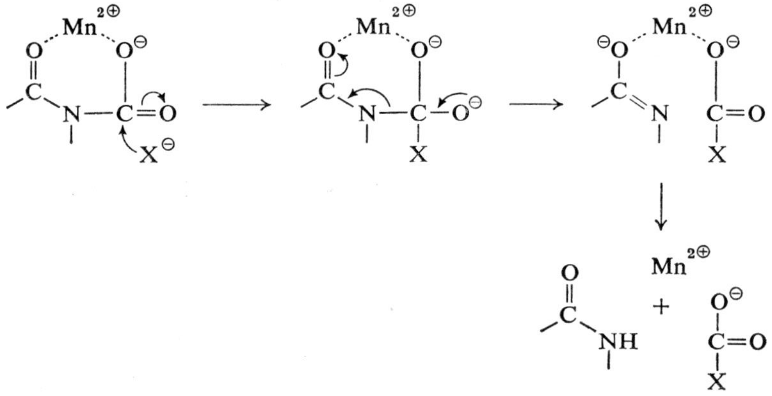

similar to that described in chapter 1. Presumably the enzyme itself provides the basic group which performs this function.

Kaziro *et al.* [24] have obtained some information about the first step in the carboxylase process, that is the ATP-dependent carboxylation of the biotin-enzyme, for propionyl-CoA carboxylase. The bicarbonate ion, $HCO_3^{\ominus}$ is the reactive species and one of the three oxygen atoms of this bicarbonate species appears in the liberated inorganic phosphate ion. This indicates that, at some stage, a nucleophilic attack by the oxygen of the bicarbonate on to the terminal phosphate group of the ATP takes place. The magnesium ions which are essential for this stage of the enzymic process could enhance such a nucleophilic attack by forming a chelate as shown below. Such a system has been suggested to account for the metal-ion catalysed hydrolysis of certain pyrophosphates and triphosphates [25].

In this case also the metal-ion plays a dual role. By complexing with the terminal phosphate group it renders that group more susceptible to nucleophilic attack. In addition its interaction with the penultimate phosphate residue will enhance the leaving-group tendencies of that residue.

Whether the carboxylation of the biotin residue in the system takes place in two distinct stages or as a concerted process is not known. Caplow [22] has discussed the possibility of a carbonyl phosphate intermediate; a two-stage process would be involved in such a case.

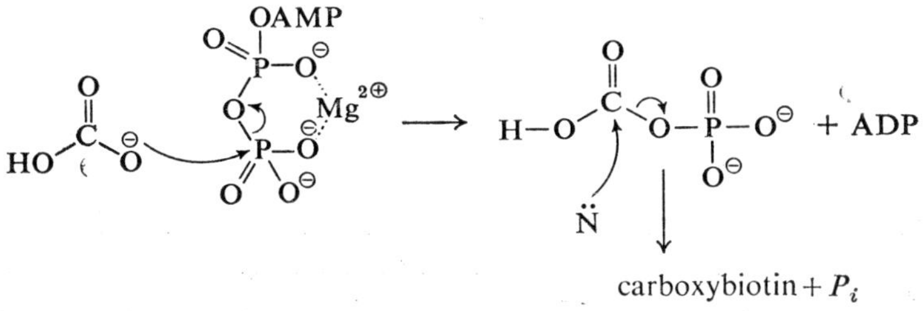

Carbonyl phosphate has not, however, been identified as an intermediate.

A single-stage, concerted process as proposed by Ochoa and Kaziro [16] would also account for the observed appearance of an oxygen atom from the bicarbonate in the phosphate ion.

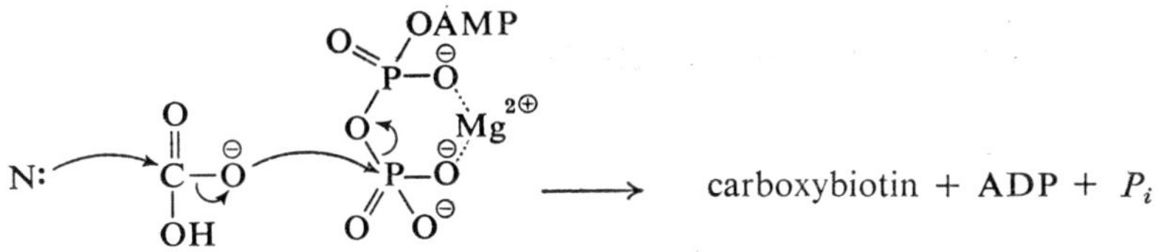

In the two possible mechanisms shown above the biotin nitrogen is represented simply as N:. It is very likely that the ability of the 1-N atom in the biotin to participate in this way is enhanced by general-base catalysis, with the enzyme providing an appropriate basic group.

## THE ACYLTRANSFERASES

It is clear that most of the enzymes which catalyse the hydrolysis of carboxylic acid derivatives are able also to catalyse acyl-transfer reactions, that is, molecules other than water may function as acceptor species (see chapter 4). However, these are classified as hydrolases rather than transferases mainly because it is believed that their hydrolytic properties are of primary importance and the group transfer reactions incidental. Those enzymes which are classified as acyltransferases catalyse acyl-transfer (almost invariably from an acyl-CoA derivative, see below) as their main function.

CoA is the abbreviation usually used for 'coenzyme A' which is a complex molecule whose most important feature is a thiol group (VII). This thiol group may be acylated so that a thiol ester is formed, the product being known as an 'acyl-CoA'.

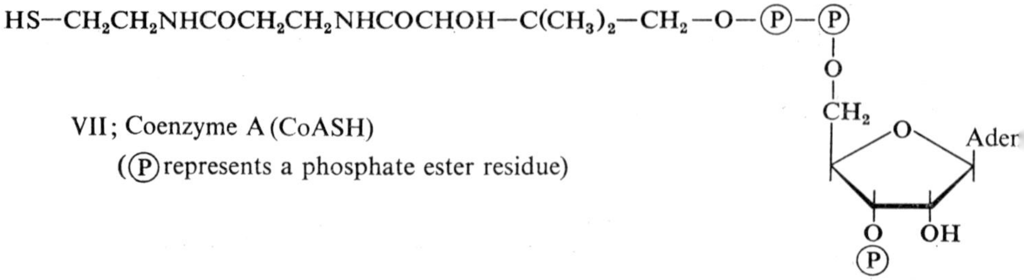


VII; Coenzyme A (CoASH)
(Ⓟ represents a phosphate ester residue)

Acyl-CoA derivatives are on the whole better donors of acyl groups than are oxy-esters, because thio-esters are more reactive. A thiol ester is regarded as an activated ester (indeed acetyl-CoA has often been referred to as 'activated acetate'), and acyl transfer from such an ester takes place by the normal acylation process (equation 6.7, where $N^{\ominus}$ represents some nucleophile).

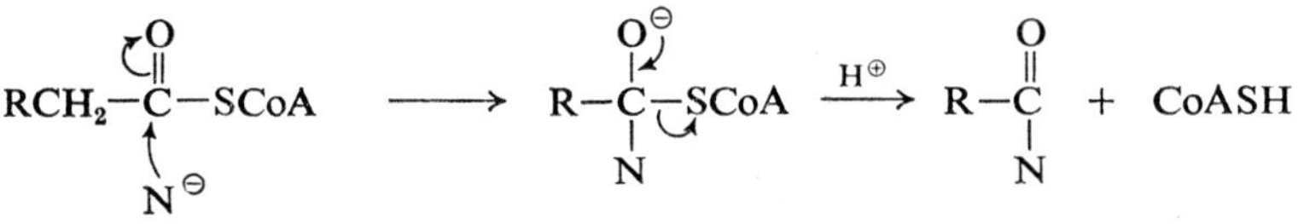


(6.7)

The enzymes in the group acyltransferases (2.3.1.–)may be specific for both the acyl donor and the acceptor. The reaction catalysed by acetyl-CoA: L-glutamate N-acetyltransferase (2.3.1.1) is the acetylation of L-glutamate. The enzyme also acts with L-aspartate and, more slowly, with some other amino-acids [26]. We may therefore assume that this enzyme possesses a binding site specific for acceptors which are amino-acids and especially for amino-acids which carry negatively-charged groups on their side-chains. Perhaps this site incorporates a protonated lysine or arginine side-chain to interact with the carboxylate group in the substrate. Such a situation has been shown to exist in the case of carboxypeptidase A (*q.v.*).

Other acyltransferases on the other hand show a breadth of specificity for the group which is transferred, that is, the acyl-group. Imidazole acetyltransferase (2.3.1.2), which catalyses the transfer of acetyl from acetyl-CoA to imidazole, also acts with propionyl-CoA. Similarly, glycine acyltransferase acts with the CoA derivatives of a range of aliphatic and aromatic acids.

It is interesting to note the specificity of these enzymes with respect to the coenzyme A portion of the acyl donors. Mechanistically, the coenzyme A functions because it is a thiol and this part of the molecule is clearly essential. However, not all thiols can replace coenzyme A. The minimum requirement in the structure of a CoA substitute appears to be an N-acylated thioethanolamine unit. Thus, N-acetyl cysteamine (VIII) is enzymically active, as are more complex molecules which possess more features of the coenzyme structure [27]. These substances, however, appear to be

$$\text{H S—CH}_2\text{CH}_2\text{—NHCO—CH}_3 \text{ (VIII; N-acetylcysteamine)}$$

bound less firmly to the enzymes than the intact CoA and therefore it is probable that the acyl-donor binding sites are complementary to most of the features of the coenzyme A structure.

It seems probable that all the enzyme-catalysed reactions in which an acyl-group is transferred from or to an acyl-CoA proceed via pathways which involve the formation of acyl-enzyme intermediates (equations 6.8 and 6.9). In the case of arylamine acetyltransferase (2.3.1.5), Bessman and Lipmann [28] showed that radioactively-labelled arylamine exchanged

$$\text{Acyl-CoA} + \text{enzyme} \rightleftharpoons \text{Acyl-enzyme} + \text{CoASH} \qquad (6.8)$$

$$\text{Acyl-enzyme} + \text{acceptor} \rightleftharpoons \text{Acyl acceptor} + \text{enzyme} \qquad (6.9)$$

rapidly with acetylarylamine in the presence of the enzyme (reaction 6.9 and its reverse; arylamine being the acceptor species).

An exchange reaction involving the acylation portion of the reaction mechanism (6.8) was demonstrated by Lynean [29,30] for the enzyme $\beta$-ketothiolase (acetyl-CoA acylthiolase, 2.3.1.16), which catalyses the transfer of acyl-groups such as propionyl. The formation of a propionyl-enzyme intermediate would mean that exchange between CoASH labelled with $^{35}$S and propionyl-CoA should be catalysed. This was shown to be the case.

Although little evidence is available concerning the nature of the covalent linkage between the enzymes and the acyl-groups, the observations reported [30,33] that many of these enzymes are inactivated by thiol-group reagents led to the suggestion that acyl-enzymes may be thio-esters formed from the side-chain thiol groups of cysteinyl residues.

With these factors in mind it is possible for us to postulate a mechanism for the action of one acyl-transfer enzyme, acetyl-CoA: acetyl-CoA C-acyltransferase (2.3.1.9) which catalyses the transformation shown in equation (6.10). In this reaction an acetyl group is transferred to the

$$\text{CH}_3\text{ CO—SCoA} + \text{CH}_3\text{CO—SCoA} \rightleftharpoons \text{CH}_3\text{COCH}_2\text{COSCoA} + \text{CoASH} \qquad (6.10)$$

2-carbon of another molecule of acetyl-CoA. The reaction is of interest because two types of chemical activation by the thiol ester function of an acyl-CoA are in evidence. In the acetyl donor the ester group is activated and is susceptible to nucleophilic attack. A thiol group in the enzyme acts as nucleophile and the acyl-enzyme is formed. In addition the methylene group of the acceptor molecule is activated (see page 239) and the reaction occurs to transfer the acyl-group from the acyl-enzyme. The postulated mechanism is shown in Fig. 6.2. In this mechanism it is suggested that a basic group $B_1$ enhances the nucleophilicity of the enzyme thiol group, and an acid $HA_1$ helps removal of the CoA leaving group. Similarly base-catalysis by $B_2$ may activate the methyl-carbon of the acetyl acceptor and the acid $HA_2$ may assist the leaving of the enzyme thiol group. At this

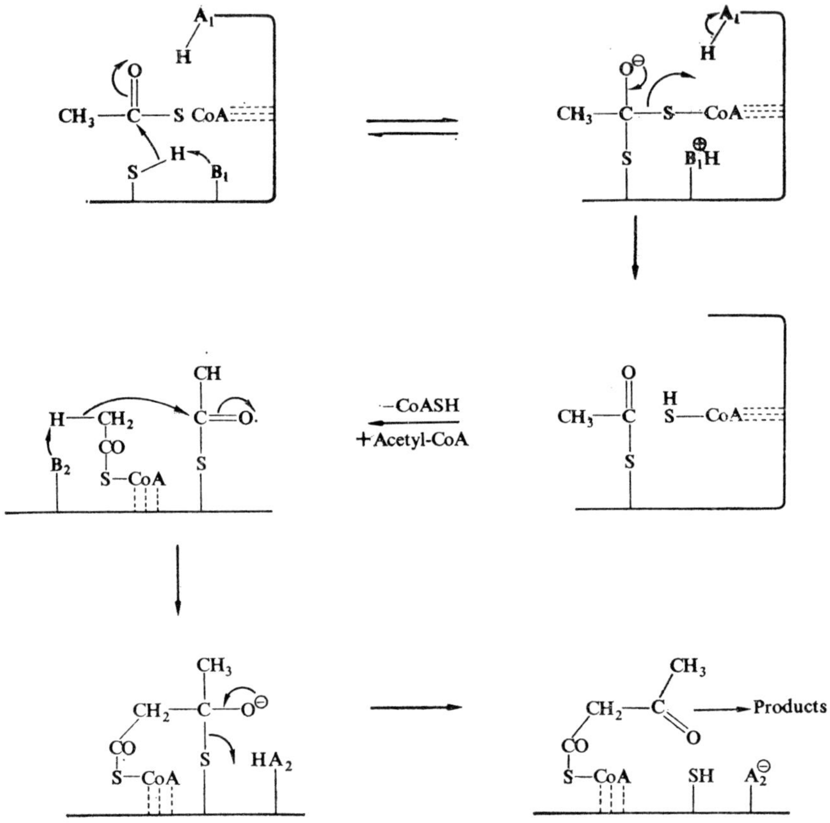


FIG. 6.2. Postulated mechanism of action for the enzyme acetyl-CoA: acetyl-CoA C-acetyltransferase (2.3.1.9)

stage the existence, identity and conformation of these groups $A_1$, $A_2$, $B_1$ and $B_2$ require confirmation.

Folk *et al.* [31, 32] have studied the aminoacyltransferase 'transglutaminase' (2.3.2.–) and have obtained strong and direct evidence for this type of mechanism. This enzyme catalyses the replacement by primary amines of amide groups or of peptides or proteins in bound glutamine residues. For example, the α-glutaminyl group is transferred from benzyloxycarbonylglutaminylglycine to hydroxylamine (equation 6.11). In addition, the enzyme catalyses the hydrolysis of esters such as p-nitrophenyl

$$\text{Z}-\overset{\displaystyle NH_2}{\overset{|}{\text{Glu}}}.\text{Gly} + NH_2OH \rightleftharpoons \text{Z}-\overset{\displaystyle NH_2}{\overset{|}{\text{Glu}}}-\text{NHOH} + \text{Gly} \qquad (6.11)$$

acetate and p-nitrophenyl trimethylacetate. Since it has been shown that the hydrolysis of these non-physiological substrates occurs at the same active site as does the normal transferase reaction, we are justified in using these compounds to obtain information about the active site and hence about the mechanism of the transfer process.

When p-nitrophenyl trimethylacetate, in which the trimethylacetate group was radioactively labelled, was used as the substrate, a labelled protein derivative was obtained. One nzymic hydrolysis a labelled peptide could be isolated and Folk and his co-workers were able to show that in this peptide the trimethylacetyl group was carried as a thio-ester on the side-chain of a cysteinyl residue (IX). This is the same cysteine residue which reacts with iodoacetamide to give an inactive protein.

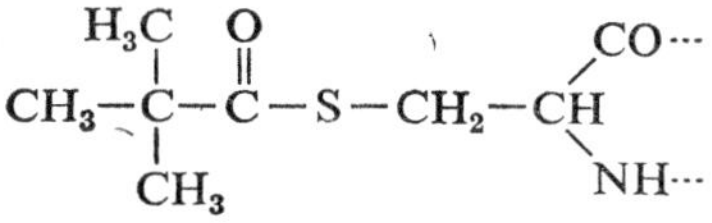

(IX)

We notice that here the situation appears to be very similar to that which obtains in the case of papain and the other active-cysteine hydrolases (see chapter 4).

## GLYCOSYL TRANSFER

Equation 6.12 is a representation of the general reaction catalysed by the glycosyl-transferase enzymes (2.4.–.–) [33, 34]. The donor is a glycoside in which the group —OX is attached to the anomeric carbon of a carbohydrate residue. This carbohydrate residue is the glycosyl group which is transferred to the acceptor.

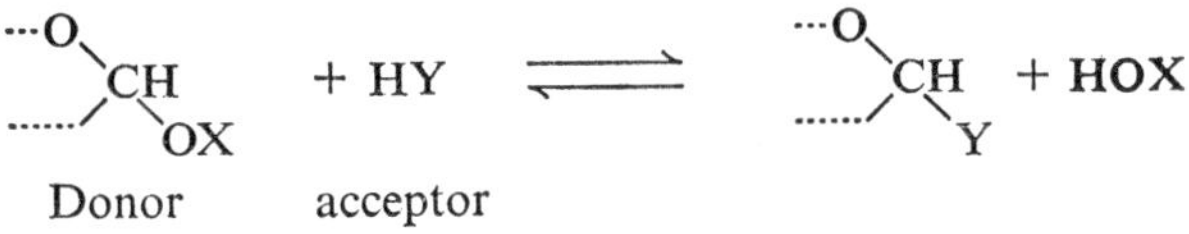


(6.12)

The specificities of these enzymes relate especially to the nature of the glycosyl group which is transferred, although the enzymes vary according to the nature of species X and Y whose reactions they catalyse. For example, X and Y may be phosphate, pyrophosphate, nucleotide diphosphate, or sugar molecules. The most intensely-studied groups are those in which $XO^{\ominus}$ is phosphate and HY is a sugar unit, and in such a case the reverse direction of equation (6.12) is known as phosphorolysis, the enzymes being named phosphorylases. Phosphorolysis bears a formal relationship to hydrolysis, in that a cleavage reaction occurs with the incorporation of phosphate into one species, in contrast to hydrolysis where a water molecule becomes incorporated.

With reference to the mechanism of the glycosyl-transferase reactions, an important point is that in each case studied bond cleavage occurs between the anomeric carbon atom and the oxygen atom of the —OX group [35, 36]. This has been taken to imply that a displacement reaction occurs at the anomeric carbon and that as a result $^{\ominus}OX$ departs as a leaving group. An $S_N2$-type displacement reaction always gives rise to inversion of configuration and this leads us to another important property of the enzyme-catalysed reactions—the type of linkage at the anomeric carbon atom and the possibility of retention or inversion of configuration at this carbon in the product. This point has been discussed by Koshland [37], who suggested that where inversion of configuration is observed on conversion of substrate into product a single nucleophilic displacement process is probably involved, and that retention of configuration implies two successive nucleophilic substitutions. (Cohn [34] has pointed out that retention of configuration could result from an abnormal, front-side nucleophilic attack.) On this basis two possible mechanisms for glycosyl-transferase enzymes have been considered. In the first an intermediate glycosyl-enzyme derivative is postulated, and the group transfer therefore proceeds in two steps. If these two steps are $S_N2$-reactions then the product will possess the same configuration as the starting material.

The second possible mechanism consists of only one step, and therefore should result in inversion of configuration. Experiment has shown that while some enzymic reactions give products where the original configurations are retained, others lead to inversion, and therefore it is clear that at least two different mechanisms may be utilized by the glycosyl-transfer enzymes.

### Sucrose Glucosyltransferase (Sucrose Phosphorylase, 2.4.1.7)

This enzyme catalyses the phosphorolysis of sucrose, giving glucose-1-phosphate and fructose (6.13). Although the enzyme is specific for the

$$\underset{\text{sucrose}}{\alpha\text{-glucose-1,2-fructose}} + P_i \rightleftharpoons \alpha\text{-glucose-1-phosphate} + \text{fructose} \qquad (6.13)$$

group transferred, that is, for the D-glucosyl group, it is relatively non-specific as far as the other species are concerned. Thus, in addition to fructose or phosphate, certain other groups can function as acceptors. These include sorbose, xylulose, arabinose, arsenate and water [38].

Sucrose glucosyltransferase functions with retention of configuration [35] and, in addition to this evidence, other information has been obtained in support of a two-stage mechanism involving a glucosyl-enzyme intermediate (eq. 6.14).

$$\begin{array}{c}\text{Sucrose} + \text{enzyme} \rightleftharpoons \text{glucosyl-enzyme} + \text{fructose} \\ \text{Glucosyl-enzyme} + P_i \rightleftharpoons \text{glucose-1-phosphate} + \text{enzyme}\end{array} \quad (6.14)$$

This information includes the results of isotope exchange studies. For example, in the absence of fructose the enzyme catalyses the incorporation of labelled phosphate into glucose-1-phosphate [40]:

$$\alpha\text{-glucose-1-phosphate} + {}^{32}P_i \rightleftharpoons \alpha\text{-glucose-1-phosphate } ({}^{32}P) + P_i$$

In addition, in the absence of phosphate the exchange of labelled D-fructose with sucrose is catalysed [41]. This latter process involves the first reversible step in the mechanism (6.14).

It has already been mentioned that arsenate can replace phosphate as a glycosyl acceptor. In fact glycosyl-arsenates are extremely unstable, being hydrolysed spontaneously in water. Sucrose phosphorylase catalyses the following two reactions involving arsenate, the first in the absence of

$$\alpha\text{-glucose-1-phosphate} \xrightarrow[+H_2O]{\text{Arsenate}} \text{glucose} + P_i$$

$$\text{sucrose} \xrightarrow[+H_2O]{\text{Arsenate}} \text{glucose} + \text{fructose}$$

fructose and the second in the absence of phosphate. These two reactions support the glucosyl-enzyme concept in that they each represent one of the steps of mechanism (6.14) (either in the forward or reverse direction) in leading to the glucosyl-enzyme intermediate. From this intermediate the glucosyl group is then transferred to arsenate and the resulting glucose-1-arsenate decomposes.

Very direct evidence for the involvement of a glycosyl-enzyme complex has been obtained recently by Silverstein *et al.* [39], who examined two aspects of the reaction. In the first place they studied the reaction kinetically. The two possible mechanisms, which may be stated as follows, give two different sets of rate equations [42] and by studying the rates of the reactions under conditions of varying concentrations of fructose, phosphate, sucrose, etc., these workers were able to show that the reaction corresponded most closely with mechanism (a), the glycosyl-enzyme mechanism (see chapter 2).

(a)

$$\begin{array}{l}G\text{—}X + E \rightleftharpoons GXE \\ GXE \rightleftharpoons G\text{—}E + X \\ GE + Y \rightleftharpoons GEY \\ GEY \rightleftharpoons GY + E\end{array}$$

(b)
$$\begin{aligned} GX + E &\rightleftharpoons GXE \\ GXE + Y &\rightleftharpoons GYE + X \\ GYE &\rightleftharpoons GY + E \end{aligned}$$

(G—X is the glycosyl-donor, GXE is the enzyme–substrate complex, G—E is the glycosyl-enzyme, Y is the glycosyl acceptor and GY and X are the products of the reaction.)

Undoubtedly the best evidence for a covalently-bound enzyme-derived intermediate is the isolation and identification of that intermediate. In the case of sucrose phosphorylase this was not an easy task. This is because water itself can function as a glycosyl acceptor and therefore the glycosyl-enzyme is unstable, being rapidly hydrolysed when attempts are made to isolate it. Silverstein *et al.*, however, went some way towards this objective by isolating the *denatured* enzyme from a reaction mixture containing sucrose, radioactively labelled in the glucose portion. The denatured glycosyl-enzyme would not be expected to be unstable. In fact the denatured protein was radioactive, containing one glucosyl residue per 80,000 molecular weight (in good agreement with the molecular weight estimated by sedimentation studies). The radioactivity was released from the protein by dilute alkali and the fragment liberated confirmed as glucose.

When sucrose labelled in the fructose portion was used the protein obtained carried no activity. Although ideally, it should be shown that the intermediate complex is able to undergo the sequence of reactions required by the mechanism, this isolation of an inactive glycosyl-enzyme compound leaves little doubt that the mechanism (a) is the correct one.

The pH-rate profile for the phosphorolysis shows a maximum at about 7, and the curve suggests that a group with $pK$ 5–6 is required in the basic form. It may be this group in this protein which is the original nucleophile bringing about the first displacement. This could be the side-chain carboxyl of a glutamic or aspartic acid residue, in which case the glycosyl-enzyme would be an ester. Alkaline hydrolysis of a denatured complex involving an ester would liberate the glucose easily. The pH-rate profile also suggests the necessity of the acidic form of a group with $pK$ about 8. This may function as an acidic catalyst assisting the departure of the leaving group. The reaction scheme (6.14) may therefore be expanded into a possible mechanism as shown in Fig. 6.3.

In certain other cases where reaction takes place with retention of configuration it is probable that a similar mechanism applies. Sucrose 6-fructosyl transferase (levansucrase, 2.4.1.10) catalyses the transfer of fructosyl residues from fructose oligosaccharides to various suitable acceptors. It catalyses exchange of labelled glucose into sucrose and various other reactions which lead to the implication of a fructosyl-enzyme intermediate for the enzymic reaction [43].

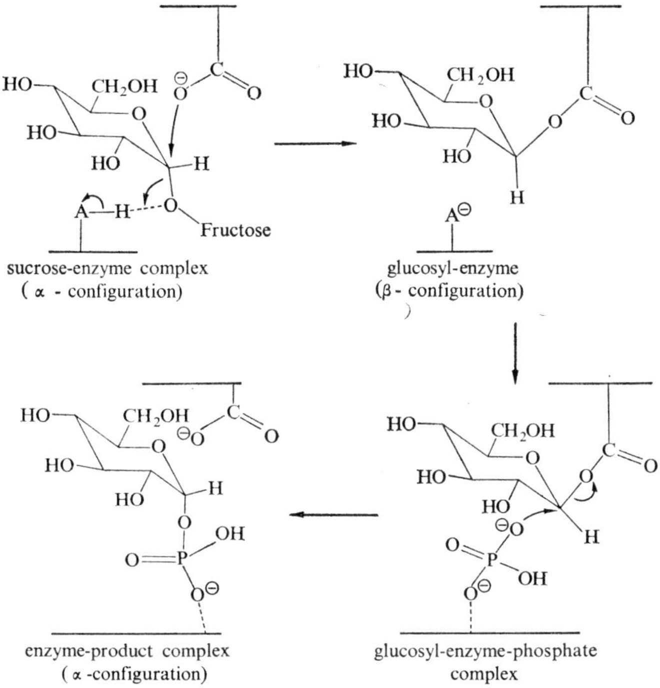


FIG. 6.3. Possible mechanism for sucrose phosphorylase

**Maltose Phosphorylase**

When this enzyme (2.4.1.8) catalyses the phosphorolysis of maltose, giving D-glucose-1-phosphate and glucose, inversion occurs at the anomeric carbon. Thus β-D-glucose-1-phosphate is the product (compare sucrose phosphorylase which gives the α-isomer although both enzymes act on disaccharides in which glucose is α-linked). Therefore Koshland's theory

$$\underset{\text{maltose}}{\alpha\text{-1,4-D-glucosyl-D-glucose}} + P_i \longrightarrow \beta\text{-D-glucose-1-phosphate} + \text{D-glucose}$$

[37] would suggest that a single displacement reaction occurs. If this is the case then a glucosyl-enzyme cannot be formed, and therefore none of the exchange reactions observed for sucrose phosphorylase, for example, would be detected. For maltose phosphorylase [44] the absence of such exchange reactions, and of arsenolysis, is consistent with a one-stage mechanism which may be represented as shown below:

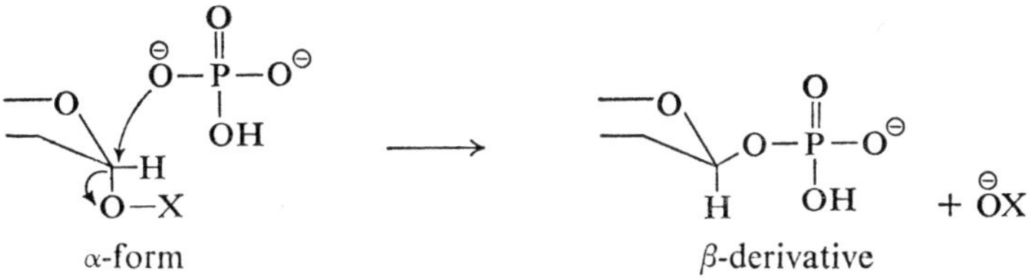

Other enzymes which give products with inverted configurations such as cellobiose phosphorylase (2.4.1.20) and purine nucleoside phosphorylase (2.4.2.1) [33, 34] presumably belong to the same category.

### α-1,4-Glucan: Orthophosphate Glucosyltransferase (2.4.1.1)

This enzyme, which for many years has been known as glycogen phosphorylase or simply as phosphorylase *a*, has been the subject of intensive study, but because of certain complicating factors little definite information has been obtained about its mechanism.

The reaction catalysed by phosphorylase involves the degradation of polysaccharides such as glycogen to monosaccharide units (equation 6.15), a process which is extremely important for the mobilization of carbohydrates in living systems.

$$\alpha\text{-1,4-Glycosyl-}(\alpha\text{-1,4-glucosyl})_n + P_i \rightarrow \alpha\text{-glucose-1-phosphate} + (\alpha\text{-1,4-glucosyl})_n \quad (6.15)$$

Phosphorylase *a* has been shown [45] to consist of four identical subunits, each unit possessing a serine residue whose hydroxyl group is involved in an ester linkage with phosphate. An interesting feature of this enzyme is that it contains pyridoxal phosphate (X), which is essential for

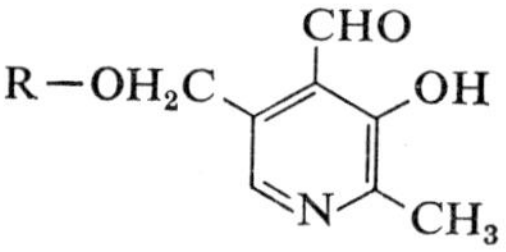


(X; (a) R = H, pyridoxal
(b) R = $PO_3\overset{\ominus}{H}$, pyridoxal phosphate)

enzymic activity [46]. No role has yet been definitely assigned to this factor. Various studies have shown that carbohydrate substrates do not become covalently bound to the pyridoxal phosphate [47], neither does inorganic phosphate exchange with its phosphate residue [48]. It is therefore unlikely that it plays a direct role in the catalytic reaction. This is in marked contrast to the situation which exists for other pyridoxal phosphate enzymes, such as the aminotransferases (see later in this chapter) where direct interactions occur. Glaser has suggested [33] that it may fulfil a function involving aggregation and disaggregation of the protein molecule.

The enzyme phosphorylase phosphatase (3.1.3.17) brings about the hydrolysis of the serine phospho-ester linkages in phosphorylase and converting it into a form known as phosphorylase *b*. This form is inactive, except in the presence of adenosine monophosphate (AMP), which functions as an allosteric activator to it (see chapter 2) [49]. ADP and ATP are, however, inhibitory [50].

Although adenosine monophosphate is not essential for the activity of phosphorylase *a*, it does cause an enhancement of catalytic activity. Recent studies have shown that both phosphorylase *a* and phosphorylase *b* give very complex kinetic patterns. Some of the complexities may be due to association and dissociation involving dimeric and tetrameric forms which have different activities and to various salt effects which also appear to influence the overall enzyme structures [51, 52].

As equation (6.15) indicates, phosphorylation catalysed by this enzyme leads to retention of configuration, a phenomenon which, for other enzymes, has suggested a two-step mechanism involving a covalent intermediate. However, in this case, the enzyme does not catalyse exchange between glucose-1-phosphate and labelled phosphate [53]. This negative result in an important experimental test of the two-step mechanism probably means either that there is no glucosyl-enzyme intermediate or that the first step cannot proceed unless the phosphate (the reactant in the second step) is present. Evidence has been obtained which tends to eliminate the second of these alternatives [34, 54].

Therefore a one-step mechanism involving a retention of configuration, i.e. a 'front-side' nucleophilic attack, may take place [34].

## PYRIDOXAL ENZYMES

### Animo Group Transfer

Transamination reactions are normally considered to include only reactions which involve transfer of an amino-group from one molecule to another without the production of free ammonia as an intermediate. The most common enzymic transaminations involve amino-acids and keto-acids, and equation (6.16) shows the general reaction.

$$\mathrm{R.CHNH_2.CO_2H + R^1.CO.CO_2H \rightleftharpoons R.CO.CO_2H + R^1.CHNH_2.CO_2H} \qquad (6.16)$$

Direct transaminations, uncatalysed either by enzymes or by any other species, were observed by Herbst and Engell [55] and Nakada and Weinhouse [56] and the mechanism of this direct process was formulated as

shown below. The first step in the reaction between an amino-acid (XI) with glyoxylic acid (XII) is the formation of a Schiff's base (XIII). A rearrangement then follows involving a prototropic shift and a new Schiff's base is produced (XIV) which on hydrolysis yields a new keto-acid and glycine.

In many other cases, where keto-acids other than glyoxylic acid were used, decarboxylation might also be observed.

Enzymatic transaminations were shown by Braunstein [57], Cohen [58] and Snell [59] to require pyridoxal phosphate (Xb) or pyridoxal (Xa) as a coenzyme, and it was therefore of great significance when Snell, Metzler and their co-workers [59] reported that pyridoxal phosphate, alone or in the presence of metal-ions, could catalyse transaminations. The reaction between an amino-acid and pyridoxal phosphate leads to the formation of the corresponding keto-acid and pyridoxamine phosphate (XV). The

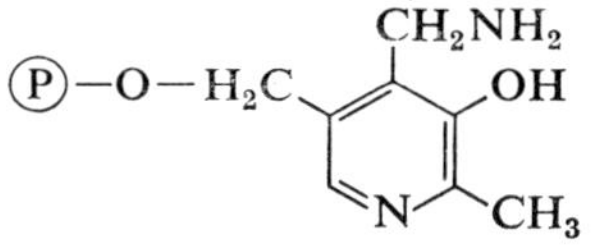


(XV)

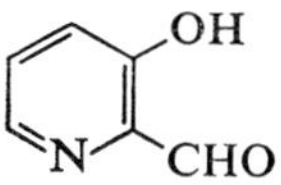


(XVIII)

mechanism of the transamination reaction brought about by pyridoxal and metal-ion was suggested [60] to be essentially similar to that suggested for the direct transaminations, that is, proceeding via a Schiff's base and involving a prototropic shift (Fig. 6.4). The highly electron-withdrawing nature of the pyridine ring, especially when protonated, in conjugation with the imine double bond in XVI, renders the $\alpha$-hydrogen of the amino-acid residue very labile and the prototropic shift can occur readily. The metal-ion is viewed as stabilizing the imine (XVI) and ensuring a planar conjugated system [61, 62].

Since the reactions in the scheme are reversible the pyridoxamine formed can react with a different keto-acid to give a new amino-acid, regenerating the pyridoxal; in this way a transamination between an amino-acid and a keto-acid is brought about.

On the basis of this type of mechanism, it is possible to explain the significance of the various structural features of the pyridoxal molecule with respect to its ability to catalyse the transamination process in the presence of metal-ions. Snell and his co-workers [59, 63] investigated different structural analogues of pyridoxal, and showed that 3-O-methyl pyridoxal was not effective. Thus, a free hydroxyl group is necessary *ortho*- to the formyl group, which is itself, of course, essential. 2-Formyl-3-hydroxy-pyridine (XVIII) was active as a catalyst, its structure being electronically very similar to that of pyridoxal. The methyl and the hydroxymethyl groups are not important, at least as far as the non-enzymic reactions are concerned. In certain other reactions catalysed by

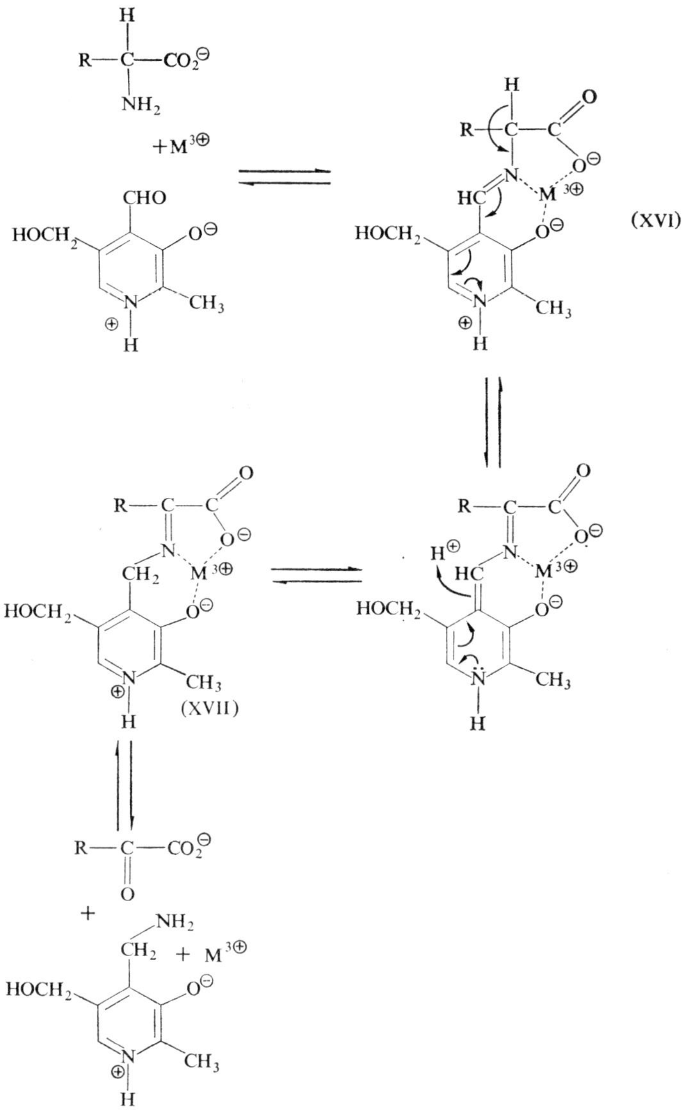


FIG. 6.4. Transamination between an amino-acid and pyridoxal

pyridoxal, which involve a similar mechanism, the pyridine ring may be replaced by a benzene ring carrying a nitro-group. This emphasizes the role of the heterocyclic ring as an electron-attracting species (see page 254).

Vernon and his co-workers [64] have confirmed this mechanism for a pyridoxal-catalysed transamination by a kinetic study, showing that the process involves the rapid production of a Schiff's base, followed by a rate-determining tautomerism (a prototropic shift) between XVI and

XVII. Because the interconversion of the two imines is the rate-determining step, its mechanism is clearly of great importance. With this point in

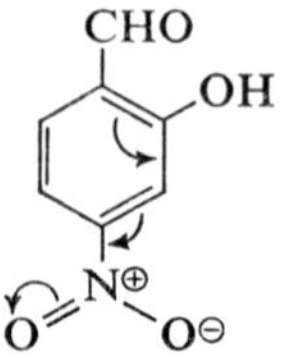

mind, Bruice and Topping [65] examined the effect of certain acids and bases on the metal-independent transamination of pyridoxal and α-aminophenylacetic acid. They found that imidazole is an extremely effective catalyst of the reaction, and that both the protonated and unprotonated imidazole species are involved in catalysing the imine interconversion. It was therefore suggested that the process could best be expressed by a path in which a complex between imine, imidazole and imidazolium ion is formed, followed by a prototropic shift in which the imidazole and the imidazolium ion function as basic and acidic catalysts respectively, by a concerted mechanism (Fig. 6.5). In that it is a concerted process, this

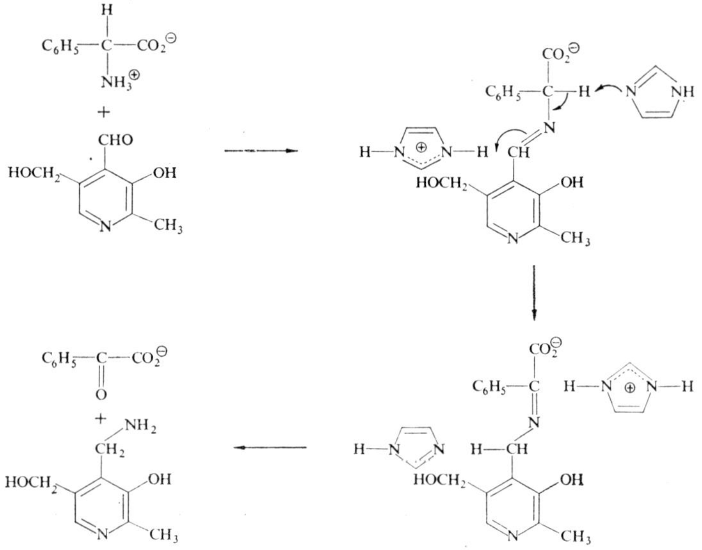

FIG. 6.5. Imidazole-catalysis of transamination (after Bruice and Topping, [65])

mechanism bears close resemblance to that suggested for the catalysis of the mutarotation of tetramethylglucose by α-hydroxypyridine (see chapter 1, page 37).

If the tautomerism step in the non-enzymic reaction of pyridoxal is rate-determining, it therefore follows that the first stage, the production of imine from the pyridoxal and the amino-acid, is rapid. We have already seen (chapter 1, page 47) that the formation of imines or similar carbonyl derivatives is catalysed by primary amines (that is, that transimination takes place more rapidly than does imine formation from carbonyl compounds). Jencks and Cordes [66] have shown that this is also the case for the production of imines from pyridoxal and amino-acids, and the point may be of importance in the enzyme-catalysed transaminations.

Having discussed very briefly the non-enzymic transaminations which may be brought about by pyridoxal and its phosphate, we may now consider some of the properties of a representative enzyme, aspartate aminotransferase (L-aspartate: 2-oxoglutarate aminotransferase, 2.6.1.1). Like all of the aminotransferases, this requires pyridoxal phosphate as a coenzyme (some enzymes require pyridoxal itself). It catalyses the reaction shown in equation (6.17).

$$\text{L-aspartate} + \text{2-oxoglutarate} \rightleftharpoons \text{oxaloacetate} + \text{L-glutamate} \qquad (6.17)$$

All of the kinetic and chemical data available suggest that the mechanism of action of this enzyme is essentially similar to the non-enzymic pyridoxal-catalysed transaminations [67, 68, 69, 70, 71]. Thus, an imine (XVIII) between enzyme-bound pyridoxal and the aspartate is formed and this tautomerizes to give a new imine (XX) which breaks down to yield oxaloacetate and enzyme-bound pyridoxamine. The reverse process takes place, involving amination of 2-oxoglutarate to L-glutamate.

In addition to the coenzyme which is involved directly in the reaction, certain amino-acid side-chains have been identified as being in the vicinity of the active site of the enzyme. The coenzyme does not exist as the aldehyde in the enzyme–coenzyme complex but rather as a Schiff's base involving the ε-amino-group of a lysyl residue. This has been shown by spectroscopic studies [72] and by experiments in which the holo-enzyme has been reduced with sodium borohydride [73, 74], which reduces imines to stable secondary amines. From the reduced holo-enzyme it has been possible to isolate peptides containing ε-pyridoxyl-lysyl residues. Thus, the first step in the enzymic reaction must involve transimination (see Fig. 6.6). Anderson and Song [75] report that the borohydride reduction is extremely rapid, suggesting that the Schiff's base is activated either by protonation or by hydrogen bonding. Such a situation will naturally favour the transamination process too.

Other studies have indicated that a second lysyl residue is close to the active site [76, 77] and Turano *et al.* [78] have suggested the close proximity of a histidyl residue. It was possible to link one of the lysine residues with a histidine by means of the bifunctional 3-bromopropionyl chloride [78]. The identification of a histidyl residue at the active site is important, since catalytic activity is lost when a histidine residue in the enzyme is destroyed by photo-oxidation [79]. It is possible that this histidine residue functions as an acidic or a basic catalyst in the tautomerism process. For many pyridoxal-dependent enzymes, however, ultra-violet absorption at about 500 nm during the reaction has led to the conclusion that the quinonoid form (XIX) of the coenzyme is involved (see chapter 7). If this is the case then the role of the histidine may not be precisely that proposed by Bruice and Topping [65] for the non-enzymic process.

Figure 6.6 shows a possible mechanism for the first half of the aspartate aminotransferase reaction. Since the enzyme is stereospecific, it may be

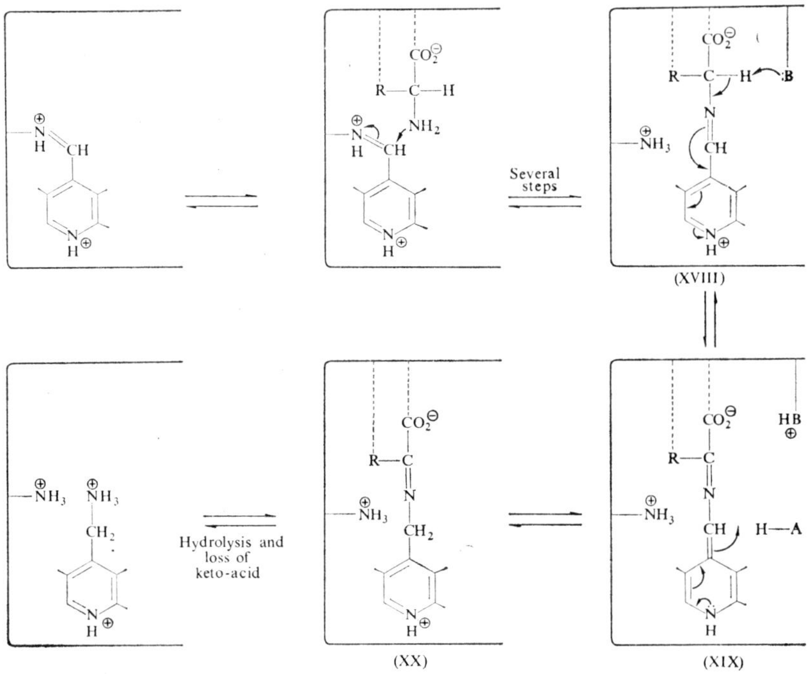


FIG. 6.6. Possible mechanism for the aspartate aminotransferase reaction. B and H—A are unspecified basic and acidic groups, respectively

assumed that the amino-acid is bound to the enzyme by both the carboxyl- and the R-group. The binding between the pyridoxamine and the enzyme is very strong ($K_m = 5{\cdot}3 \times 10^{-7}$M for the enzyme from a plant source [80]). Very little is known about the mechanism of the hydrolysis of the Schiff's base (XX) which leads to the liberation of the keto-acid.

**Serine Hydroxymethyltransferase (2.1.2.1)**

At this stage it is appropriate for us to discuss another transferase which requires pyridoxal phosphate for its activity. This enzyme, which plays an important role in the metabolism of serine, catalyses the transfer of a hydroxymethyl group between glycine and tetrahydrofolate (equation 6.18).

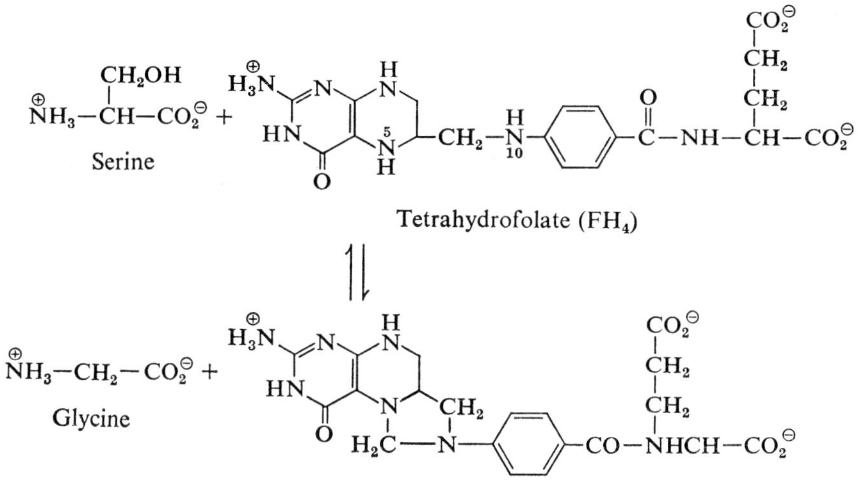


$N^5$, $N^{10}$-methylene-tetrahydrofolate
($N^5$, $N^{10}$-methylene $FH_4$) (6.18)

In $N^5$,$N^{10}$-methylene-$FH_4$, the one-carbon fragment added is at the oxidation level of formaldehyde. A similar reaction has been demonstrated in a non-enzymic system. In the presence of pyridoxal and metal-ions, serine undergoes decomposition to glycine and formaldehyde [81], and the suggested mechanism for the reaction involves an intermediate Schiff's base (Fig. 6.7) between the amino-acid and the aldehyde. This imine

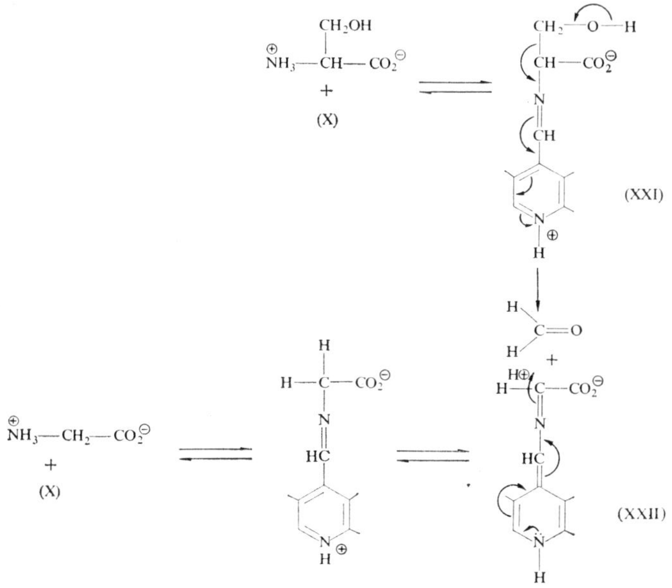


FIG. 6.7. Pyridoxal-catalysed decomposition of serine into glycine and formaldehyde

(XXI) breaks down by means of the electronic changes shown, again the strong electron-withdrawing powers of the heterocyclic ring being the driving force for the reaction.

In the other direction, that is, the direction of serine synthesis, the imine formed from glycine and pyridoxal is envisaged as undergoing the

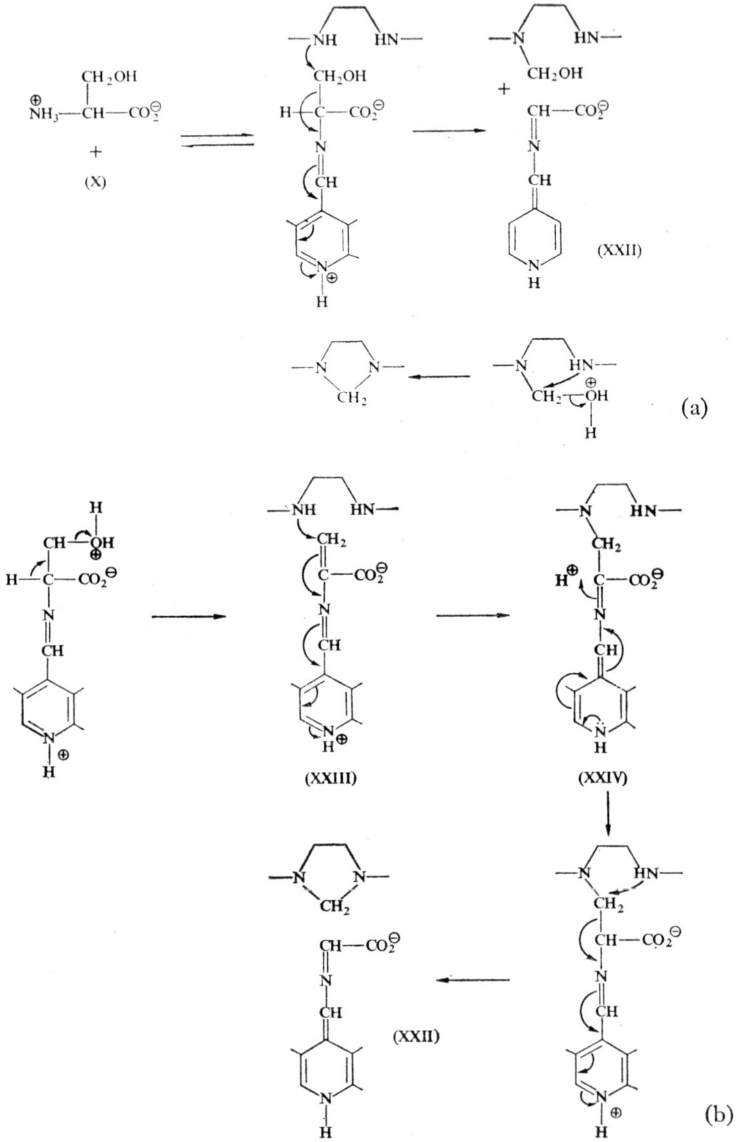


FIG. 6.8. Alternative mechanisms for the transfer of a methylene group to THF. Only the $N^5$- and $N^{10}$-positions of the THF are shown

prototropic shift giving XXII. Then follows a nucleophilic attack on the carbonyl carbon of the formaldehyde. Notice how similar this reaction is to the catalysis of aldol condensations by primary amines, discussed in chapter 1.

Figures 6.8a and 6.8b show two hypothetical pathways for the serine hydroxymethyltransferase reaction. They differ mainly in the stage at which the hydroxyl group is lost to produce the methylene group. In Figure 6.8a a direct nucleophilic attack on the hydroxymethyl group of the serine-pyridoxal Schiff's base gives $N^5$- (or $N^{10}$)-hydromethyl $FH_4$. Attack by the other nitrogen atom with loss of protonated hydroxyl would yield the $N^5$,$N^{10}$-methylene $FH_4$. The imine (XXII) would rearrange to give the glycine–pyridoxal complex and hence liberate free glycine.

Alternatively, the serine–pyridoxal Schiff's base may lose water. The electron-attracting conjugated system activates the proton on the serine $\alpha$-carbon and this, together with some acidic catalysis for the loss of the hydroxyl group, may lead to dehydration to the dehydroalanine system (XXIII). Nucleophilic attack by the $N^5$ (or $N^{10}$) nitrogen of tetrahydrofolate gives XXIV which, by taking up a proton in the manner shown, regenerates a system in which the methylene group is again susceptible to nucleophilic substitution, leading to $N^5$,$N^{10}$-methylene $FH_4$.

## TRANSFER OF PHOSPHORUS-CONTAINING GROUPS

The phosphate group plays a vital role in biological chemistry, in the form of phosphate esters, amides and anhydrides. These include simple phosphate esters such as sugar phosphates, nucleotides, amides such as creatine phosphate, pyrophosphates and mixed carboxylic-phosphate anhydrides. To a considerable extent the metabolism of these derivatives involves enzymes which catalyse the transfer of phosphorus-containing groups from donors to appropriate acceptors. Where a phosphate group itself is transferred from a nucleoside triphosphate (usually ATP) the enzyme is known as a *kinase*, the kinases being classified according to the nature of the acceptor species.

Phosphoglucomutase (2.7.5.1) is a representative of a small but interesting group of enzymes which catalyse the transfer of phosphate groups; in a reaction catalysed by such an enzyme the donor is regenerated and the overall process appears to be an intramolecular transfer.

The most important member of the *nucleotidyl* transferases is ribonuclease (2.7.7.16), whose structure and function have been studied intensively in the past few years.

### Kinases

These are enzymes which catalyse the transfer of a phosphate group from a nucleoside triphosphate (ATP) to a suitable acceptor, with production of the acceptor-phosphate and the nucleoside diphosphate (ADP). When the acceptors are alcohols, the products are phosphate esters. For example, hexokinase (2.7.1.1) catalyses the production of a hexose-6-phosphate

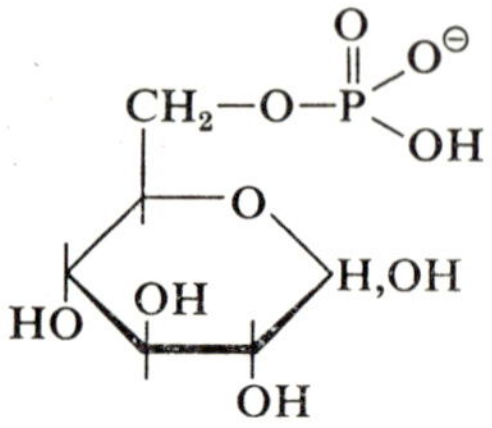

(XXV; D-glucose-6-phosphate)

(e.g. glucose-6-phosphate, XXV) from a hexose and ATP. Another example is phosphorylase kinase (2.7.1.38) by whose action the hydroxyl group of a serine residue in the enzyme phosphorylase becomes phosphorylated. In the case of pyruvate kinase (2.7.1.40) the hydroxyl group which accepts the phosphate residue is derived from the *enol* form of pyruvate (equation 6.19).

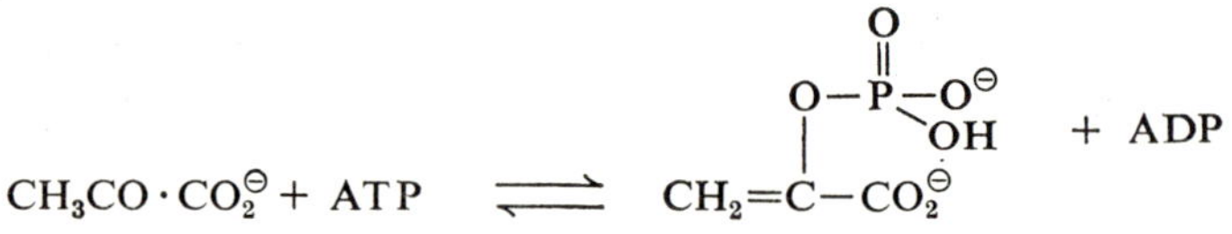

pyruvate phosphoenol pyruvate

(6.19)

When a carboxyl group functions as acceptor, as in the reaction of acetate kinase (2.7.2.1), the product is a mixed anhydride (XXVI), while in phosphocreatine (XXVII) which is formed by the action of creatine kinase (2:7.3.2), a P—N bond is present.

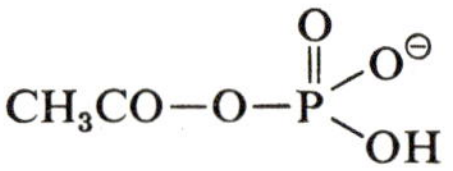

(XXVI)

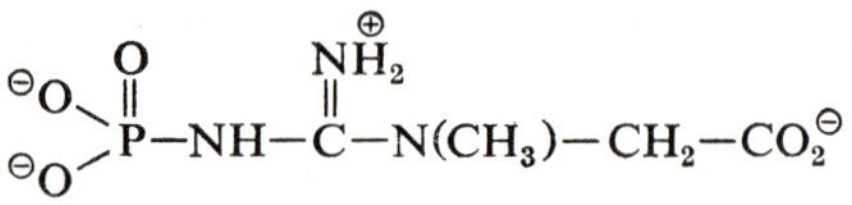

(XXVII)

The first question that arises when the mechanism of a transferase is under discussion is whether the transfer is direct or where an intermediate covalent enzyme derivative is involved. In Chapter 2 the kinetic patterns obtained for certain mechanistic pathways were discussed. For a mechanism in which the transferred group is first transferred to the enzyme giving a relatively stable covalent complex, a 'ping-pong' mechanism is usually implied, with appropriate kinetic results. However, where the group transfer is direct between one substrate and the other, ternary complexes are formed and the group transfer reaction takes place as an interconversion of one ternary complex into another. Thus if A is the acceptor,

B the donor and E the enzyme then such a situation is described by equation (6.20), P and Q being the products.

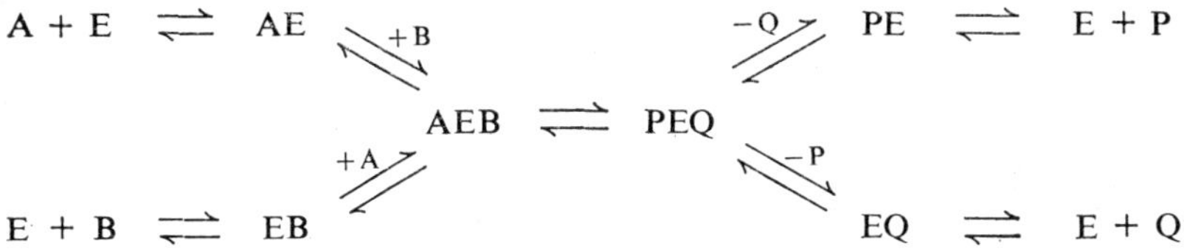


(6.20)

For almost all the kinases studied, kinetic analyses indicate random order mechanisms, in which the interconversion of the ternary complexes is rate-determining [82]. The absence of phosphoenzyme intermediate has been confirmed in some cases by experiments using labelled substrates [83].

In the general case, the ternary complex AEB may be formulated as in Fig. 6.9 where, for hexokinase, A is a hexose (glucose, mannose or

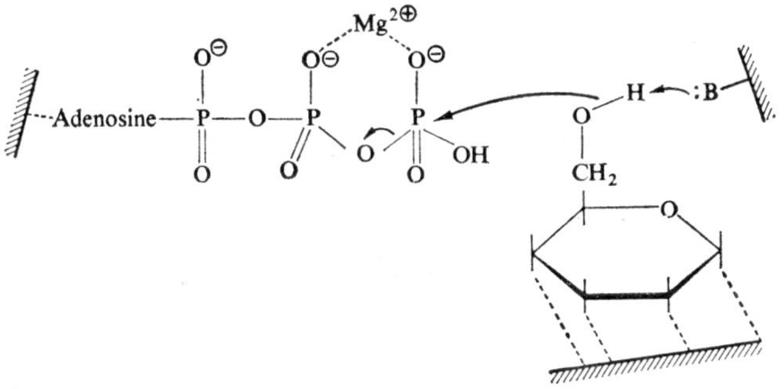


FIG. 6.9. Possible mechanism for hexokinase action

fructose can act as acceptor). B, the phosphate donor, is a complex between the nucleoside triphosphate (ATP, deoxy ATP can also function) and a magnesium ion. Magnesium ions are essential for all kinase reactions. As in the case of the carboxylase reaction discussed earlier (page 240) the metal-ion enhances the susceptibility of the terminal phosphate group to nucleophilic attack and in addition probably improves the ability of the penultimate residue to function as a leaving group. A basic group B: on the enzyme may act as a basic catalyst, as shown (see also Chapter 5).

For creatine kinase, Watts and Rabin [82] proposed a mechanism in which the function of the basic catalyst was carried out by a thiol group hydrogen-bonded to an imidazole ring. Since most, if not all, kinases are inactivated by thiol-group reagents, this possibility may apply to all of them. It is interesting that a similar system, a thiol-group hydrogen bonded to an imidazole ring, has been postulated to exist in the active site of certain peptidases such as papain (see Chapter 4).

The enzyme phosphoenol pyruvate synthase (as yet unclassified) functions by a different mechanism. The overall reaction catalysed (equation 6.21) involves the production of AMP and phosphate and is therefore not the same as that catalysed by pyruvate kinase (see equation 6.19).

$$\text{Pyruvate} + \text{ATP} \rightleftharpoons \text{phosphoenol pyruvate} + \text{AMP} + \text{P}_i \qquad (6.21)$$

In the case of the pyruvate kinase reaction the phosphate residue appearing in the phosphoenol pyruvate originated as the terminal (or $\gamma$-) phosphate in the ATP. However, when phosphoenol pyruvate is obtained by the action of phosphoenol pyruvate synthase, the phosphate residue is derived from the penultimate (or $\beta$-) position of ATP [84]. Cooper and Kornberg, and Berman *et al.* [84], have described experiments with labelled substrates which indicate that the reaction actually proceeds in several stages (equations 6.22, 6.23, 6.24).

$$\text{E} + \text{ATP} \rightleftharpoons \text{E—P—P} + \text{AMP} \qquad (6.22)$$

$$\text{E—P—P} + \text{H}_2\text{O} \rightleftharpoons \text{E—P} + \text{P} \qquad (6.23)$$

$$\text{E—P} + \text{pyruvate} \rightleftharpoons \text{E} + \text{phosphoenol pyruvate} \qquad (6.24)$$

In this proposed pathway the first step is the transfer of a pyrophosphoryl residue to the enzyme. The pyrophosphoryl-enzyme (E—P—P) is then hydrolysed to a phosphoryl-enzyme species which is able to donate the phosphate group to the acceptor. This scheme is supported by the following experimental results. An enzyme-catalysed incorporation of $^{14}$C-8 labelled AMP into ATP in phosphate buffer, in the absence of pyruvate, was observed (equation 6.22). $^{32}$P-Labelled phosphate ($P_i$) was incorporated into ATP in the presence of the enzyme, $Mg^{2\oplus}$ ions, ATP and AMP and in the absence of pyruvate, but *not* in the absence of AMP (equations 6.22 and 6.23). The enzyme also catalysed the incorporation of $^{14}$C-labelled pyruvate into phosphoenol pyruvate in the absence of phosphate (equation 6.24).

**Phosphoglucomutase (2.7.5.1)**

The net overall reaction catalysed by this enzyme appears at first sight to be an isomerization (equation 6.25), in which a phosphate group is transferred from one position in a glucose molecule to another. However,

$$\alpha\text{-D-glucose-1-phosphate} \rightleftharpoons \alpha\text{-D-glucose-6-phosphate} \qquad (6.25)$$

detailed studies have shown that this reaction is not simply an intramolecular group transfer process. Leloir *et al.* [85] found that glucose-1,6-diphosphate played an important role in the reaction which they formulated as shown:

$$\alpha\text{-D-glucose-1-phosphate} + \alpha\text{-D-glucose-1,6-diphosphate} \rightleftharpoons$$
$$\alpha\text{-D-glucose-1,6-diphosphate} + \alpha\text{-D-glucose-6-phosphate} \qquad (6.26)$$

Thus, the reaction involves the intermolecular transfer of a phosphate group from the 1-position of the diphosphate to the 6-position of glucose 1-phosphate. Support for this scheme was obtained from experiments in which labelled substrates were used [86]. When the enzyme was incubated with α-D-glucose-1-phosphate-labelled in both the carbohydrate ($^{14}C$) and the phosphate ($^{32}P$) portions and α-D-glucose-1,6-diphosphate, at equilibrium the isotopes were found to be distributed among mono- and di-phosphate esters in equilibrium amounts, i.e., there was incorporation of $^{14}C$ into both glucose-1,6-diphosphate and glucose-6-phosphate.

The transfer of the phosphate group in reaction (6.26) occurs via a phospho-enzyme intermediate. Najjar *et al.* [87] were able to isolate the enzyme in the phosphorylated and dephosphorylated forms, the phospho-enzyme being quite stable. The phosphate group is bound to the enzyme through a phosphoester link with the side-chain of a serine residue in the protein [88]; enzymic digestion has afforded peptides containing phosphoseryl residues [89].

The phosphorylated enzyme undergoes all the reactions expected of it as an intermediate. For example, α-D-glucose-1-phosphate can be converted into the 1,6-diphosphate according to equation (6.27), which is freely reversible [87]. Therefore, although under normal circumstances

$$\alpha\text{-D-glucose-1-phosphate} + \text{phosphoenzyme} \rightleftharpoons \alpha\text{-D-glucose-1,6-diphosphate} + \text{enzyme} \qquad (6.27)$$

glucose-1,6-diphosphate is an essential co-factor for the phosphoglucomutase reaction, if greater than catalytic amounts of phosphorylated enzyme are present, then added glucose-1,6-diphosphate is not necessary.

The dephosphorylated enzyme undergoes no reaction with either glucose-1-phosphate or glucose-6-phosphate. In addition glucose-1,6-diphosphate does not react with the fully phosphorylated enzyme [89].

The overall phosphoglucomutase (α-D-glucose-1,6-diphosphate: α-D-glucose-1-phosphate phosphotransferase) reaction therefore conforms to he path shown below.

α-D-glucose-1-phosphate + phosphoenzyme

⇌ α-D-glucose-1,6-diphosphate : enzyme ⇌

α-D-glucose-6-phosphate + phosphoenzyme

The precise nature of the glucose diphosphate : enzyme complex is very difficult to specify. Koshland and his colleagues [90] measured the equilibrium constant of reaction (6.27) and found it to be far lower than expected from calculations by other workers. They concluded that phosphoglucomutase catalysed the reversible transfer of phosphate without the obligatory formation of *free* α-D-glucose-1,6-diphosphate.

One further aspect of the phosphoglucomutase reaction deserves comment. Only one serine residue in the enzyme molecule becomes phosphorylated during the catalytic process, and it appears therefore that the hydroxyl group of this particular serine residue possesses greater-than-normal nucleophilic activity. The pH–activity curve of the enzyme [89] shows a peak at about pH 7·5 and the reaction appears to depend on an ionizable group with a $pK_a$ of 6–7, this group being required in the basic form. It may be possible to assign this group as the side-chain imidazole of a histidine residue, especially since Ray and Koshland [91] have shown that destruction of a histidine residue by photo-oxidation causes loss of catalytic activity. It is possible that the side-chain imidazole of a histidine residue functions as a basic catalyst increasing the reactivity of the serine hydroxyl group in the active site. Such a situation appears to exist in the case of the active serine hydrolases, and is discussed in greater detail for chymotrypsin.

**Ribonuclease (2.7.7.16) 92a**

The ribonucleases from various sources bring about the degradation of ribonucleic acid to oligonucleotides through the hydrolysis of the phosphodiester linkages. Detailed studies of the reaction have shown that in general the action of these enzymes is better described as transferase rather than hydrolase [92]. The ribonuclease-catalysed degradation of a ribonucleic acid involves firstly transfer of a phosphate group from the 5′-position of one nucleotide residue to the 2′-position of the neighbouring residue so that a 2′,3′-cyclic phosphate is produced. This step cleaves the polynucleotide chain (Fig. 6.10). The cyclic phosphate is subsequently cleaved by hydrolysis giving a 3′-phosphate mono-ester. Both stages of the reaction, the formation and the opening of the cyclic derivative, are catalysed by ribonuclease. The first stage being reversible, it is possible

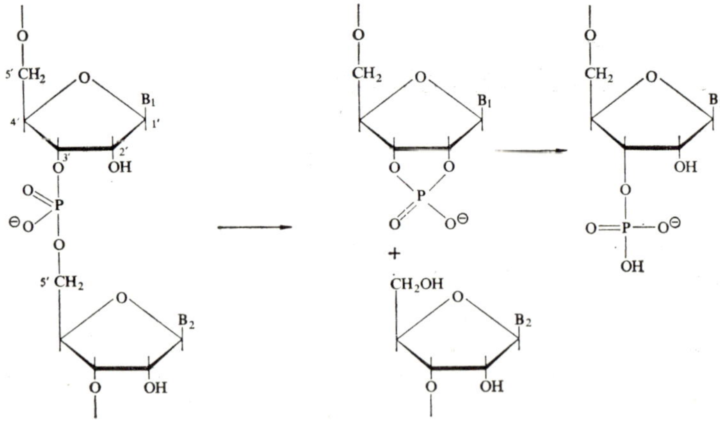

FIG. 6.10. Pathway for ribonuclease-catalysed hydrolysis of RNA

to synthesize various nucleotide derivatives from a cyclic phosphate and an alcohol.

Bovine pancreatic ribonuclease A has been studied far more intensively than any other. Its specificity is such that the phosphate residue of a pyrimidine nucleotide is transferred to the 2′-position. In Fig. 6.10 this means that the base $B_1$ must be pyrimidine.

Ribonuclease A is a single-chain protein containing 124 amino-acid residues, with a molecular weight of 13,800. The amino-acid sequence was determined mainly by Hirs, Moore, Smyth and Stein [93] and is shown in Fig. 6.11. The molecule possesses four intrachain cysteine

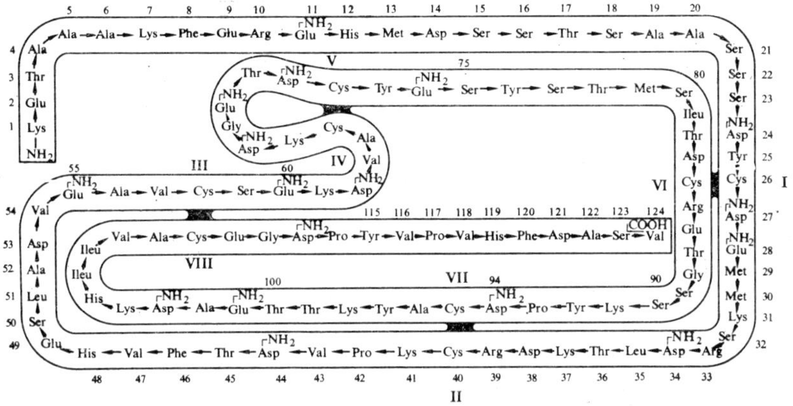

FIG. 6.11. The amino-acid sequence of the enzyme ribonuclease A. As reported by Smyth, Stein and Moore [93]. (By courtesy of the American Society of Biological Chemists)

bridges; these may be broken by reduction, a process which causes loss of catalytic activity. This loss is presumably due to the destruction of the tertiary structure of the protein. On reoxidation most of the activity is recovered, suggesting that the protein possesses within itself the ability to fold itself into the correct conformation [94].

Of special interest are the histidine residues in positions 12 and 119. When treated with iodoacetate ribonuclease gave two products [95]. The major product possessed a carboxymethyl group on histidine-119, and the minor product was carboxymethylated on histidine-12. No material was obtained in which both histidines in the same molecule had reacted. This experiment gives rise to several interesting and important points. Under the conditions used (pH 5·5) the side-chain imidazole groups of histidine residues do not normally react with iodoacetate. Therefore the environment of each of the histidines in ribonuclease must be a unique one. The fact that both histidines in one molecule are never carboxymethylated suggests that reaction of one residue prevents the reaction of the other and therefore that these two residues, although widely separated in the primary amino-acid sequence, are in fact very close together in the natural state of the enzyme. Iodoacetamide does not react under the same conditions; hence the negative charge on the iodoacetate molecule is clearly important. It is likely that the carboxylate ion becomes bound to one imidazole group (protonated) increasing the probability of the reaction, as shown. Based on this idea, the distance between the two imidazole rings may be calculated as being approximately 5 Å.

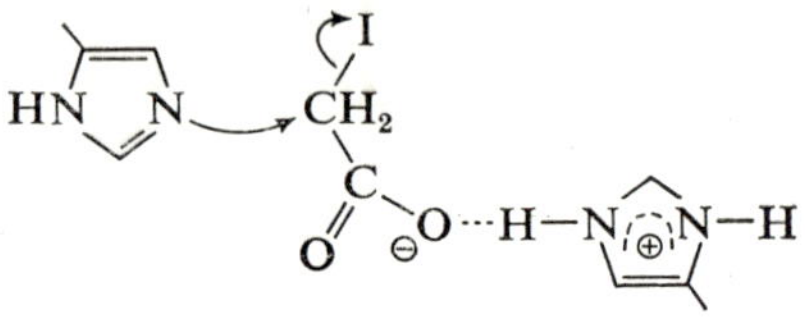

Both of the products from the iodoacetate reaction were inactive and therefore the region of the enzyme which contains these two histidine residues is the active site and either one or both of them is likely to be essential for the activity. Competitive inhibitors of the enzyme slowed down the reaction with iodoacetate, confirming that the reaction takes place at the active site. Destruction of histidine residues 12 and 119 by photo-oxidation caused loss of activity [96].

Other amino-acid residues of importance to the catalytic activity of ribonuclease are the aspartic acid at position 121, and lysine at 41. Removal of the aspartic acid by carboxypeptidase causes loss of activity [97], as does blockage of the ε-amino-group at position 41 with fluorodinitrobenzene [98].

The product obtained when a ribonuclease A was treated with the proteolytic enzyme subtilopeptidase A (3.4.4.16), retained its catalytic activity [99]. This product, known as ribonuclease-S, can be separated into two fragments one of which is the N-terminal peptide containing 20 amino-acid residues (S-peptide) the other being the remainder of the protein (S-protein) containing 104 residues. Each of these fragments, S-peptide and S-protein, is totally inactive. However, when the two are mixed in 1 : 1 molar proportions full catalytic activity is recovered. The reconstituted enzyme is known as ribonuclease-S′. This result suggests that each of the fragments possesses some of the factors necessary for the catalytic activity of the enzyme, and that combination of the fragments rebuilds the active site. Any binding between the S-peptide and the S-protein must be non-covalent (see the amino-acid sequence); this binding must, however, be extremely firm since 1 : 1 recombination regenerates virtually total activity. Thus the equilibrium (6.28) favours combination to a very high degree.

$$\text{S-peptide} + \text{S-protein} \rightleftharpoons \text{ribonuclease-S}' \qquad (6.28)$$

Ribonuclease-S provides a unique system for the study of two important aspects of enzyme chemistry, these being the roles of certain amino-acid residues in the catalytic process and the nature of the protein–peptide interaction. Hofmann and his co-workers [100] and, later, Scoffone *et al.* [101] synthesized various peptides related to S-peptide in order to study both of these aspects. Of the very large number of peptides synthesized by Hofmann *et al.*, the following examples indicate the type of result which could be obtained. The fully synthetic S-peptide, that is, the one containing the 1–20 sequence of the enzyme, gave full activity when combined in 1 : 1 molar proportions with S-protein. Similarly the peptide corresponding to 1–14 of the S-peptide sequence gave full activity at a 1 : 1 ratio. This indicates that the residues 15–20 are necessary for neither the peptide–protein binding nor the catalytic activity. However, the peptide corresponding to 1–13 gave only 50% activity at a peptide–protein ratio of 3 : 1. Full activity could be obtained only when very high peptide–protein ratios were used. This suggests that the aspartyl residue in position-14 does not contribute to the catalytic activity of the enzyme but that it does contribute significantly to the binding between the peptide and the protein in ribonuclease-S′ (and presumably also to the intramolecular forces in the intact enzyme).

The dodecapeptide with the 1–12 sequence gave 50% activation at a peptide–protein ratio of 88, a result which suggests that the methionine-13 residue also contributes to the protein–peptide binding. When histidine-12 is absent, as in the sequence 1–11, no catalytic activity could be detected on combination of peptide with S-protein. This very strong evidence for the essential nature of the histidine-12 residue in the catalytic process was confirmed when Hofmann and his group [102] prepared the peptide corresponding to the 1–14 sequence but in which the histidine-12 residue had been replaced by a 3-pyrazolylalanine residue. This possesses a pyrazolyl ring on its side-chain, which is similar in size and shape to the imidazole ring of a histidine, but has very different chemical properties. This new peptide was not only incapable of activating the S-protein, but also could function as a competitive inhibitor for the interaction between normal S-peptide and S-protein.

Other results have indicated that the glutamyl residue in position-2 is also important for the binding of the peptide to the protein.

Rabin and his co-workers have carried out some important kinetic studies on ribonuclease A [103]. These showed that the active site possessed two ionizing groups, which in the free enzyme had $pK_a$ values 5·22 and 6·78, one being in the acidic form. A study of the catalysis process in media containing organic solvents indicated that both of these groups gave cationic acids, and that therefore they were most probably histidyl residues. These workers also deduced that a binding site existed on the enzyme for a water molecule (during the hydrolysis step) and that this probably involved a hydrogen bond between the water and the histidine side-chain in the basic form. Similarly they observed shifts in the $pK_a$'s of the active site groups on binding of the substrate which suggested that the histidine in the acidic form interacted in some way with the substrate.

More recent kinetic studies, some involving high-speed techniques [104, 105], have shown that in detail the mechanism is very complex. Hammes and co-workers [105] found that at least three ionizable groups might be present on the enzyme ($pK$'s 5, 6 and 6·7), and that a minimum of five intermediates are involved in the hydrolysis of the cytidine-2′,3′-cyclic phosphate. For the hydrolysis of the dinucleotide derivative cytidyl-3′,5′-cytidine a minimum of seven states of the enzyme–substrate complex are required. Free cystidine is released from this dinucleotide phosphate after the rate-determining step has taken place and it is assumed that the rupture of the 5′-O-phosphate bond and the formation of the cyclic phosphate occur in the rate-determining step. The reaction may therefore be viewed as being very similar to that of the kinases in that a direct transfer of phosphate takes place at the rate-determining stage.

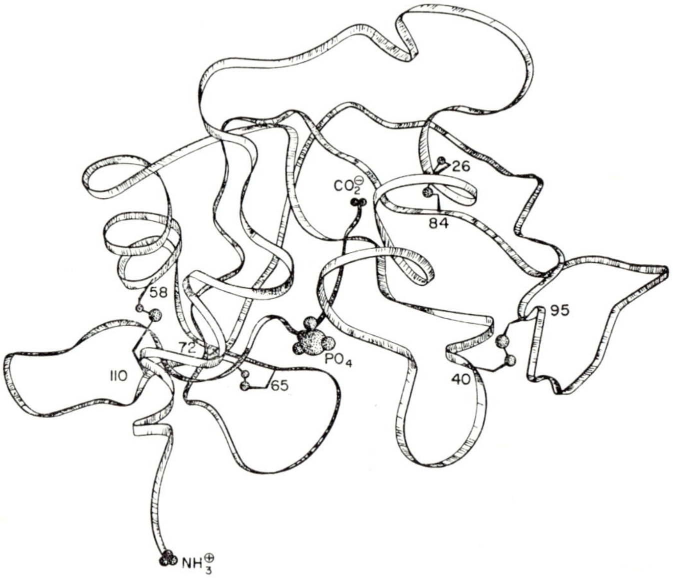


FIG. 6.12. The ribonuclease molecule (G. Kartha). The folding of the main chain of bovine pancreatic ribonuclease molecule. This protein is a single chain of 124 amino-acid residues starting at the amino end marked $NH_3^+$ and ending at the carboxy terminal marked $CO_2^-$. The main chain is cross-linked at four places by disulphide bridges from half cystine residues. The disulphide pairing for these bridges are 26–84; 40–95; 58–110 and 65–72 in the sequence. The region of the site of enzyme activity is indicated by the binding of the phosphate ion ($PO_4$) and nucleotide inhibitors in the cleft of the molecule. This model was obtained by X-ray diffraction studies of crystalline bovine pancreatic ribonuclease (By courtesy of G. Kartha)

The results of X-ray crystallographic studies confirm those obtained by chemical methods. Figure 6.12 shows the way in which the peptide chain of the intact ribonuclease molecule is folded [106]. The N-terminal 1–20 portion (S-peptide) lies on the front surface and the 20–21 peptide bond is in the loop at the top of the molecule. This explains the action of subtilopeptidase A since the peptide bond cleaved is clearly accessible. (In fact the action of this enzyme on ribonuclease is not as specific as was at first thought. Several bonds in the vicinity of the residue–21 are cleaved and the resulting S-peptide is not homogeneous [107]. Since the residues in this region are involved in neither the catalytic process nor the peptide–protein binding, this effect is relatively unimportant to the mechanism.)

Figure 6.13 shows the three-dimensional structure of ribonuclease-S as obtained by Richards and his group [108]. The active site of the enzyme

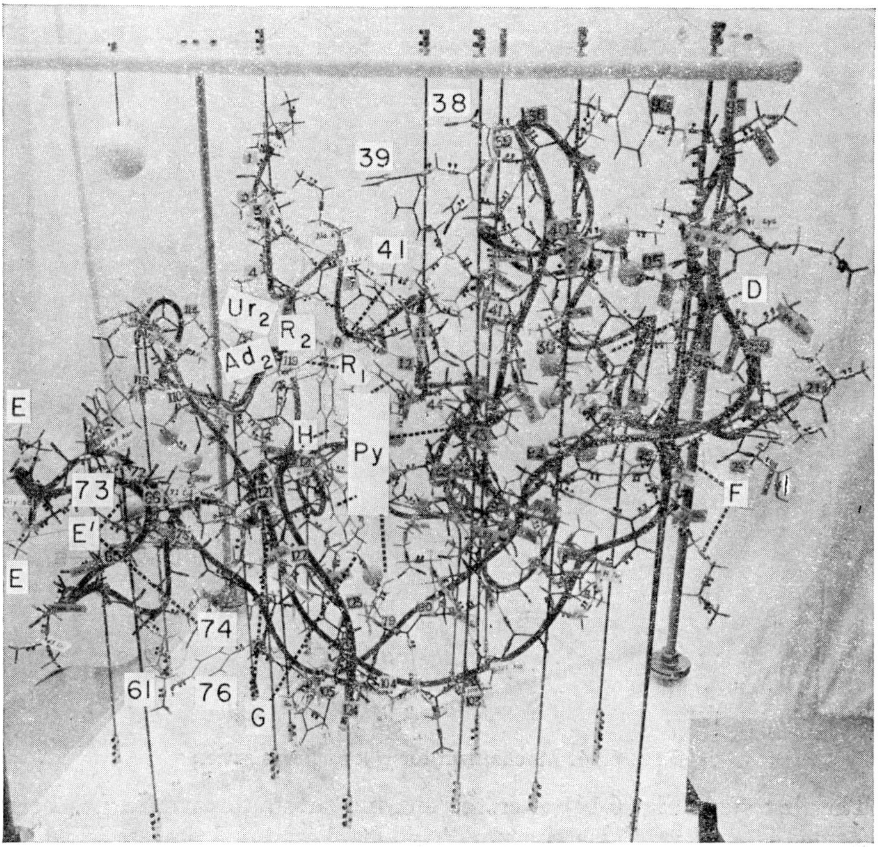


FIG. 6.13. Three-dimensional structure of ribonuclease-S [108]. (By courtesy of F. M. Richards)

was indicated by the position of binding of uridylic and iodo-uridylic acids. Histidines 12 and 119 are present at this site, indeed histidine-12 has been indicated as being very close to the 2′-position of a bound nucleoside (in agreement with deductions by Rabin based on ORD and other measurements during substrate and inhibitor binding [103]). Lysine-41 is also in the vicinity of the active site, as is aspartic acid-121 [97, 98]. In confirmation of the results obtained with synthetic peptides, aspartic acid-14 contributes to the peptide–protein binding by forming an ion pair with the side-chain guanidinium group of arginine-33. In addition, residues 15–20 and 21–23 could be removed without effect on the remainder of the system.

A mechanism first suggested by Rabin *et al.* [103] appears to be in agreement with most, if not all, of the data available for ribonuclease A.

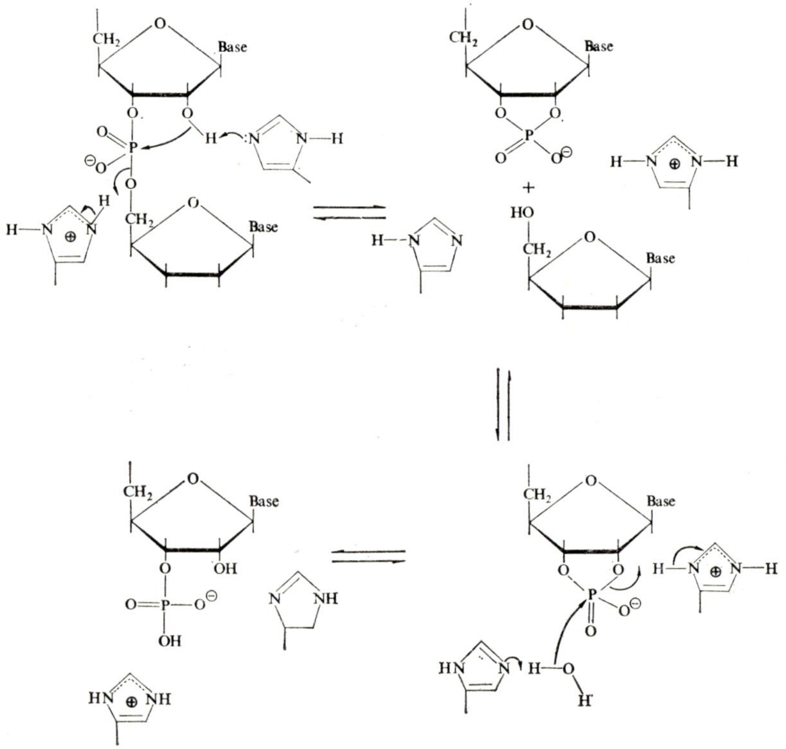

FIG. 6.14. Mechanism for ribonuclease action

The first step (Fig. 6.14) after the enzyme–substrate complex has been formed is envisaged as a concerted acid–base-catalysed process in which the basic imidazole group interacts with the 2′-hydroxyl group and the imidazolium group assists the leaving group. When the released nucleoside has dissociated away from the enzyme product complex, a molecule of water (or possibly an alcohol) becomes bound and the cleavage of the cyclic phosphate group proceeds in the reverse manner, again by a concerted acid–base process. This step is a transfer of a group to a water molecule and may be considered as a hydrolysis. In each case the transfer step has a symmetrical transition state (XXVIII) in which the phosphorus atom has a trigonal bipyramid electronic configuration.

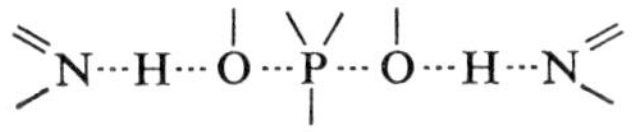


(XXVIII)

For a brief discussion of another aspect of ribonucleic acid hydrolysis, see Chapter 5.

## REFERENCES


1. HODGKIN, D., KAMPER, J., LINDSEY, J., MACKAY, M., PICKWORTH, J., ROBERTSON, J. H., SHOEMAKER, C. B., WHITE, J. G., PROSEN, R. J. and TRUEBLOOD, K. N., *Proc. R. Soc.* **242A,** 228 (1957)
2. GUEST, J. R., FRIEDMAN, S. and WOODS, D. D., *Nature, Lond.* **194,** 340 (1962)
3. BRADY, R. O. and BARKER, H. A., *Biochim. biophys. Res. Commun.* **4,** 464 (1961); VITOLS, E., WALKER, G. and HUENNEKENS, F. M., *Biochem. biophys. Res. Commun.* **15,** 372 (1964)
4. SMITH, E. L., MERVYN, L., JOHNSON, A. W. and SHAW, N., *Nature, Lond.* **194,** 1175 (1962)
5. BERNHAUER, K. and IRION, E., *Biochem. Z.* **339,** 530 (1964)
6. TAYLOR, R. T. and WEISSBACH, H., *J. biol. Chem.* **242,** 1509 (1967)
7. WOOD, J. M. and WOLFE, R. S., *Biochemistry* **5,** 3598 (1966)
8. WEISSBACH, H., PETERKOVSKY, A. and BARKER, H. A., *Comprehensive Biochemistry*, Eds. M. Florkin and E. H. Stotz, Vol. 16, p. 189 (Elsevier, 1965)
9. WEISSBACH, H. and TAYLOR, R., *Fedn. Proc.* **25,** 1649 (1966)
10. WEISSBACH, H., PETERKOVSKY, A., REDFIELD, B. G. and DICKERMAN, H., *J. biol. Chem.* **238,** 3318 (1963)
11. TAYLOR, R. T. and WEISSBACH, H., *Archs biochem. Biophys.* **119,** 572 (1967)
12. STADTMAN, T. C. and BLAYLOCK, B. A., *Fedn Proc.* **25,** 1657 (1966)
13. KAZIRO, Y., LEONE, E. and OCHOA, S., *Proc. natn. Acad. Sci. U.S.A.* **46,** 1319 (1960)
14. LANE, M. D., HALENZ, D. R., KOSOW, D. P. and HEGREE CARMAN, S., *J. biol. Chem.* **235,** 3082 (1960); HALENZ, D. R. and LANE, M. D., *Biochim. biophys. Acta* **48,** 425 (1961)
15. KAZIRO, Y. and OCHOA, S., *J. biol. Chem.* **236,** 3131 (1961)
16. OCHOA, S. and KAZIRO, Y., *Comprehensive Biochemistry* Vol. 16, p. 210 (Elsevier, 1965)
17. LANE, M. D. and LYNEN, F., *Proc. natn. Acad. Sci. U.S.A.* **49,** 379 (1963)

18. Kosow, D. P. and Lane, M. D., *Biochem. biophys. Res. Commun.* **7,** 439 (1963)
19. Knappe, J., Wenger, B. and Wiegand, U., *Biochem. Z.* **337,** 232 (1963); Knappe, J., Biederbick, K. H. and Brummer, W., *Angew. Chem.* **74,** 432 (1962)
20. Wood, H. G., Lochmuller, H., Riepertinger, C. and Lynen, F., *Biochem. Z.* **337,** 247 (1963)
21. Knappe, J., Ringelmann, E. and Lynen, F., *Biochem. Z.* **335,** 168 (1961)
22. Caplow, M., *J. Am. chem. Soc.* **87,** 5774 (1965)
23. Mildvan, A. S., Scrutton, M. C. and Utter, M. F., *J. biol. Chem.* **241,** 3488 (1966)
24. Kaziro, Y., Hass, L. F., Boyer, P. D. and Ochoa, S., *J. biol. Chem.* **237,** 1460 (1962)
25. Koshland, D. E., jr., *Phosphorus Metabolism,* Eds. W. D. McElroy and B. Glass, Vol. 1, p. 536 (Johns Hopkins Press, Baltimore, 1951)
26. *Enzyme Nomenclature,* p. 97 (Elsevier, Amsterdam, 1965)
27. Jaenicke, L. and Lynen, F., *The Enzymes* Vol. 3, p. 3 (Academic Press, New York, 1960)
28. Bessman, S. P. and Lipmann, F., *Archs biochem. Biophys.* **46,** 252 (1953)
29. Lynen, F., Henning, U., Bublitz, C., Sorbo, B. and Kroplin-Rueff, L., *Biochem. Z.* **330,** 269 (1953)
30. Lynen, F., *Fedn Proc.* **12,** 683 (1953)
31. Folk, J. K., Mullooly, J. P. and Cole, P. W., *J. biol. Chem.* **242,** 1838 (1967); Folk, J. K., Cole, P. W. and Mullooly, J. P., *J. biol. Chem.* **242,** 2615, 4329 (1967)
32. Folk, J. K. and Cole, P. W., *J. biol. Chem.* **240,** 2951 (1965)
33. Glaser, L., *Comprehensive Biochemistry* Vol. 15, p. 93 (Elsevier, 1964)
34. Cohn, M., *The Enzymes* Vol. 5, p. 179 (Academic Press, New York, 1961); Brown. D. H. and Cori, C. F., *The Enzymes* Vol. 5, p. 207 (Academic Press, New York, 1961)
35. Cohn, M., *J. cell. comp. Physiol.* **54,** Suppl. 1, 17 (1959); *J. biol. Chem.* **180,** 771 (1949)
36. Hestrin, S., *First IUB/IUBS Symposium; Biological Structure and Function* Vol. 1, p. 315 (1961)
37. Koshland, D. E., jr., *The Mechanism of Enzyme Action,* Eds. W. D. McElroy and B. Glass, p. 608 (Johns Hopkins Press, 1954)
38. Doudoroff, M., *The Enzymes* Vol. 5, p. 229 (Academic Press, New York, 1961)
39. Silverstein, R., Voet, J., Reed, D. and Abeles, R. H., *J. biol. Chem.,* **242,** 1338 (1967)
40. Doudoroff, M., Barker, H. A. and Hassid, W. Z., *J. biol. Chem.* **168,** 725 (1947)
41. Woloshow, H., Putman, E. W., Doudoroff, M., Hassid, W. Z. and Barker, H. A., *J. biol. Chem.,* **180,** 1238 (1949)
42. Alberty, R. A., *Adv. Enzymol.* **17,** 1 (1956)
43. Dedonder, R. A., *Bull. soc. chim. biol.* **42,** 1748 (1960); Joyeux, Y. and Dedonder, R. A., *Compt. rend.* **255,** 608 (1962)
44. Fitting, C. and Doudoroff, M., *J. biol. Chem.* **199,** 153 (1952)

45. MADSEN, N. B. and CORI, C. F., *J. biol. Chem.* **223,** 1055 (1956); MADSEN, N. B., *J. biol. Chem.* **223,** 1067 (1956)
46. BARANOWSKI, T., ILLINGWORTH, B., BROWN, D. H. and CORI, C. F., *Biochim. biophys. Acta* **25,** 16 (1957); CORI, C. F. and ILLINGWORTH, B., *Proc. natn. Acad. Sci. U.S.A.* **43,** 547 (1957)
47. ILLINGWORTH, B., JANSZ, H. S., BROWN, D. H. and CORI, C. F., *Proc. natn. Acad. Sci. U.S.A.* **44,** 1180 (1958)
48. FISCHER, E. H., KENT, K. B., SNYDER, E. R. and KREBS, E. G., *J. Am. chem. Soc.* **80,** 2906 (1958)
49. SEALOCK, R. W. and GRAVES, D. J., *Biochemistry* **6,** 201 (1967)
50. MADSEN, N. B. and SHECHOSKY, S., *J. biol. Chem.* **242,** 3301 (1967)
51. WANG, J. H. and GRAVES, D. J., *Biochemistry* **3,** 1437 (1964); WANG, J. H., SHONKA, M. L. and GRAVES, D. J., *Biochemistry* **4,** 229 (1965)
52. LOWRY, O. H., SCHULZ, D. W. and PASSONEAU, J. V., *J. biol. Chem.* **242,** 271 (1967)
53. COHN, M. and CORI, C. F., *J. biol. Chem.* **175,** 89 (1948)
54. KREBS, E. G. and FISCHER, E. H., *Adv. Enzymol.* **24,** 263 (1962)
55. HERBST, R. M. and ENGELL, L. L., *J. biol. Chem.* **107,** 505 (1934); HERBST, R. M., *J. Am. chem. Soc.* **58,** 2239 (1936); HERBST, R. M., *Adv. Enzymol.* **4,** 75 (1944)
56. NAKADA, H. I. and WEINHOUSE, S., *J. biol. Chem.* **204,** 831 (1953)
57. BRAUNSTEIN, A. E., *Enzymologia* **7,** 25 (1939)
58. COHEN, P. P., *Biochem. J.* **33,** 1478 (1939)
59. SNELL, E. E., *J. Am. chem. Soc.* **67,** 194 (1945); METZLER, D. E. and SNELL, E. E., *J. Am. chem. Soc.* **74,** 979 (1952); METZLER, D. E., OLIVARD, J. and SNELL, E. E., *J. Am. chem. Soc.* **76,** 644 (1954); LONGENECKER, J. B. and SNELL, E. E., *J. Am. chem. Soc.* **79,** 142 (1957)
60. METZLER, D. E., IKAWA, M. and SNELL, E. E., *J. Am. chem. Soc.* **76,** 648 (1954)
61. SNELL, E. E., *Chemical and Biological Aspects of Pyridoxal Catalysis*, Eds. E. E. Snell, F. M. Fasella, A. E. Braunstein and A. R. Faneli, p. 1 (Pergamon, London, 1963)
62. BRAUNSTEIN, A. E. and SHEMYAKIN, M. M., *Doklady Akad. Nauk S.S.R.* **85,** 1115 (1952); *Biokhimiya* **18,** 393 (1953), quoted in ref. [61]
63. IKAWA, M. and SNELL, E. E., *J. Am. chem. Soc.* **76,** 653 (1954)
64. BANKS, B. E. C., DIAMANTIS, A. A. and VERNON, C. A., *J. chem. Soc.* 4253 (1961)
65. BRUICE, T. C. and TOPPING, R. M., *Chemical and Biological Aspects of Pyridoxal Catalysis* p. 29 (Pergamon, London, 1963)
66. JENCKS, W. P. and CORDES, E., *Chemical and Biological Aspects of Pyridoxal Catalysis* p. 57 (Pergamon, London, 1963)
67. VELICK, S. F. and VAVRA, J., *J. biol. Chem.* **237,** 2109 (1962)
68. HAMMES, G. G. and FASELLA, P., *Chemical and Biological Aspects of Pyridoxal Catalysis*, p. 185 (Pergamon, London, 1963)
69. BANKS, E. C., LAURENCE, A. C., VERNON, C. A. and WOOTTON, J. F., *Chemical and Biological Aspects of Pyridoxal Catalysis* p. 197 (Pergamon, London, 1963)

70. TURANO, C., FASELLA, P. and GIARTOSIO, A., *Biochim. biophys. Acta* **58,** 255 (1962)
71. JENKINS, W. T. and SIZER, I. W., *J. biol. Chem.* **235,** 620 (1960)
72. JENKINS, W. T. and SIZER, I. W., *J. biol. Chem.* **234,** 1179 (1959)
73. FISCHER, E. H. and KREBS, E. G., *Abst. Nat. Meeting A.C.S.* 24C (1959); HUGHES, R. C., JENKINS, W. T. and FISCHER, E. H., *Proc. natn. Acad. Sci. U.S.A.* **48,** 1615 (1957)
74. TURANO, C., FASELLA, P. and GIARTOSIO, A., *Proc. Vth Int. Cong. Biochem.* Abstracts, p. 128 (Moscow, 1961)
75. ANDERSON, J. A. and SONG, P.-S., *Archs biochem. Biophys.* **122,** 224 (1967)
76. POLYANOVKSY, O. L. and KEIL, B. A., *Biokhimiya* **28,** 375 (1963)
77. TURANO, C., *Abst. Meeting Fed. Eur. Biochem. Soc.* 7 (1964)
78. TURANO, C., GIARTOSIO, A., RIVA, F., BARONCELLI, V. and ROSSA, F., *Archs biochem. Biophys.* **117,** 678 (1966)
79. MARTINEZ-CARRION, M., TURANO, C., RIVA, F. and FASELLA, P., *J. biol. Chem.* **242,** 1426 (1967)
80. WONG, K. F. and COSSINS, E. A., *Biochem. biophys. Res. Commun.* **25,** 651 (1966)
81. METZLER, D. E., LONGENECKER, J. B. and SNELL, E. E., *J. Am. chem. Soc.* **75,** 2786 (1953); METZLER, D. E., LONGENECKER, J. B. and SNELL, E. E., *J. Am. chem. Soc.* **76,** 639 (1954)
82. APPS, D. K., *Biochem. J.* **104,** 35P (1967); FROMM, H. J. and ZEWE, V., *J. biol. Chem.* **237,** 3027 (1962); PARRY, M. J. and WALKER, D. G., *Biochem. J.* **105,** 473 (1967); LARSSON-RAZNIKIEWISZ, M., *Biochim. biophys. Acta* **132,** 33 (1967); WATTS, D. C. and RABIN, B. R., *Biochem. J.* **85,** 507 (1962); KERSON, L. A., GARFINKEL, D. and MILDVAN, A. S., *J. biol. Chem.* **242,** 2124 (1967) and other references quoted in these papers.
83. GAMBLE, J. L. and NAJJAR, V. A., *Science* **120,** 1023 (1954); KAUFMAN, S. *J. biol. Chem.* **216,** 153 (1955)
84. COOPER, R. A. and KORNBERG, H. L., *Biochim. biophys. Acta* **141,** 211 (1967); COOPER, R. A. and KORNBERG, H. L., *Biochim. biophys. Acta* **104,** 618 (1965); BERMAN, K., ITADA, N. and COHN, M., *Biochim. biophys. Acta* **141,** 214 (1967)
85. LELOIR, L. F., TRUCCO, R. E., CARDINI, C. E., PALADINI, A. and CAPUTTO, R., *Archs Biochem.* **19,** 339 (1948); CARDINI, C. E., PALADINI, A. C., CAPUTTO, R., LELOIR, L. F. and TRUCCO, R. E., *Archs Biochem.* **22,** 87 (1949)
86. SUTHERLAND, E. W., COHN, M., POSTERNAK, T. and CORI, C. F., *J. biol. Chem.* **180,** 1285 (1949)
87. NAJJAR, V. A. and PULLMAN, M. E., *Science* **119,** 631 (1954); NAJJAR, V. A., *The Enzymes* Vol. 6, p. 161 (Academic Press, New York, 1962)
88. ANDERSON, L. and JOLLÉS, G. R., *Archs biochem. Biophys.* **71,** 127 (1957); MILSTEIN, C. and SANGER, F., *Biochem. J.* **79,** 456 (1949); HARSHMAN, S. and NAJJAR, V. A., *Fedn Proc.* **21,** 233 (1962)
89. CORI, C. F. and BROWN, D. H., *Comprehensive Biochemistry* Vol. 15, p. 212 (Elsevier, 1964)

90. GOUNARIS, A. D., HORTON, H. R. and KOSHLAND, D. E., JR., *Biochim. biophys. Acta* **132,** 41 (1967)
91. RAY, W. J., JR., and KOSHLAND, D. E., JR., *J. biol. Chem.* **237,** 2493 (1962)
92. ANFINSEN, C. B. and WHITE, F. H., *The Enzymes* Vol. 5, p. 95 (Academic Press, New York, 1961)
92a. BARNARD, E. A., *Ann. rev. Biochem.* **38,** 677 (1969)
93. SMYTH, D. G., STEIN, W. H. and MOORE, S., *J. biol. Chem.* **238,** 227 (1963); HIRS, C. H. W., MOORE, S. and STEIN, W. H., *J. biol. Chem.* **235,** 633 (1960)
94. WHITE, F. H., *J. biol. Chem.* **235,** 383 (1960)
95. CRESTFIELD, A. M., STEIN, W. H. and MOORE, S., *J. biol. Chem.* **238,** 2413 (1963)
96. KENKARE, U. W. and RICHARDS, F. M., *J. biol. Chem.* **241,** 3197 (1966)
97. POTTS, J. T., YOUNG, D. M., ANFINSEN, C. B. and SANDOVAL, A., *J. biol. Chem.*, **239,** 3781 (1964)
98. HIRS, C. H. W., *Brookhaven Symp. Biol.* **15,** 154 (1962); MARFEY, P. S., UZIEL, M. and LITTLE, J., *J. biol. Chem.* **240,** 3270 (1960)
99. RICHARDS, F. M. and VITHAYATHIL, P. J., *J. biol. Chem.* **234,** 1459 (1959); VITHAYATHIL, P. J. and RICHARDS, F. M., *J. biol. Chem.* **235,** 2343 (1960)
100. HOFMANN, K., HAAS, W., SMITHERS, M. J., WELLS, R. D., WOLMAN, Y., YANAIHARA, N. and ZANETTI, G., *J. Am. chem. Soc.* **87,** 620 (1965); FINN, F. M. and HOFMANN, K., *J. Am. chem. Soc.* **89,** 5298 (1967); FINN, F. M., VISSA, J. P. and HOFMANN, K., *Proc. IXth Eur. Peptide Symposium* (1968)
101. SCOFFONE, E., MARCHIORI, F., MORODER, L., ROCCHI, R. and SCATTURIN, A., *Proc. IXth Eur. Peptide Symposium* (1968) and earlier references
102. HOFMANN, K., FINN, F., LIMETTI, M., MONTIBELLER, J. and ZANETTI, G., *J. Am. chem. Soc.* **88,** 3633 (1966)
103. HERRIES, D. G., MATHIAS, A. P. and RABIN, R. B., *Biochem. J.* **85,** 127 (1962); FINDLAY, D., MATHIAS, A. P. and RABIN, B. R., *Biochem. J.* **85,** 134, 139 (1962); ROSS, C. A., MATHIAS, A. P. and RABIN, B. R., *Biochem. J.* **85,** 145 (1962); FINDLAY, D., HERRIES, D. G., MATHIAS, A. P., RABIN, B. R. and ROSS, C. A., *Biochem. J.* **85,** 152 (1962); DEAVIN, A., MATHIAS, A. P. and RABIN, B. R., *Biochem. J.* **101,** 14C (1966); MATHIAS, A. P. and RABIN, B. R., *Biochem. J.* **103,** 62P (1967)
104. RAMSDEN, E. N. and LAIDLER, K. J., *Can. J. Chem.* **44,** 2597 (1966)
105. HAMMES, G. G. and SCHIMMEL, P. R., *J. Am. chem. Soc.* **87,** 4665 (1965); CATHOU, R. E. and HAMMES, G. G., *J. Am. chem. Soc.* **87,** 4674 (1965)
106. KARTHA, G., BELLO, J. and HARKER, D., *Nature, Lond.* **213,** 862 (1967)
107. DOSCHER, M. S. and HIRS, C. H. W., *Biochemistry* **6,** 304 (1967)
108. WYCKOFF, H. W., HARDMAN, K. D., ALLEWELL, N. M., INAGAMI, T. JOHNSON, L. N. and RICHARDS, F. M., *J. biol. Chem.* **242,** 3984 (1967)

# CHAPTER 7

# LYASES

The reactions catalysed by the enzymes in Class 4 (the lyases) may on the whole be regarded as eliminations, in that the enzymes remove groups from substrates non-hydrolytically, usually with the formation of double bonds. Thus, the hydro-lyases, or dehydratases, bring about the elimination of water:

$$H-X-\overset{|}{\underset{|}{C}}-OH \longrightarrow X=\overset{|}{\underset{|}{C}} + H_2O$$

and ammonia-lyases cause loss of ammonia; histidine ammonia-lyase (4.3.1.3) converts histidine into urocanate (equation 7.1.). In the case of a

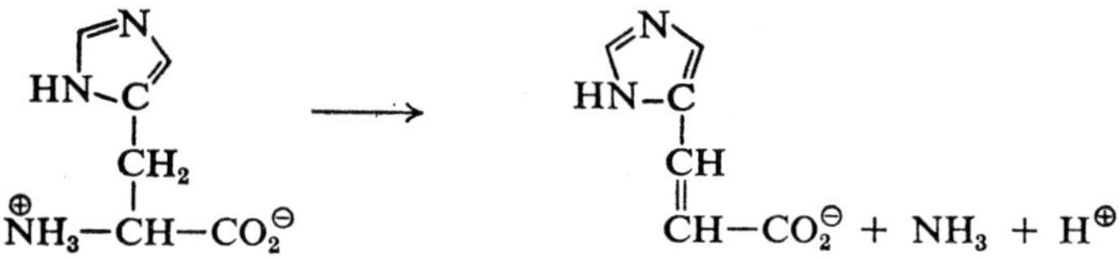

(7.1)

decarboxylation the double bond appears in the carbon dioxide which is eliminated.

$$R—CO_2H \rightarrow R—H + CO_2$$

A less simple example of a lyase-catalysed reaction is that catalysed by ketose-1-phosphate aldolase (4.1.2.7) in which dihydroxyacetone phosphate is eliminated, a double bond being formed, not between two carbon atoms but between a carbon and an oxygen (equation 7.2).

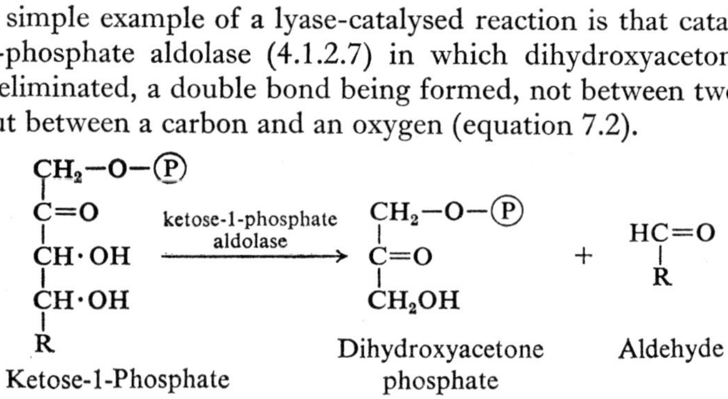


(7.2)

Many lyases are dependent on the presence of one or more co-factors for their activity, and on this basis they fall into five groups; (a) those which are pyridoxal (or pyridoxal phosphate)-dependent, (b) those which require thiamine pyrophosphate and a metal-ion, (c) those requiring no coenzyme (with active lysyl residues), (d) metal-dependent lyases, (e) lyases which require a vitamin $B_{12}$-derived coenzyme. This chapter deals with enzymes in the first four of these categories. Propanediol dehydratase (4.2.1.28), which requires the vitamin $B_{12}$ coenzyme, is mentioned in the next chapter, along with others which require this coenzyme.

## PYRIDOXAL PHOSPHATE ENZYMES

Where the lyase reaction involves an amino-acid as substrate the enzyme is usually found to be dependent on pyridoxal or its phosphate.

### Non-enzymic Reactions of Pyridoxal

In addition to the transaminations mentioned earlier, pyridoxal is able to catalyse, non-enzymically, a series of reactions which are entirely analogous to those catalysed by certain lyases [1]. Table 7.1 gives a short list of some of these reactions.

Table 7.1. *Reactions of amino-acids catalysed by pyridoxal phosphate* (from ref. [1])

1. Decarboxylation to primary amines.
2. Degradation of $\beta$-hydroxy-amino-acids to glycine and an aldehyde.
3. Dehydration of $\beta$-hydroxy-amino-acids (with loss of ammonia).
4. Synthesis of sulphur-containing amino-acids from serine.
5. Synthesis of tryptophan from serine.
6. Desulphurization of cysteine (with loss of ammonia).
7. $\beta,\gamma$-Elimination of homocysteine (with loss of ammonia).

The proposed mechanisms by which the pyridoxal is able to bring about these chemical conversions all involve the prior formation of pyridoxal-amino-acid Schiff-bases as in the case of the transamination reactions. In such a compound (I), the presence of the pyridine ring (especially when protonated) activates or labilizes the carboxyl group, the $\alpha$-hydrogen or the R group of the amino-acid species. In most of the reactions metal-ions have been found to enhance the effect of the catalyst, in a manner similar to that described in chapter 6.

Figure 7.1 depicts the process of decarboxylation of an $\alpha$-amino-acid. After the elimination of the carbon dioxide, the molecule takes up a proton from the solvent as shown (II). The decarboxylation process differs from most of the other pyridoxal-catalysed reactions in that the $\alpha$-hydrogen of the amino-acid species remains attached to the carbon throughout.

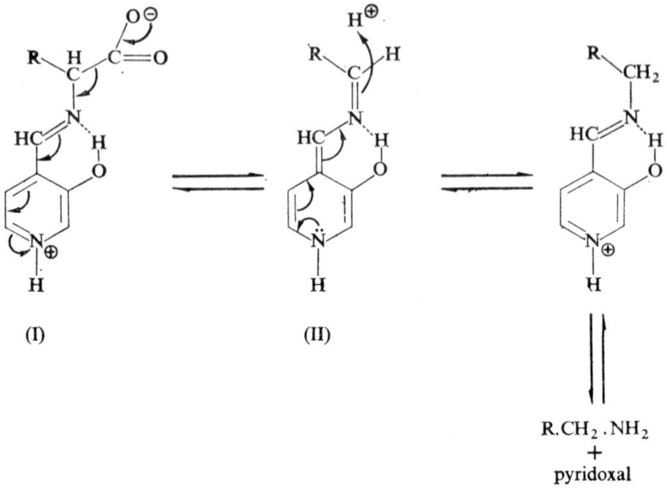


FIG. 7.1. Pyridoxal-catalysed decarboxylation of an α-amino-acid. The methyl and hydroxymethyl groups of the pyridoxal molecule are omitted in this and subsequent figures

The absence of any necessary reaction in which this α-hydrogen is lost, is shown by the observation that amino-acids carrying substituents other than hydrogen undergo decarboxylation readily [4].

The presence of metal-ions, which enhances the rates of other reactions, actually slows down the decarboxylation process [4]. In a metal-complex such as III, the coordination of the carboxylate group to the metal-ion would tend to decrease the ability of the carboxyl group to eliminate as carbon dioxide. It is most probable then that, even when metal-ions are present, decarboxylation occurs in the metal-free complex (I) [3, 4].

The second reaction listed in Table 7.1, degradation of β-hydroxy-amino-acids, such as serine or threonine to glycine and an aldehyde, has already been discussed in chapter 6 (see Fig. 6.7), with particular reference to serine.

In Fig. 7.2, several reactions involving the serine-pyridoxal Schiff-base are represented. The first step is the breakage of the α-carbon–hydrogen bond, giving the species (IV, V) which may be regarded as a carbanion stabilized by resonance. The activation of this α-hydrogen assists the loss of the hydroxyl group (protonated or solvated as shown), the result being the α,β-elimination of water to give the imine (VI) which is derived from α-amino-acrylic acid. This is the key intermediate in reactions 3, 4, 5 and 6 of Table 7.1. Hydrolysis gives α-amino-acrylate which spontaneously reacts to give pyruvate and ammonia and this sequence of events brings about the deaminative dehydration of serine.

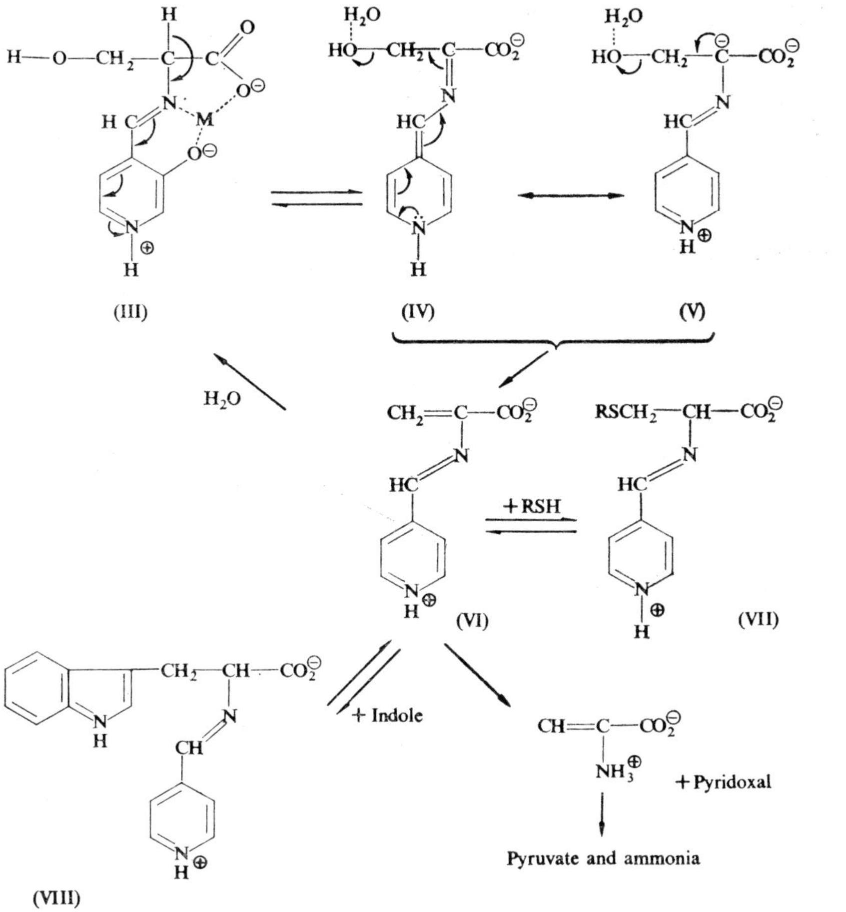


FIG. 7.2. Mechanisms for the deaminative dehydration of serine, synthesis of sulphur-containing amino-acids (VII) and tryptophan (VIII) from serine, and desulphurization of cysteine catalysed by pyridoxal and a metal-ion [14]. The metal-ion and the phenolic oxygen of the pyridoxal are omitted from all the structures except III

Addition of various thiols (RSH) to the double bond of VI gives compound (VII) and hence various sulphur-containing amino-acids including cysteine (R = H), methyl cysteine (R = $CH_3$) and cystathionine (R = homocysteinyl). Reversal of this process (VII → VI) followed by hydration of the unsaturated compound (VI) gives the serine derivative (III) and achieves the desulphurization of cysteine and its derivatives. Tryptophan (VIII) is produced when indole adds to the double bond in VI. The reactions of VI giving III, VII and VIII, may involve nucleophilic attack on the methylene carbon followed by uptake of a proton by the α-carbon.

A rather more complex process must be envisaged for the $\beta,\gamma$-elimination which effects the desulphurization of homocysteine with loss of ammonia (reaction 6, Table 7.1). The usual loss of the $\alpha$-hydrogen gives IX (Fig. 7.3), in which the $\beta$-hydrogen may be regarded as being labilized due to the presence of the imine-group (in which the nitrogen atom is coordinated on to the metal-ion not shown in Fig. 7.3). The carbanion is

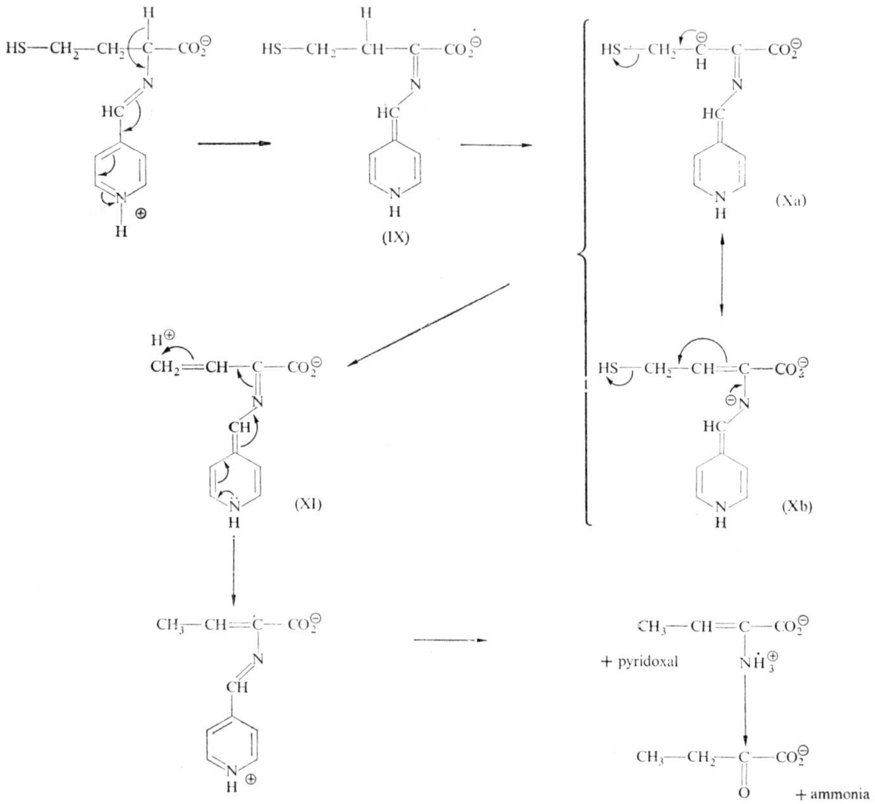


FIG. 7.3. Mechanism for the deaminative desulphurization of homocysteine

stabilized by resonance (Xa, Xb). The $\beta,\gamma$-elimination of $H_2S$ is therefore assisted. Uptake of a proton by the methylene carbon (XI) followed by hydrolysis of the imine gives $\alpha$-amino-$\beta$-methylacrylate which yields 2-ketobutyrate spontaneously. (For this, and alternative representations of this reaction, see ref. [5].)

## Enzyme-catalysed Reactions

L-Glutamate decarboxylase (4.1.1.15) catalyses the production of 4-aminobutyrate from L-glutamate (equation 7.3), that is, the $\alpha$-carboxyl group is

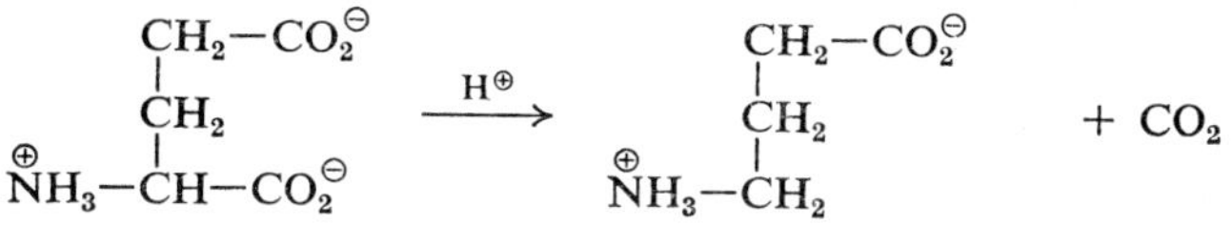

Glutamate 4-aminobutyrate

(7.3)

lost. The *E. coli* enzyme contains two tightly-bound molecules of pyridoxal phosphate per molecule of protein (molecular weight, 300,000) [6]. Ultraviolet spectral data indicate that, as in the transaminases, the pyridoxal phosphate is bound to the enzyme via an aldimine link, probably with the ε-amino group of a lysyl residue. The initial enzyme–substrate complex must therefore be formed by a transimination. The subsequent reactions almost certainly proceed by a mechanism similar to the non-enzymic pyridoxal-catalysed process.

When the enzyme-catalysed reaction was carried out in $^{2}H_2O$ the product contained one atom of deuterium [2]. This shows that the decarboxylation does not involve labilization of the α-hydrogen. If this had been the case two deuterium atoms would be incorporated. Since the enzyme is specific for the L-isomer of glutamate it might be expected that the deuterium atom taken up by the α-carbon (which is the centre of asymmetry) would be taken up stereospecifically. Indeed this is the case [7]. The two enantiomers of 4-monodeutero-4-amino butyrate could be prepared [7]. One isomer was prepared by the decarboxylation of L-α-deuteroglutamate. In this isomer the deuterium atom is *not* the atom taken up from the solvent. The other isomer was obtained when L-glutamate was decarboxylated enzymically in the presence of $^{2}H_2O$. Here the deuterium atom is derived from the solvent, and in this isomer exchange of the isotope with water was catalysed by the enzyme. In addition to emphasizing the stereospecificity of the process, this result is in agreement with a mechanistic scheme such as that in Fig. 7.1, which predicts that one α-hydrogen in the amine product would undergo exchange with solvent.

The two analogous isomers of monodeuterotyramine have been demonstrated with the enzyme tyrosine decarboxylase (4.1.1.25) [8].

In the case of aspartate-4-decarboxylase (4.1.1.2), the situation is more complex. This enzyme brings about the elimination of the β-carboxyl group of aspartate giving alanine (equation 7.4). Although in this enzyme

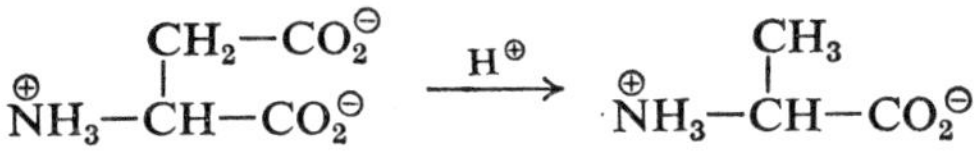

(7.4)

(from bacterial sources) [9, 10, 11] the pyridoxal phosphate appears to be bound to the enzyme via an aldimine link [12], the ultraviolet spectrum suggests that it is not bound in exactly the same way as in most pyridoxal-enzymes [11, 12]. A mechanism for the $\beta$-decarboxylation is given in Fig. 7.4. The initial imine loses its $\alpha$-hydrogen giving XII. In this compound the $\beta$-carboxyl group is activated by the presence of the nitrogen

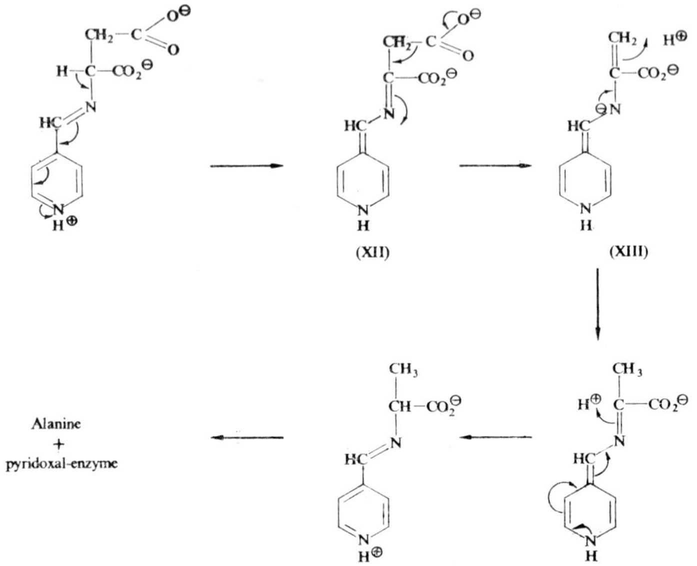


FIG. 7.4. Mechanism for aspartate 4-decarboxylase

atom (just as the $\beta$-hydrogen is activated in IX). In the enzyme–substrate complex this nitrogen is probably protonated or at least hydrogen bonded. Loss of carbon dioxide from XII gives XIII which takes up a proton at the $\beta$-carbon. Subsequent reactions give alanine.

On incubation with substrate, or various other amino-acids, the enzyme loses activity [3, 9]. The activity may be regenerated by addition of pyridoxal phosphate or a keto-acid. Meister and his co-workers [9, 10] were able to show that this behaviour was due to the ability of the enzyme to catalyse relatively slow, non-specific transaminations. In the presence of an amino-acid this causes the enzyme to be converted into the pyridoxamine form, which is inactive in decarboxylation. The active pyridoxal form is produced again by transamination when pyridoxal phosphate or keto-acids are added.

Snell and co-workers [13] have shown that the enzyme tryptophanase (4.2.1.20) catalyses a series of amino-acid interconversions including the $\alpha,\beta$-eliminations shown in equations (7.5), (7.6), (7.7) and (7.8), and the $\beta$-replacements (7.9), (7.10) and (7.11).

$$\text{Tryptophan} + H_2O \rightarrow \text{pyruvate} + NH_3 + \text{indole} \tag{7.5}$$

$$\text{Serine} \rightarrow \text{pyruvate} + NH_3 \tag{7.6}$$

$$\text{Cysteine} + H_2O \rightarrow \text{pyruvate} + NH_3 + H_2S \quad (7.7)$$

$$\text{S-Methylcysteine} + H_2O \rightarrow \text{pyruvate} + NH_3 + CH_3SH \quad (7.8)$$

$$\text{Serine} + \text{indole} \rightarrow \text{tryptophan} + H_2O \quad (7.9)$$

$$\text{Cysteine} + \text{indole} \rightarrow \text{tryptophan} + H_2O \quad (7.10)$$

$$\text{4,5 or 6-methyl tryptophan} + \text{indole} \rightarrow \text{4,5 or 6-methylindole} + \text{tryptophan} \quad (7.11)$$

These reactions are analogous to the non-enzymic interconversions shown in Fig. 7.2 [14], and Snell *et al.* have shown that essentially similar mechanisms apply to the enzyme-catalysed processes. The whole scheme may be summarized as in the equations (7.12), the key reaction being the production of the enzyme-bound α-amino-acrylate which is common to

$$\begin{aligned} E + R.CH_2\,CH(NH_2).CO_2H &\rightleftharpoons E.R.CH_2CH(NH_2).CO_2H \\ E.R\,CH_2CH(NH_2).CO_2H &\rightleftharpoons E.CH_2{=}C(NH_2).CO_2H + RH \quad (7.12) \\ E.CH_2{=}C(NH_2).CO_2H &\rightarrow E + CH_3.CO.CO_2H + NH_3 \end{aligned}$$

all the interconversions. The α-amino-acrylate–enzyme complex may break down irreversibly to give pyruvate and ammonia, or may react with a new RH species to give the new amino-acid.

A detailed kinetic study allowed comparisons of the rate constants and $K_m$ values to be made [13], and the results were in complete agreement with the mechanism proposed. They also showed that the rate-limiting step in each case was the α,β-elimination leading to the production of the α-amino-acrylate complex.

In the holo-enzyme, the pyridoxal phosphate is bound to the protein by an aldimine link to the ε-amino-group of a lysyl residue. An ultraviolet absorption band at 420 nm is attributed to the hydrogen-bonded form (XIV) [13, 14, 15, 16], while a band at 337 nm which predominates at higher pH values probably represents (XV). Reduction of the holo-enzyme with sodium borohydride causes inactivation. The holo-enzyme reacts with amino-acids, substrates and non-substrates such as alanine, giving the new imine (XVI). Borohydride reduction would not be expected to inactivate this complex since the resulting covalent bonds would be between substrate and coenzyme, and the free apo-enzyme would be liberated. The experiments of Morino and Snell [13] showed this to be the case.

On addition of a substrate or alanine to the holo-enzyme, a new ultraviolet absorption band at 500 nm could be observed. In the case of all substrates this band disappeared as the reaction proceeded to give pyruvate and ammonia. When alanine was used the band remained, no product being formed. Snell *et al.* [13] concluded that this ultraviolet band was due to the quinonoid form (XVIIb), which is formed when the α-hydrogen is lost from the amino-acid. They provided evidence for this conclusion by showing that the exchange of the α-hydrogen with tritiated water was closely related to the appearance of this band. Similar enzyme–substrate complexes absorbing at 500 nm have been reported for other pyridoxal-dependent enzymes [17]. Fig. 7.5 shows the mechanistic scheme proposed by Snell *et al.* for the key α,β-elimination in tryptophanase.

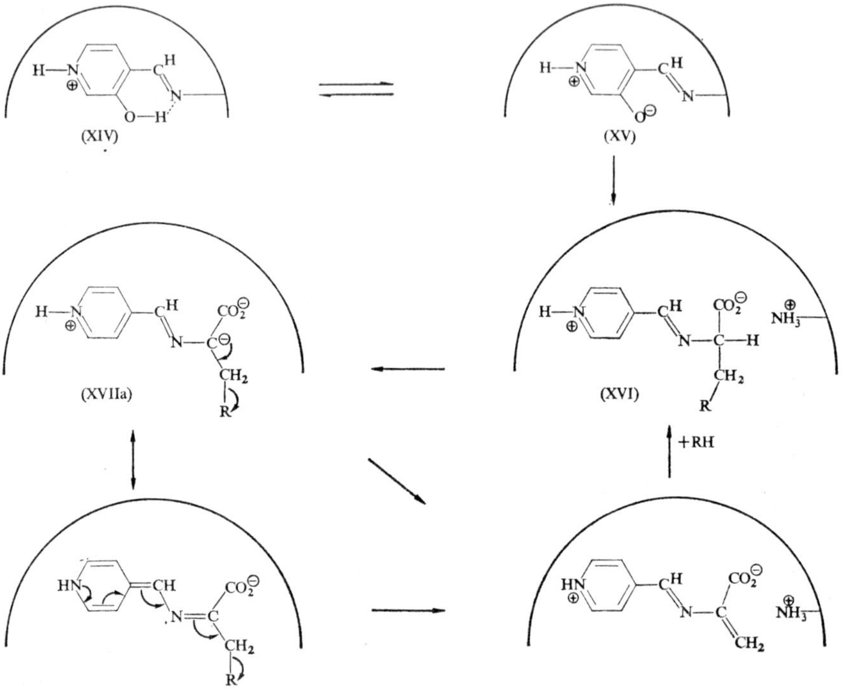


FIG. 7.5. Mechanism for the α,β-elimination reactions of tryptophanase (Snell *et al.* [13])

Histidine deaminase catalyses the elimination of ammonia from histidine (see equation 7.1). Abeles and his co-workers [18] have reported that the enzyme is inactivated by carbonyl reagents (sodium borohydride, cyanide ion, bisulphite, phenylhydrazine and hydroxylamine), indicating that the

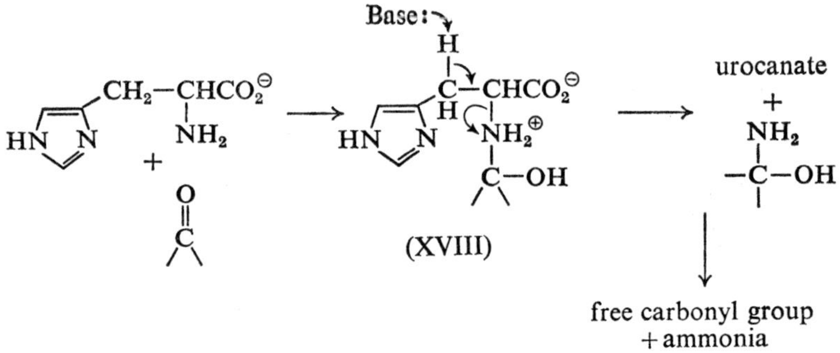

enzyme contains a carbonyl group essential for activity. This type of evidence is often diagnostic of a pyridoxal phosphate enzyme. However, these workers report that pyridoxal phosphate is absent. They have proposed a mechanism for the reaction in which interaction between the histidine amino-group and the (as yet undefined) carbonyl group produces a carbinolamine (XVIII). Since carbinolamines possess $pK_a$ values 2–3 units lower than those of amines [19], they suggest that the carbinolamine provides a better leaving group for the elimination. The elimination gives urocanate and an ammonia–enzyme complex which can break down to give ammonia and the free carbonyl. Dissociation of ammonia, the last stage, appears to be rate-determining.

## THIAMINE PYROPHOSPHATE-DEPENDENT ENZYMES

2-Oxoacid carboxy-lyase (pyruvate decarboxylase, carboxylase, 4.1.1.1) was the first enzyme for which thiamine pyrophosphate (XIX) was found to be a coenzyme [20]; hence the original name for this coenzyme, cocarboxylase. Thiamine possesses a thiazolium group, linked by a methylene

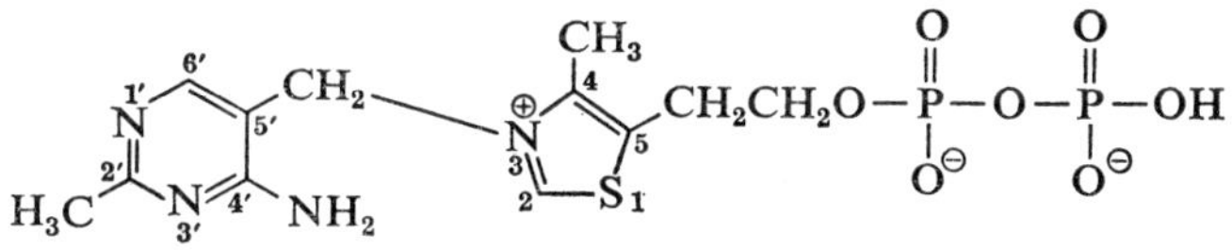


(XIX; thiamine pyrophosphate)

bridge to a substituted pyrimidine ring. It functions as a coenzyme for various enzymes only as the pyrophosphate derivative. All enzymes which require thiamine pyrophosphate as a co-factor also require magnesium ions; in the absence of either of these factors no enzymic activity can be observed.

The various reactions catalysed by the thiamine pyrophosphate-dependent enzymes are basically similar, involving cleavage of a carbon–carbon bond adjacent to the keto-group in the substrate. Thus, pyruvate decarboxylase catalyses the degradation of pyruvate, or other keto-acids, to carbon dioxide and acetaldehyde (or the corresponding aldehyde (equation 7.13)). An oxidative decarboxylation is brought about by pyruvate dehydrogenase (1.2.2.2, 1.2.4.1) (7.14). Oxoacid carboxy-lyases from various

$$R-\overset{\overset{O}{\|}}{C}-CO_2H \longrightarrow R-\overset{\overset{O}{\|}}{C}H + CO_2 \qquad (7.13)$$

$$\text{Pyruvate} \longrightarrow \text{acetate} + CO_2 \qquad (7.14)$$

sources also catalyse the synthesis of acyloins, for example, acetoin may be produced from pyruvate plus acetaldehyde, or from two moles of pyruvate.

Other reactions in this category catalysed by thiamine pyrophosphate-enzymes are the transketolase (equation 7.16) and phosphoketolase (equation 7.17) reactions.

**Non-enzymic Thiamine-catalysed Reactions**

As in the case of the pyridoxal-containing enzymes, much of our present understanding of the reaction pathways of the thiamine pyrophosphate-enzymes arises from studies of non-enzymically catalysed reactions. The reactions which are catalysed by thiamine and its derivatives are essentially similar to those catalysed by the enzymes. Thus, thiamine and thiazolium compounds catalyse the acyloin condensation of furfural to furoin [21] (equation 7.15). In the presence of thiamine the decarboxylation of pyru-

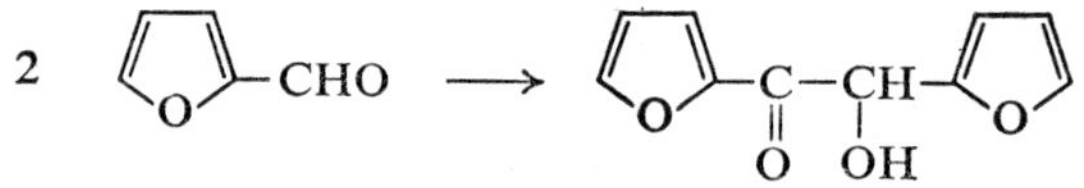


(7.15)

vate and the synthesis of acetoin from pyruvate and acetaldehyde are effected [22]. In addition thiamine and pyruvate, in the absence of acetaldehyde, give α-acetolactate (XXV), which under certain conditions loses carbon dioxide rapidly giving acetoin [23].

These reactions involving pyruvate, which may be summed up as shown, suggested [24] that the pyruvate during the course of the reaction was

Pyruvate ⟶ Intermediate —Pyruvate⟶ acetolactate ⟶ acetoin

Intermediate —acetaldehyde↘ acetoin

converted into a so-called 'active acetaldehyde' intermediate which is able to react in several ways: (a) it can condense with more pyruvate giving acetolactate, (b) it can react with acetaldehyde to give acetoin, (c) it can decompose (interact with water) giving acetaldehyde. In the non-enzymic reactions (c) does not occur readily and acetaldehyde is rarely formed.

When pyruvate and thiamine were incubated at pH 8·8, evolution of carbon dioxide was rapid, while acetoin or acetolactate production was delayed, subject to a pronounced lag [25]. This implies that the initial decarboxylation is rapid, giving the 'active aldehyde' intermediate which must be relatively stable and therefore accumulates.

The nature of this intermediate was explained by Breslow [26, 27], who showed, first, that the hydrogen atom on the 2-position of 3,4-dimethylthiazolium bromide (XX) exchanged very rapidly with deuterium in $D_2O$. The lability of this hydrogen means that the anion (XXI) must be relatively stable, the stability being, in the main, due to the presence of the positively-charged nitrogen atom adjacent to the anionic carbon. This

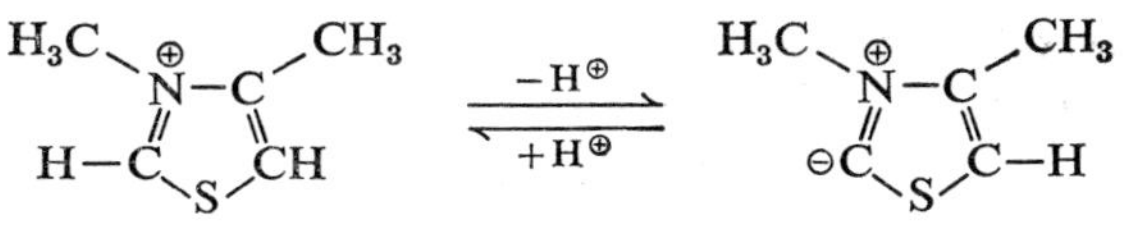


(XX) (XXI)

thiazolium anion would have a pronounced nucleophilic character, and Breslow proposed that in the mechanism for the conversion of pyruvate to acetoin, acetolactate and acetaldehyde, the first step was a nucleophilic attack on the carbonyl group of the pyruvate molecule (Fig. 7.6).

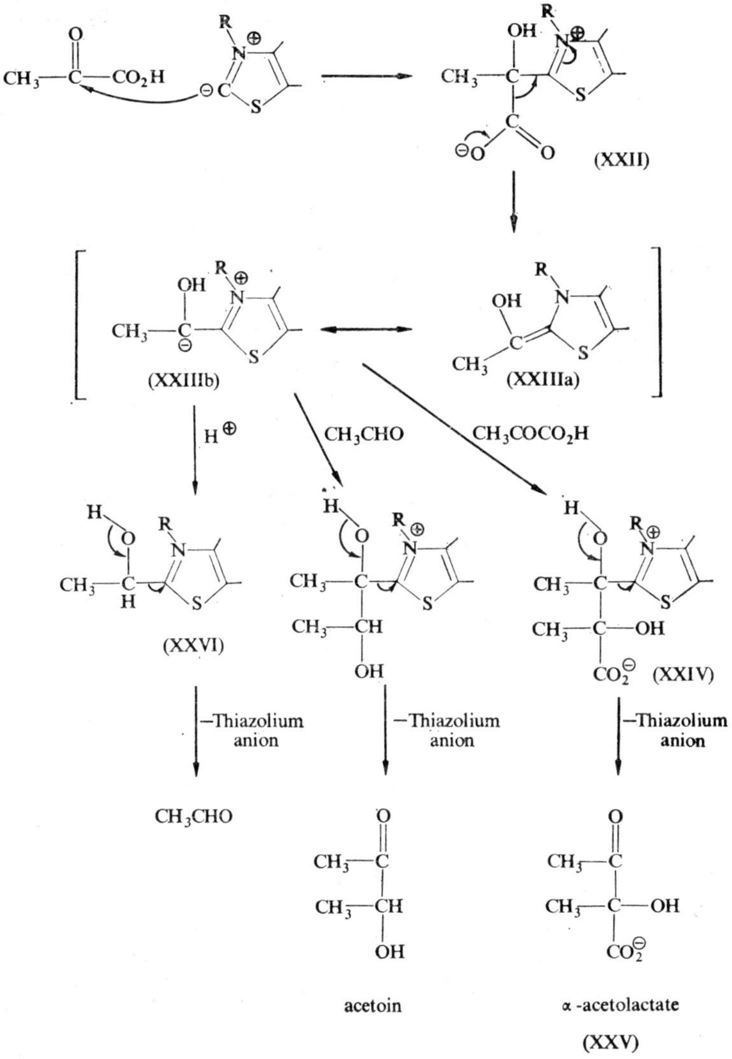


FIG. 7.6. Pathways for the thiazolium-catalysed reactions of pyruvate [26, 27]

The decarboxylation of the resultant species (XXII) is assisted by the positive charge of the thiazolium ring, and this process gives a species (XXIIIa) which may be regarded as a carbanion (XXIIIb) stabilized by resonance as shown. This carbanion will also be nucleophilic and may attack pyruvate, to give the complex (XXIV) which breaks down to give α-acetolactate (XXV), or acetaldehyde giving eventually acetoin. If the carbanion (XXIIIb) takes up a proton from the solvent then a 2-(1′-hydroxyethyl)-thiazolium derivative (XXVI) is produced, which may break down to give acetaldehyde.

A 2-(1′-hydroxyethyl)-thiazolium derivative has been synthesized and it has been shown to exchange the hydrogen on the 1-carbon in water [28]. Such an exchange confirms the stability of the anion (XXIIIb).

At this point it is worth emphasizing the role played by the thiazolium species in both the decarboxylation and the acyloin reactions (see also references [24, 29]). Similar roles are played by pyridoxal, cyanide ion and certain amines in similar reactions. We have seen that the decarboxylation of the thiazolium derivative (XXII) is assisted by the presence of the positively-charged nitrogen atom, the result being a carbanion stabilized by resonance involving that nitrogen atom. In chapter 1 the cyanide-ion-catalysed decarboxylation of α-ketoacids was discussed and the effect of the cyanide ion was there attributed to its ability to stabilize an analogous carbanion (XXVII, chapter 1). The stability of the carbanion, and its nucleophilic character, was also shown to be the important factor in the cyanide-ion-catalysed acyloin reactions.

Just as the positively-charged nitrogen in the thiazolium group, and nitrogen in the cyanide ion, assists the decarboxylation by withdrawing electrons and thus stabilizing the resultant carbanions, so also does the hetero-cyclic ring of pyridoxal (in compounds such as I, this chapter) assist in the decarboxylation of amino-acids. In Fig. 7.2 the stabilization of a carbanion by resonance involving the heterocyclic ring is indicated for the α,β-eliminations. β,γ-Eliminations may be brought about when a carbanion (Xa, Fig. 7.3) is stabilized by an imino-nitrogen.

Decarboxylations, and aldol and acyloin condensations, may also be catalysed by amines, by mechanisms which involve similar principles to those described above. It is of great interest that amino-groups derived from the proteins appear to be implicated in the reaction mechanisms of certain decarboxylases and aldolases. These are discussed later in the chapter.

Considerable evidence has been obtained to justify the mechanistic pathways for thiazolium-catalysed reactions shown in Fig. 7.6. Thus, 2-(1′-hydroxyethyl)-thiamine pyrophosphate has been shown to be produced from thiamine pyrophosphate and acetaldehyde [30], and to react with acetaldehyde to yield acetoin via the acetoin–thiamine pyrophosphate complex [30, 31]. Synthetic 2-(1′-hydroxyethyl)-derivatives of 3,4-dimethyl thiazolium and 3-benzyl-4-methyl thiazolium were shown to be very efficient catalysts of the acyloin reactions [32].

**Enzyme-catalysed Reactions**

Once a mechanistic scheme has been established for the catalysis of reactions by a coenzyme in the absence of its apo-enzyme, it is natural that an essentially similar mechanism should be proposed for the enzymic reaction. That the mechanistic pathway shown in Fig. 7.6 applies for the reactions of the thiamine pyrophosphate enzymes is now supported by a considerable body of experimental evidence.

Pyruvate decarboxylase, as stated earlier, catalyses the decarboxylation of pyruvate, yielding acetaldehyde. The reaction proceeds via the intermediate formation of 'active acetaldehyde' bound to the apo-enzyme. 2-(1′-Hydroxyethyl)-thiamine pyrophosphate was isolated from a reaction mixture containing pyruvate and the enzyme by Holzer and Beauchamp [33] and incubation of the enzyme (from which the coenzyme had been removed) with $^{14}$C-labelled hydroxyethylthiamine pyrophosphate gave $^{14}$C-labelled acetaldehyde [30]. In this latter experiment Krampitz *et al.* observed a rapid initial release of acetaldehyde followed by a much slower release. The exchange of thiamine pyrophosphate for the hydroxyethyl derivative must therefore be slow, showing that the apo-enzyme–coenzyme binding is quite strong.

A major difference exists between the enzymic and non-enzymic reactions of thiamine and pyruvate. In the case of the non-enzymic process acyloin production predominates, very little acetaldehyde being detected. By contrast acetaldehyde is the major product of the pyruvate decarboxylase reaction. Juni [34] has suggested that this may be because the acetaldehyde–thiamine pyrophosphate complex is relatively stable and that the loss of free acetaldehyde in the enzymic reaction is brought about by an extra step in the mechanism. The decarboxylation of pyruvate is inhibited by acetaldehyde and it was proposed that the active site of the enzyme possessed, in addition to a suitable locus for pyruvate, an extra acetaldehyde-binding site. The dissociation of acetaldehyde from the enzyme complex was envisaged as involving transfer of the 'active aldehyde' fragment to this latter site from which dissociation was more rapid. Added acetaldehyde would bind to this site and thus prevent the overall reaction. Transfer of the aldehyde group to this site is apparently not necessary for all substrates of the enzyme. Acetaldehyde does not inhibit the decarboxylation of larger keto-acids. Aldehyde–thiamine pyrophosphate complexes from longer-chain aldehydes are far less stable than that of acetaldehyde and require no extra dissociation mechanism.

The suggestion of an extra site which accepts the 'active aldehyde' fragment is of interest in view of the work carried out by Schellenberger and his co-workers [35, 36] on the coenzyme activity of various analogues of thiamine pyrophosphate. Although the thiazolium group itself is able to catalyse certain reactions of pyruvate, the structural requirements for coenzyme activity with the enzyme are more strict [24]. Thus, the pyrophosphate and 4′-amino groups, which can have no role in the non-enzymic reactions, appear to be essential for the activity of pyruvate decarboxylase. In general it has been assumed that structural features in the coenzyme apart from the thiazolium group are involved in the coenzyme–apoenzyme binding. This is almost certainly the case for the pyrophosphate part of the molecule. However, the 4′-amino-group may have a different role.

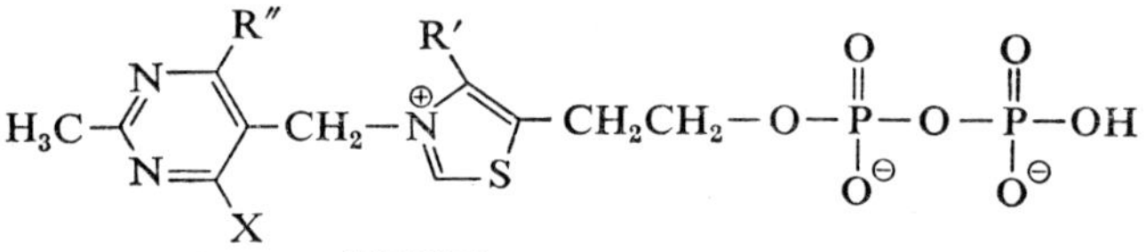


(XXVII)

4′-Hydroxy-4′-desaminothiamine pyrophosphate (XXVII; X = OH R′ = $CH_3$, R″ = H) binds to the active site of the apo-enzyme almost exactly as strongly as does thiamine pyrophosphate itself, and yet is inactive as a coenzyme. When this hydroxy compound is used in place of thiamine pyrophosphate, the 'active aldehyde' appears to be formed, but it is unable to lose acetaldehyde and regenerate active enzyme. The 4′-amino-group appears therefore to have a role in the process of dissociation of the acetaldehyde from the 2-position of the thiazolium portion of the coenzyme. Schellenberger [35] suggests that an intramolecular transfer of the aldehyde species from the 2- to the 4′-position takes place. The acetaldehyde acceptor site proposed by Juni [34] is therefore, according to this hypothesis, the 4′-amino group. Schellenberger *et al.* demonstrated the feasibility of such a process by showing that certain nucleophilic compounds such as 4-amino-pyrimidines catalysed the release of acetaldehyde from hydroxyethylthiazolium salts. In addition, 6′-methylthiamine pyrophosphate (XXVII; X = $NH_2$, R′ = $CH_3$, R″ = $CH_3$) was found to be inactive as a coenzyme, although it did bind to the apo-enzyme to a certain extent, while 6′-methyl-4-nor-thiamine pyrophosphate (XXVII; X = $NH_2$, R′ = H, R″ = $CH_3$) was active. It was suggested that where the coenzyme analogue carried a methyl group in each of 6′ and 4-positions, interaction between these methyl groups caused the coplanarity of the two rings to be lost to such an extent that the 4′-amino-group was no longer suitably positioned for the intramolecular transfer.

A kinetic scheme in which the transfer of the aldehyde group between

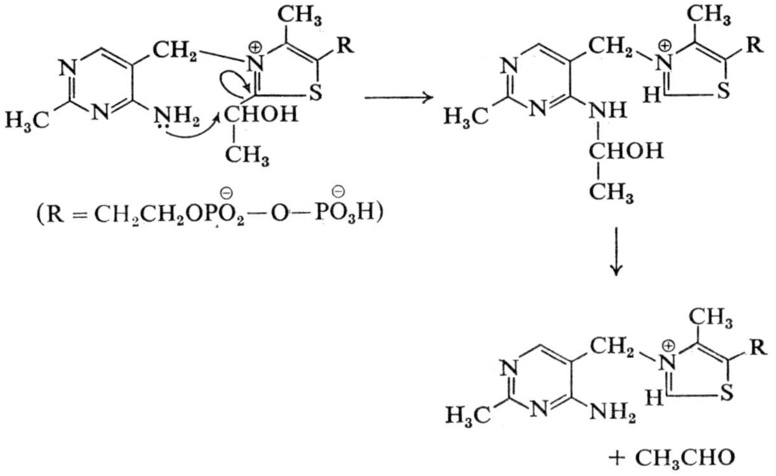


the 2- and 4′-positions of the coenzyme is rate-determining is compatible with results obtained from the enzymic reaction [36].

The magnesium ions required for the activity of pyruvate decarboxylase may be replaced by various other metal ions. The role of the metal-ions appears to be in ensuring the binding between the coenzyme and the apo-enzyme. The rate of combination of the components of the enzyme (apo-enzyme, coenzyme and metal-ion) varied according to the nature of the metal-ion used, but once the enzyme was formed, its activity was independent of the metal-ion which it contained [36].

The oxidative decarboxylation of pyruvate requires an enzyme possessing two coenzymes. The first of these is thiamine pyrophosphate while the second is a coenzyme involved in the oxidative process. In the presence of a suitable electron acceptor, pyruvate is converted into acetate plus carbon dioxide. In the absence of an oxidizing agent the enzyme catalyses the acyloin condensation via the same intermediate as that which is involved in the simple decarboxylase reaction. Thus, porcine *pyruvic oxidase* catalysed the synthesis of α-acetolactate from pyruvate and the hydroxyethylthiamine pyrophosphate [37] and a bacterial preparation gave acetoin from acetaldehyde and hydroxyethylthiamine pyrophosphate [30].

The overall reaction therefore involves decarboxylation to the hydroxyethyl derivative followed by oxidation, probably to an acetyl-thiamine. Acyl thiazolium compounds are susceptible to nucleophilic attack giving acyl derivatives [38]. Attack by water gives acetate, by phosphate gives acetyl phosphate and by coenzyme A gives acetyl-CoA.

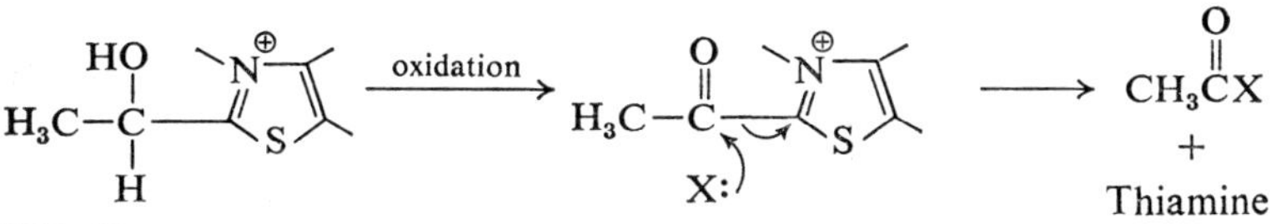

The reaction catalysed by *transketolase* (sedoheptulose-7-phosphate: D-glyceraldehyde-3-phosphate glycolaldehyde-transferase, 2.2.1.1) is a transfer process (equation 7.16) in which a two-carbon fragment (glycolaldehyde) is transferred from D-xylulose-5-phosphate to D-ribose-5-phosphate. In view of this, the enzyme is classified as a transferase; we shall consider it here because, as a thiamine pyrophosphate-dependent enzyme, its mechanism of action is similar to those discussed above. The carbon–

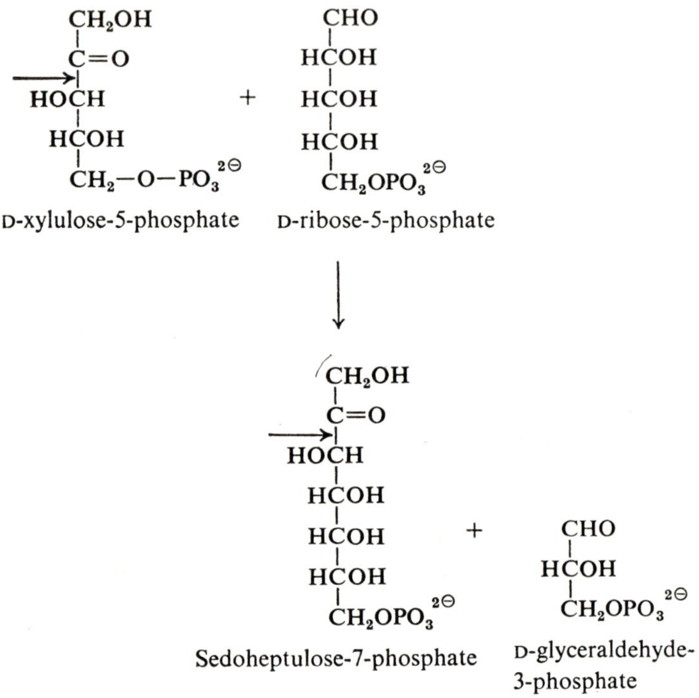


(7.16)

carbon bonds cleaved and formed in the reaction are shown, and the overall process is completely reversible (the systematic name of the enzyme is obtained by considering the reverse direction). In the forward direction of the reaction as in equation (7.16) the xylulose-5-phosphate may be regarded as a glycolaldehyde donor, and the ribose phosphate as an acceptor.

Transketolase exhibits a fairly wide specificity [39]. Thus, for example, D-fructose-6-phosphate, octulose-8-phosphate, L-erythrulose and even hydroxypyruvate can function as glycolaldehyde donors while glycolaldehyde and D-glucose-6-phosphate can act as acceptors.

The transketolase reaction occurs in two distinct stages. In the absence of an acceptor, xylulose-5-phosphate is converted into glyceraldehyde phosphate plus 'active glycolaldehyde', a two-carbon species which remains bound to the enzyme [40, 41]. This two-carbon fragment is transferred to an acceptor introduced into the system. The same 'active glycolaldehyde'

$$\text{E} + \text{xylulose-5-phosphate} \rightleftharpoons \text{E-glycolaldehyde} + \text{glyceraldehyde phosphate}$$

$$\text{E} - \text{glycolaldehyde} + \text{ribose-5-phosphate} \longrightarrow \text{E} + \text{sedoheptulose-7-phosphate}$$

was obtained when D-fructose-6-phosphate was used as donor.

When the enzyme was incubated with $^{14}C$-labelled fructose phosphate the isolated protein was shown to be radioactive [42]. The radioactivity could be transferred to an acceptor such as ribose-5-phosphate. In addition radioactively-labelled glycolaldehyde could be obtained from the protein by heating.

The nature of the 'active glycolaldehyde' species became clear when Breslow's explanation of thiamine-catalysed reactions was proposed. The reaction of transketolase is envisaged as proceeding by a nucleophilic attack on the C-2 carbonyl group of the substrate donor giving the complex (XXVIII), which eliminates the two-carbon fragment (attached to the coenzyme) liberating the aldehyde RCHO. The isolatable 'active glycolaldehyde' is dihydroxyethylthiamine pyrophosphate (XXIX), the conjugate base of which can interact with a new aldehyde species R′CHO to give the product (Fig. 7.7).

Dihydroxyethylthiamine pyrophosphate has been prepared non-enzymically from thiamine pyrophosphate and formaldehyde by a reaction which is analogous to the production of acetoin-thiamine from thiamine and acetaldehyde via hydroxyethylthiamine [30]. This synthetic compound has been shown to be active when used with transketolase, ribose-5-phosphate being converted into sedoheptulose-7-phosphate.

Since hydroxyethylthiamine pyrophosphate was obtained as an intermediate in the pyruvate oxidase reaction, Holzer *et al.* [43] reasoned that if hydroxypyruvate were incubated with this enzyme then the intermediate obtained would be the dihydroxyethylthiamine derivative, identical to that obtained in the transketolase reaction. This was shown to be the case; the coenzyme derivative obtained gave sedoheptulose phosphate when incubated with ribose-5-phosphate and transketolase. Further evidence for the structure of 'active glycolaldehyde' was obtained by degradative studies [44].

Whereas transketolase brings about the cleavage of D-xylulose-5-phosphate to triose phosphate plus a two-carbon fragment which is then transferred to a suitable acceptor, *phosphoketolase* (D-xylulose-5-phosphate D-glyceraldehyde-3-phosphate-lyase, 4.1.2.9) cleaves the same substrate

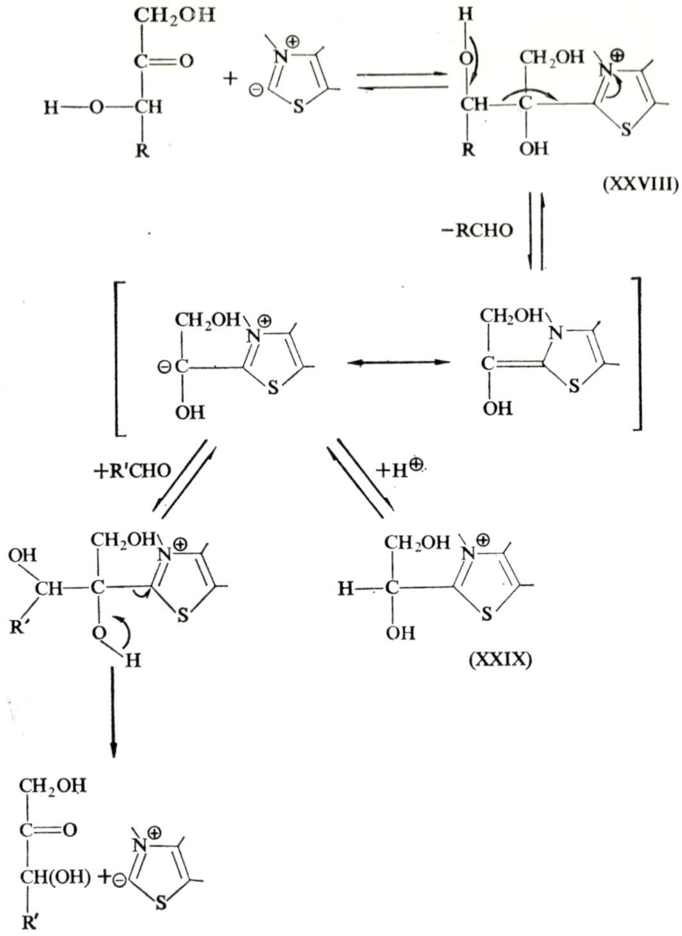


FIG. 7.7. Mechanism for the transketolase reaction

to the same triose phosphate, the two-carbon fragment being finally liberated as acetyl phosphate. The similarity between these two reactions led,

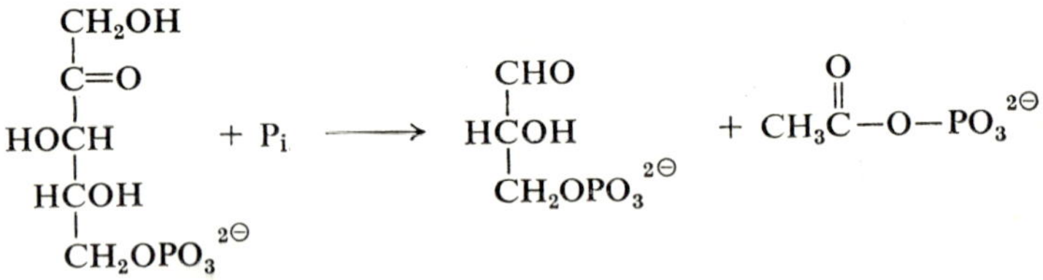

(7.17)

naturally, to similar mechanisms being proposed. Thus, a possible mechanism is as shown in Fig. 7.8, in which, as in the transketolase reaction,

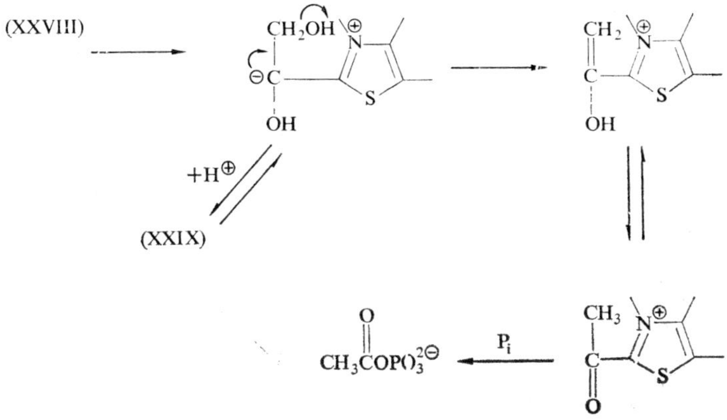


FIG. 7.8. Proposed mechanism for the phosphoketolase reaction

the first steps lead to the formation of the dihydroxyethylthiamine pyrophosphate ion, by elimination of the triose phosphate. Elimination of water then follows giving an enol which ketonizes, giving the acetyl derivative. Attack by phosphate produces acetyl phosphate. In support of this mechanism is the observation that in the presence of phosphoketolase dihydroxyethylthiamine pyrophosphate yielded acetyl phosphate [45].

The enzyme does not catalyse exchange between acetyl phosphate and phosphate [46], a result which implies that the reaction of phosphate with the acetyl derivative is, in practical terms, irreversible. Horecker [39] suggests that the phosphate ion may react with the enol form directly. The 'active glycolaldehyde' is probably a very transient intermediate in the process [39] since no exchange between xylulose-5-phosphate and the triose phosphate is catalysed by the enzyme [46].

## 'ACTIVE LYSINE' LYASES

In this chapter we have already encountered two types of enzymes which catalyse decarboxylations. The pyridoxal-enzymes function with amino-acids as their substrates and thiamine pyrophosphate-enzymes can catalyse the decarboxylation of α-ketoacids. (The reverse of the carbon dioxide ligase reaction is another possible biological mechanism for decarboxylation.) The decarboxylation of β-ketoacids provides yet another example of a decarboxylation mechanism.

### Acetoacetate Decarboxylase (4.1.1.4)

This enzyme, from *Clostridium acetobutylicum*, has been studied with great thoroughness by F. H. Westheimer and his co-workers [47, 48, 49, 50, 51]. It is normally isolated as an aggregate containing sub-units of molecular weight approximately 29,000 [51], there being two such units per active site. The variations of the kinetic parameters $V$ and $V/K_m$ with pH indicate ionizable groups with $pK_a$ values *ca.* 5·5 and 6·5. These $pK_a$'s should correspond to the ionization of two groups in the enzyme–substrate complex and in the enzyme, respectively. They are probably both cationic acids (that is, of the type $BH^{\oplus}$) because the $pK_a$ values decreased in water containing 30% ethanol (see also Chapter 6, page 267).

The decarboxylation of $\beta$-ketoacids catalysed by primary amines is discussed briefly in chapter 1 (p. 47). The reaction proceeds by formation of a Schiff-base between the amine and the $\beta$-carbonyl group, followed by the elimination of carbon dioxide which leads to the production of an enamine. The enamine tautomerizes to an imine which, on hydrolysis, yields the ketonic product. In such a mechanism the carbonyl oxygen of the keto-acid is lost in the intermediate stages and therefore exchange of this oxygen in water would be expected. Demonstration that such an exchange takes place when the decarboxylation of acetoacetate (labelled with $^{18}O$ in the carbonyl group) is catalysed by the enzyme [48] led immediately to the suggestion that a Schiff-base mechanism also applied to the enzymic process. The amino-group was envisaged as being provided by the active site of the enzyme. The pathway for such a scheme is shown in Fig. 7.9.

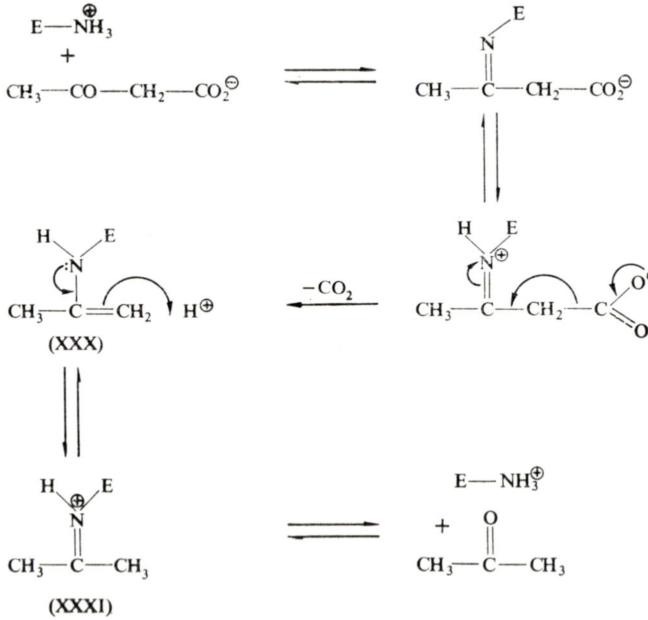


(XXX)

(XXXI)

FIG. 7.9. Schiff-base mechanism for the decarboxylation of acetoacetate in the presence of acetoacetate decarboxylase (see refs. [47, 50])

An intermediate in this pathway is the enamine (XXX), the tautomer of the acetone-amine Schiff-base (XXXI). On this basis the enzyme should catalyse the exchange of the hydrogen atoms of acetone with those from water. The demonstration of such an exchange by Westheimer and his co-workers [51] lends further support for the suggested mechanism.

However, the best evidence in favour of the Schiff-base mechanism has been obtained from experiments with sodium borohydride. This reagent inactivates acetoacetate decarboxylase in the presence of its substrate, but not in its absence. When the enzyme was treated with sodium borohydride in the presence of $^{14}C$-labelled substrate, the radioactive label became bound to the protein [47] and acid hydrolysis gave a radioactive compound identified as ε-isopropyllysine (XXXII). Therefore the amine which formed the Schiff-base was the ε-amino-group of a lysyl residue in the enzyme, and under the conditions of the experiment the substrate

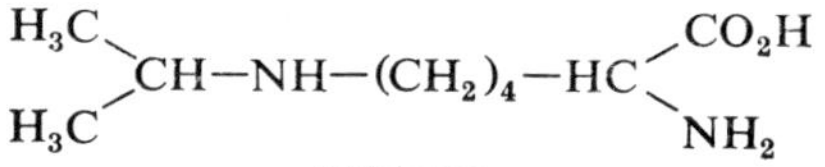

(XXXII)

had been decarboxylated before the reduction took place; only the Schiff-base of the decarboxylated molecule was detected.

The inhibition of the enzyme by cyanide is also of interest. Cyanide is not instantly effective in reducing the reaction rate; a marked induction period or lag is observed. This lag could not be eliminated by preincubation of cyanide either with the enzyme or with the substrate and therefore the cyanide presumably reacts with some kind of enzyme–substrate complex, probably the Schiff-base [47]. If this is the case then cyanide should protect the enzyme from inactivation by borohydride. Such protection has been observed [47].

Acetoacetate decarboxylase is also inhibited by acetopyruvate which interacts with the active site amine but is not decarboxylated [51]. The

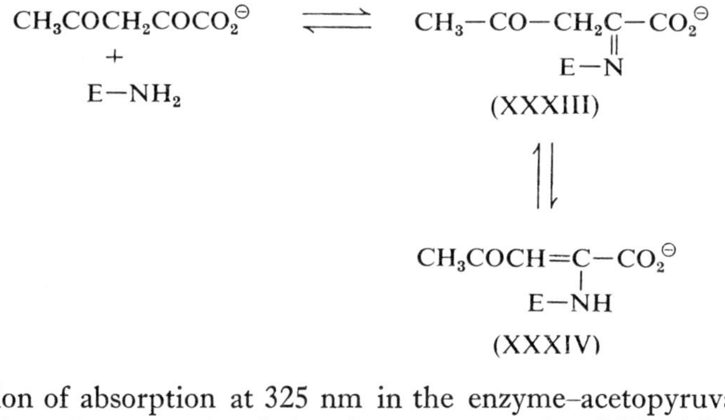

(XXXIII)

(XXXIV)

observation of absorption at 325 nm in the enzyme–acetopyruvate complex has been interpreted as being due to an enamine (XXXIV) resulting from tautomerism of the originally-formed Schiff-base (XXXIII). One acetopyruvate molecule per 60,000 molecular weight of protein becomes bound. Borohydride does not make the inhibition by acetopyruvate permanent by reducing the imine (XXXIII). Presumably the complex exists predominantly in the enamine form (XXXIV) which is not susceptible to reduction.

The 'active lysine' at the active site of the enzyme can be acetylated by either acetic anhydride or dinitrophenyl acetate, the acetylated protein being inactive enzymically [51]. One acetyl residue per active lysine is apparently introduced.

**Aldolases and Transaldolase**

Evidence similar to that obtained for acetoacetate decarboxylase above, has established that Schiff-base mechanisms also apply to many enzymes which catalyse aldol condensations and dealdolizations [52]. The best known of these are *fructosediphosphate aldolase* (fructose-1,6-diphosphate

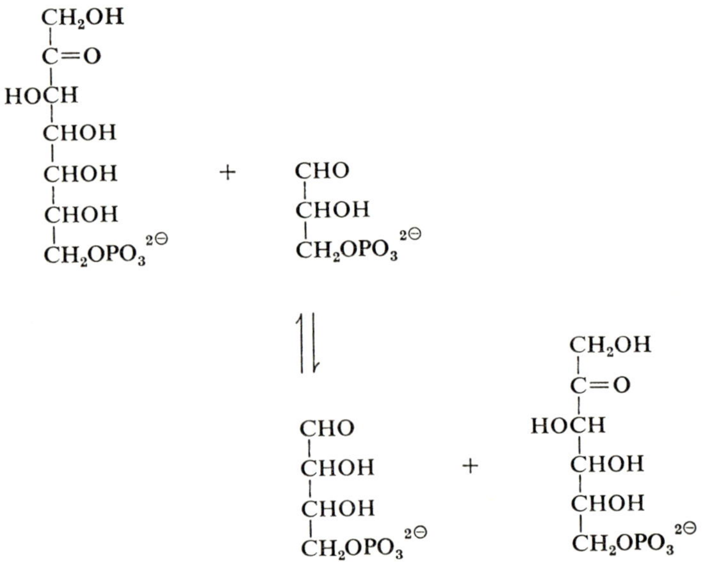

(7.18)

D-glyceraldehyde-3-phosphate-lyase, 4.1.2.13) and *transaldolase* (sedoheptulose-7-phosphate: D-glyceraldehyde-3-phosphate dihydroxyacetonetransferase, 2.2.1.2). These two enzymes are closely related in the nature of the reactions which they catalyse and in their mechanisms.

Transaldolase catalyses the transfer of a dihydroxyacetone group from a ketose to a suitable acceptor aldehyde. For example, in equation (7.18) the dihydroxyacetone group is transferred from sedoheptulose-7-phosphate to the acceptor D-glyceraldehyde-3-phosphate giving D-fructose-6-phosphate.

In the presence of a suitable substrate (a ketose which can donate a dihydroxyacetone group) and in the absence of an acceptor, an enzyme-bound $C_3$-fragment is produced, which has been designated 'active dihydroxyacetone'. This enzyme–dihydroxyacetone complex is relatively

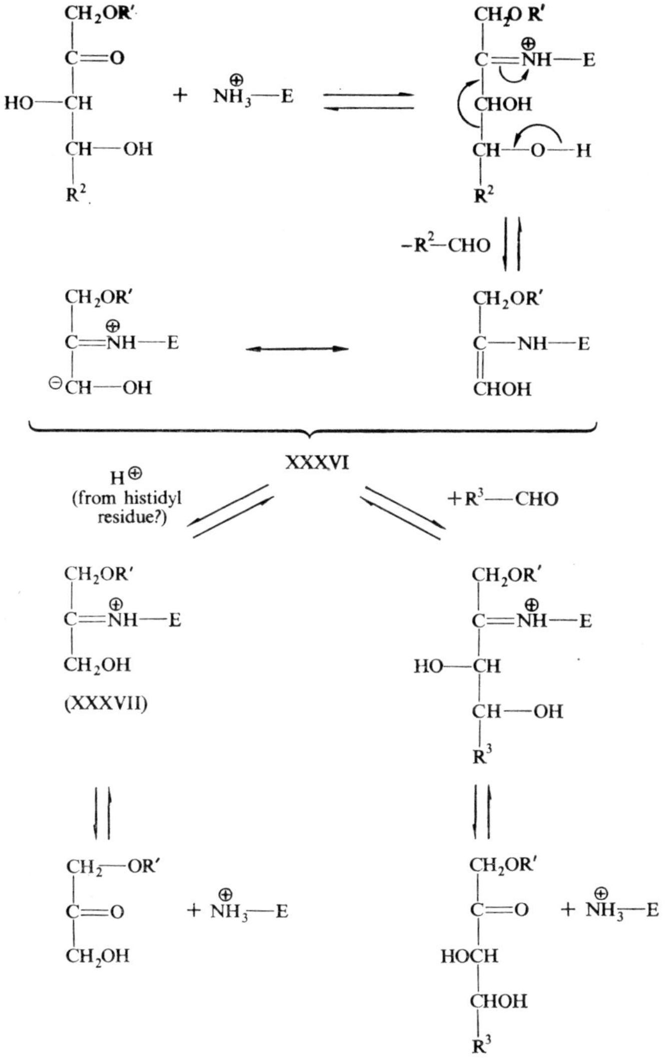


FIG. 7.10. Transaldolase (R′ = H) and aldolase (R′ = $PO_3^{2\ominus}$) mechanisms [39, 52]

stable; it can be isolated by precipitation or by ion-exchange chromotography [39, 53]. If prepared by incubation of the enzyme with fructose-6-phosphate labelled in the C-1 position with $^{14}C$, the complex can yield either radioactive dihydroxyacetone on heating or a labelled ketose phosphate on incubation with a suitable acceptor [39, 53]. Therefore, in the reaction shown in equation (7.18) the $C_3$ fragment is transferred from donor to acceptor via an enzyme-bound intermediate. As in the case of the decarboxylase the nature of the link between the enzyme and the dihydroxyacetone was established by reduction with borohydride. In the presence of a substrate, this reagent inactivates the enzyme and if the substrate is radioactively labelled, the label becomes firmly bound to the protein [53]. Hydrolysis of such a protein derivative gave labelled $N^6$-$\beta$-glyceryllysine (XXXV) indicating that the dihydroxyacetone had been linked to the enzyme via a Schiff-base with the $\varepsilon$-amino group of a lysyl residue [54, 55, 56]. One lysine per molecular weight of 57,000 appears to be involved [53].

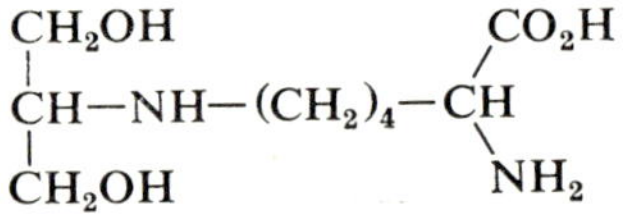


(XXXV)

In view of these results, a mechanism for the action of transaldolase similar to the non-enzymic amine-catalysed aldol condensations (see Chapter 1, p. 49) is accepted. The process (Fig. 7.10) involves the now familiar elimination assisted by the protonated imino-group, giving rise to the enamine, the 'active dihydroxyacetone' (XXXVI) which may be viewed as a resonance-stabilized carbanion. This can react with the aldehydo group of the acceptor molecule ($R^3$-CHO) giving the ketose–Schiff-base and eventually the ketose product itself. The optical purity of the product indicates that the condensation is stereospecific, the acceptor molecule being held in the correct position on the surface of the enzyme.

The reaction pathway for fructosediphosphate aldolase is essentially very similar to that of transaldolase, the main difference being that the enzyme–$C_3$ complex (active dihydroxyacetone phosphate) is more unstable and cannot be isolated. Therefore the reaction, rather than being a transfer process, is in fact a dealdolization (equation 7.19).

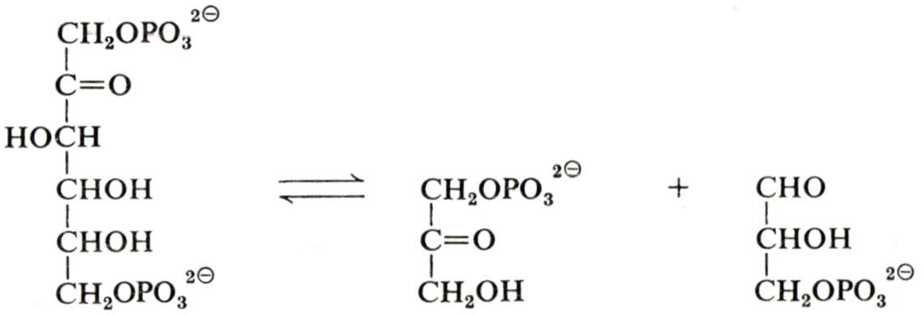


D-fructose-1,6-diphosphate — Dihydroxyacetone phosphate — D-glyceraldehyde-3-phosphate

(7.19)

Borohydride inactivates this enzyme in the presence of an excess of dihydroxyacetone phosphate [57] and the product on hydrolysis gave $N^6$-$\beta$-glyceryllysine, identical to that obtained with transaldolase (the phosphate group being lost on hydrolysis). Therefore, the dihydroxyacetone phosphate was bound to the protein via a Schiff-base. Just as the Schiff-base mechanism predicted that acetoacetate decarboxylase would catalyse exchange of hydrogen between acetone and water, so also it was expected that fructosediphosphate aldolase would catalyse exchange of the hydrogens in dihydroxyacetone phosphate. Demonstration of this exchange [58] lends support to the postulated mechanism. The exchange reaction is highly stereospecific, only one hydrogen being exchanged [58].

Some light has been thrown on the nature and importance of the proton transfer involved in this exchange (the interconversion of the dihydroxyacetone phosphate Schiff-base (XXXVII) and its conjugate base (XXXVI)) by the photo-oxidation studies carried out by Horecker and his co-workers [59]. The enzyme–$C_3$ complex in the case of fructosediphosphate aldolase has already been described as relatively unstable, dissociating rapidly while the similar complex of transaldolase is stable. Photo-oxidation of fructosediphosphate aldolase in the presence of 'rose bengal' which caused the destruction of histidine residues exclusively [60] caused loss in the enzyme's ability to catalyse the cleavage of fructose diphosphate. Ingram and Hochster [61] have also reported that photo-oxidative destruction of histidine destroys enzymic activity. In addition, Horecker *et al.* [59] reported that the photo-oxidation also reduced the rate of tritium exchange in dihydroxyacetone phosphate. Interestingly, however, the ability of the enzyme to function as a transaldolase (that is, to transfer dihydroxyacetone phosphate to an acceptor) was not impaired. Therefore the histidyl residues destroyed may be involved in the transfer of protons to the $C_3$ fragment, the transfer being essential for the dissociation of dihydroxyacetone phosphate from the protein. In the absence of this transfer process, the presence of an acceptor provides an alternative pathway for the loss of the $C_3$ fragment and the regeneration of the free enzyme. This role for histidine in the aldolase reaction is indicated in Fig. 7.10.

Fructosediphosphate aldolase does not show a very broad substrate specificity. For example, although it is able to catalyse the degradation of fructose-1-phosphate its activity towards this substrate is considerably lower than that towards the diphosphate. It is probable that a binding site exists on the enzyme for the 6-phosphate group of the substrate. Treatment of the enzyme with carboxypeptidase causes the loss of 2-3 tyrosines per molecule (the molecule probably contains four peptide chains [62, 63, 64, 65, 66, 67, 68, 69]) and the resulting protein loses activity towards the diphosphate [70, 71]. This suggests that the C-terminal tyrosine residues provide a binding site for the 6-phosphate group of the substrate. Horecker *et al.* [72] provided support for this thesis by showing that acetylation of tyrosine residues in the enzyme reduced its activity towards the diphosphate but not towards the monophosphate.

Another aldolase, ketose-1-phosphate aldolase (4.1.2.7), shows a greater degree of specificity towards the monophosphate relative to the diphosphate than does the diphosphate aldolase. In addition it appears to possess fewer C-terminal tyrosine residues per molecule [73]. It may be that these two enzymes are very similar indeed, the main difference being that the diphosphate aldolase possesses more C-terminal tyrosine residues and hence more effective binding for the 6-phosphate group of the substrate.

Many aldolases appear to function by a Schiff-base mechanism, being inactivated by borohydride in the presence of substrates. These include 2-keto-4-hydroxyglutarate aldolase (4.1.2.–) [74], deoxyribose phosphate aldolase (4.1.2.4) [75] and 2-keto-3-deoxy-6-phosphogluconate aldolase (4.1.2.14) [75, 76]. Melocke [76] has shown that this last enzyme, which catalyses the synthesis of 2-keto-3-deoxy-6-phosphogluconate from pyruvate and D-glyceraldehyde-3-phosphate, is irreversibly inactivated by bromopyruvate. Pyruvate itself forms a Schiff-base with the enzyme. The bromopyruvate presumably forms a Schiff-base with the active lysine of the enzyme and in doing so introduces the bromomethyl group into the vicinity of a nucleophilic group. This may be a histidyl residue whose normal role is proton transfer.

### 2-Keto-3-deoxy-L-arabinonate Dehydratase

Abeles and his co-workers [77] report another lyase reaction in which an enzyme–substrate Schiff-base is involved. This is a dehydration, brought about by 2-keto-3-deoxy-L-arabinonate dehydratase (4.2.1.–), producing 2,5-diketo-valerate from 2-keto-3-deoxy-L-arabinonate (equation 7.20).

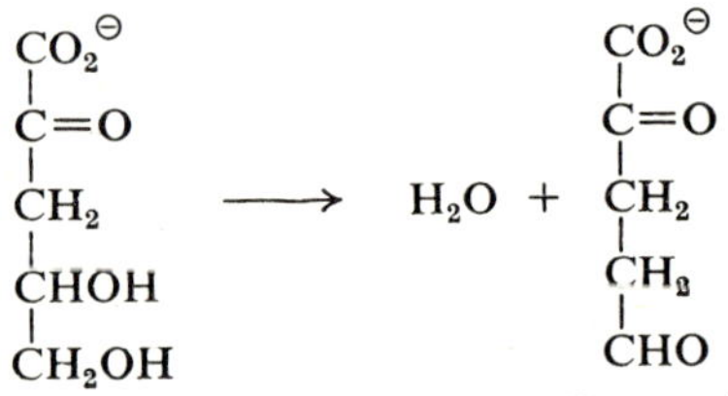

2,5-diketovalerate

(7.20)

Sodium borohydride inactivates the enzyme in the presence of substrate giving rise to the incorporation of one molecule of substrate per molecule of protein. The structure requirements for Schiff-base formation are apparently not stringent, α-ketobutyrate and 2-keto-4-hydroxyvalerate are both competitive inhibitors of the enzymic reaction, and borohydride causes the irreversible inactivation of the enzyme in the presence of either of these compounds.

When the enzymic reaction is carried out in deuterated water the product contains two deuterium atoms, one on each of the C-3 and C-4 atoms. The uptake of deuterium at these two positions is stereospecific, the product being 3(R), 4(R)-dideutero-2,5-diketovalerate.

Abeles *et al.* [77] proposed a mechanism similar to that in Fig. 7.11, to account for the experimental observations. In this mechanism the Schiff-

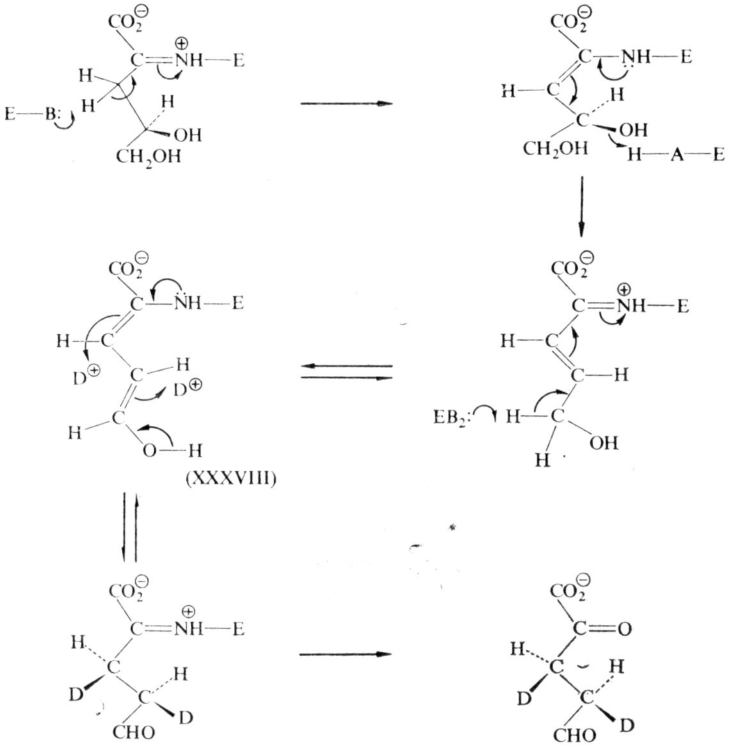

FIG. 7.11. Proposed mechanism for 2-keto-3-deoxy-L-arabinonate dehydratase (after Abeles *et al.* [77])

base first rearranges to an enamine from which the water is eliminated. A further rearrangement takes place to give the conjugated enamine (XXXVIII) which can take up deuterium stereospecifically at the C-3 and C-4 atoms.

## METAL-DEPENDENT LYASES

Aldolases, on the whole, fall into two groups, known as class I and class II aldolases. All class I aldolases are similar, forming Schiff-bases with their substrates, and Rutter [52] has suggested, on phylogenetic grounds, that they may all resemble a pre-existing enzyme perhaps similar to acetoacetate decarboxylase. The class II aldolases, which are bacterial in origin, are similar to the class I enzymes with respect to their specificities but are clearly different in their mode of action in that they require the presence of metal-ions [78]. A metal-ion may be visualized in assisting a dealdolization by virtue of its electrophilic properties. Its role here is similar to that of a Schiff-base or a thiamine group.

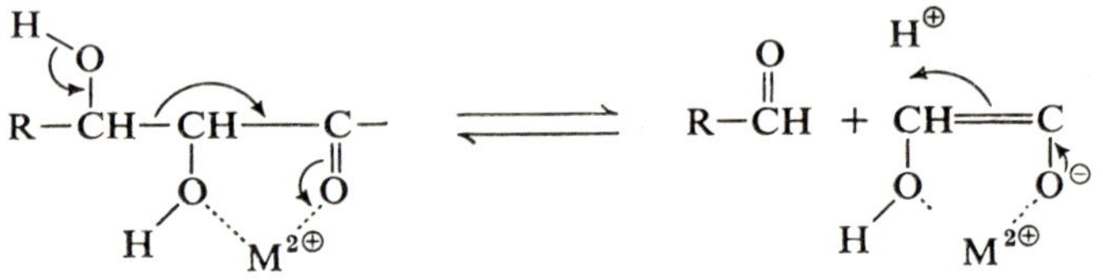

### Carbonic Anhydrase (Carbonate Hydro-lyase, 4.2.1.1)

This is a lyase of a distinctly different type. The reaction which it catalyses is of great physiological importance, the hydration of carbon dioxide or the reverse, the dehydration of carbonate.

$$CO_2 + H_2O \rightleftharpoons H\,CO_3^{\ominus} + H^{\oplus} \text{ (or } H_2CO_3\text{).}$$

The enzyme has a molecular weight of approximately 30,000 and its amino-acid composition has been reported by Nyman and Lindskog [79]. Its most important feature from the enzymic point of view is its possession of one zinc atom per molecule, removal of which causes loss of catalytic activity [80]. Metals other than zinc bind to the apo-enzyme but of these only cobalt(II) is able to restore the catalytic activity [80]. The kinetic parameters of the zinc- and cobalt-enzymes are similar, and changes in the visible absorption spectrum of the cobalt-enzyme brought about by inhibitors are paralleled by changes in activity [81], and therefore it is reasonable to assume that the zinc and cobalt ions are bound to the apo-enzyme at the same site and play similar roles in the catalytic process. Magnetic susceptibility and infra-red studies on the cobalt-enzyme [82] indicate that the arrangement of ligands around the metal-ion is probably tetrahedral.

Carbonic anhydrase is strongly inhibited by various sulphonamides, the most potent of which is acetazolamide (XXXIX). This compound is

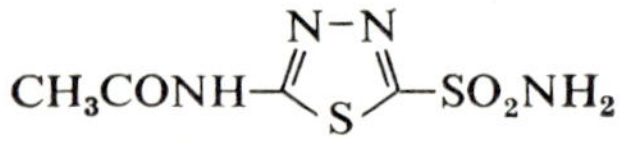

(XXXIX)

bound to the protein only in the presence of the metal-ion [83] and therefore inhibits the enzyme by coordination at the metal-ion [84], a point which has been confirmed by X-ray crystallographic studies [85, 86].

The enzyme is also inhibited by certain anions, for example azide, cyanide and sulphide, which appear to inhibit by becoming coordinated to the metal-ion [83]. These anions bring about spectral changes similar to those produced by acetazolamide [83].

The overall three-dimensional structure of the enzyme has been shown by X-ray crystallographic studies [85, 86]. The zinc atom lies near the centre of the molecule, in a large cavity. The only thiol group in the protein is at a distance of about 14 Å from the metal atom and therefore cannot be coordinated to it. It is possible that a terminal amino-acid-residue provides one ligand to the zinc since one chain end is very close to the cavity which presumably delineates the active site.

Several studies have shown that the catalytic activity of the enzyme depends on a group with $pK_a$ about 7 [87, 88, 89, 90, 91, 92]. It is possible that this group might consist of a water molecule coordinated to the zinc ion [93], although since there is at present strong evidence that a histidine residue is in the vicinity of the active site, it is perhaps more likely that the imidazole side-chain controls the pH-dependence of the reaction. Carbonic anhydrase is inactivated by bromoacetate and by iodoacetamide which have been shown to alkylate the 3-position of the imidazole ring of one histidine residue per molecule [94]. Since sulphonamide derivatives had been found to inhibit the enzyme strongly but reversibly by coordinating to the metal-ion, the effect of chloroacetylsulphonamides was tested. These gave rise to irreversible inactivation, by alkylation of one histidine per molecule. The inactivation by these compounds takes place in two steps. First, the enzyme–inhibitor complex is formed in which the sulphonamide is coordinated to the zinc ion. Then the imidazole group of the nearby histidine reacts with the chloroacetyl group producing a covalently-bound complex. In the absence of the zinc ion, the chloroacetyl derivatives do not inactivate the enzyme. Furthermore the chloroacetyl compounds do not bind to the enzyme which has been inactivated with bromoacetate [94]. Bromoacetazolamide has been shown [95] to inactivate carbonic anhydrase B (see ref. [96]).

The role of the metal-ion in the catalytic process is of course critical. It does not appear to provide a binding site for the carbon dioxide molecule [88, 97, 98]. Accurate differential infra-red spectroscopy showed that when bound to the enzyme, the carbon dioxide molecule is not distorted [99]. It appears to be situated in a hydrophobic cavity. When bound to the cobalt-enzyme, carbon dioxide did not change the visible spectrum, consistent with the point that the carbon dioxide does not bind to the metal-ion [99]. Azide, which binds to zinc ion, displaces carbon dioxide from its cavity, there being a 1 : 1 correspondence between the amount of azide bound to the zinc and the carbon dioxide displaced [99]. This implies that the cavity for the carbon dioxide is sufficiently close to the zinc atom for the azide ion to extent into it.

Infra-red studies show that bicarbonate displaces azide from the zinc and also carbon dioxide from its cavity. On the basis of these results Riepe and Wang concluded that the bicarbonate ion was coordinated to the zinc ion through its negatively-charged oxygen atom with the neutral oxygen and hydroxyl in the hydrophobic cavity. The hydration therefore takes place by the attack of metal-bound hydroxide ion (or water molecule assisted by base catalysis by the histidyl residue side-chain) on the carbon dioxide molecule (Fig. 7.12).

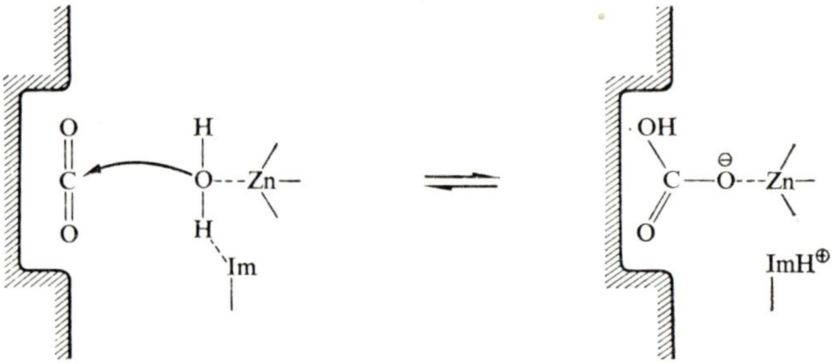


FIG. 7.12. Hydration of carbon dioxide by carbonic anhydrase, after Riepe and Wang [99]

Carbonic anhydrase is not specific for carbon dioxide. The enzyme catalyses the hydration of various aldehydes and the hydrolysis of certain esters [87, 100, 101]. Inhibition of the esterase and aldehyde hydratase activities parallels the carbonic anhydrase activity, and the same pH dependence is also observed. Therefore these reactions almost certainly occur at the same active site and by essentially similar mechanisms. All of the reactions catalysed by carbonic anhydrase are fundamentally identical, involving the hydration of a carbonyl group; in the case of an ester the resultant tetrahedral complex breaks down rapidly. (The hydrolysis of a sultone [101] involves hydration of a S═O group.) Thus, the hydrolysis of p-nitrophenyl acetate yields eventually an acetate group coordinated to the zinc ion (Fig. 7.13).

Notice that in these mechanisms for carbonic anhydrase the suggested role of the zinc ion is not the same as that proposed for metal-ions in almost every other reaction. Normally metal-ions are envisaged as coordinating to the oxygen atom of the carbonyl group and assisting the cleavage of the carbon–oxygen double bond (see Chapter 1, see also carboxypeptidase). The zinc does not appear to assist this process for carbon dioxide. Its main function appears to be that of ensuring the presence of a water molecule in the correct stereochemical arrangement for a nucleophilic attack.

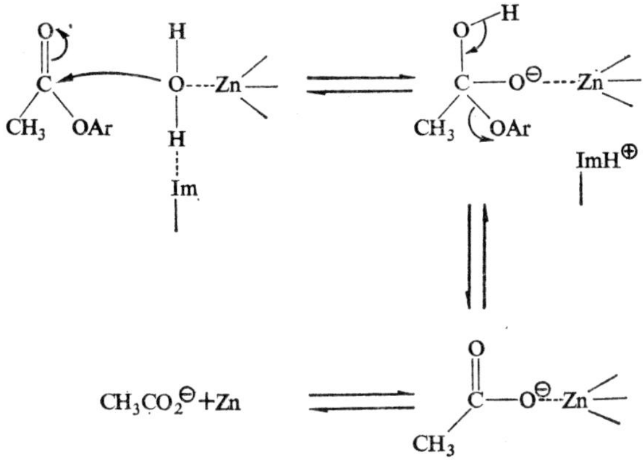


FIG. 7.13. Hydrolysis of p-nitrophenyl acetate by carbonic anhydrase, after Pocker *et al.* [87] Ar = p-nitrophenyl group

### Malate Synthase (L-Malate Glyoxylate-lyase, CoA-acetylating, 4.1.3.2)

This is an example of another metal-dependent lyase, this time a ketoacid-lyase. This enzyme catalyses the synthesis of L-malate from acetyl-CoA and glyoxylate according to equation (7.21). During the synthetic reaction the coenzyme A thioester group is hydrolysed.

$$\begin{array}{c} CH_3CO\cdot SCoA \\ \\ CHO \\ | \\ CO_2^{\ominus} \end{array} \xrightleftharpoons{+H_2O} \begin{array}{c} CO_2^{\ominus} \\ | \\ CH_2 \\ | \\ CHOH \\ | \\ CO_2^{\ominus} \end{array} + \; CoA\cdot SH \qquad (7.21)$$

Individual stages in the reaction have been examined by Eggerer and Klette [102]. The first stage is probably the enolization of acetyl-CoA; in the absence of glyoxylate the enzyme catalyses the exchange of hydrogens on the methyl group of the acetyl. The rate of enolization was estimated as about one thousand times slower than the rate of synthesis of malate. Magnesium ions were essential for the enolization, although magnesium was partially replaced by other metal-ions. The metal-ion probably assists the enolization by coordinating with the carbonyl oxygen. Various α-keto-acids enhanced the rate of enolization. They were effective in the order: pyruvate > oxalacetate > ketobutyrate > ketoglutarate > ketovalerate; thus the keto-acid which resembles glyoxylate most closely is most effective. Pyruvate, which inhibits the synthesis of malate, increased the rate of enolization of acetyl-CoA a thousand-fold to a value very similar to the rate of synthesis of malate in the presence of glyoxylate.

In view of these results Eggerer and Klette suggested that the enolization of acetyl-CoA by the enzyme involved a concerted acid–base catalysis in which the metal-ion functioned as the acid and the carboxylate group of the keto-acid operated as the base (Fig. 7.14).

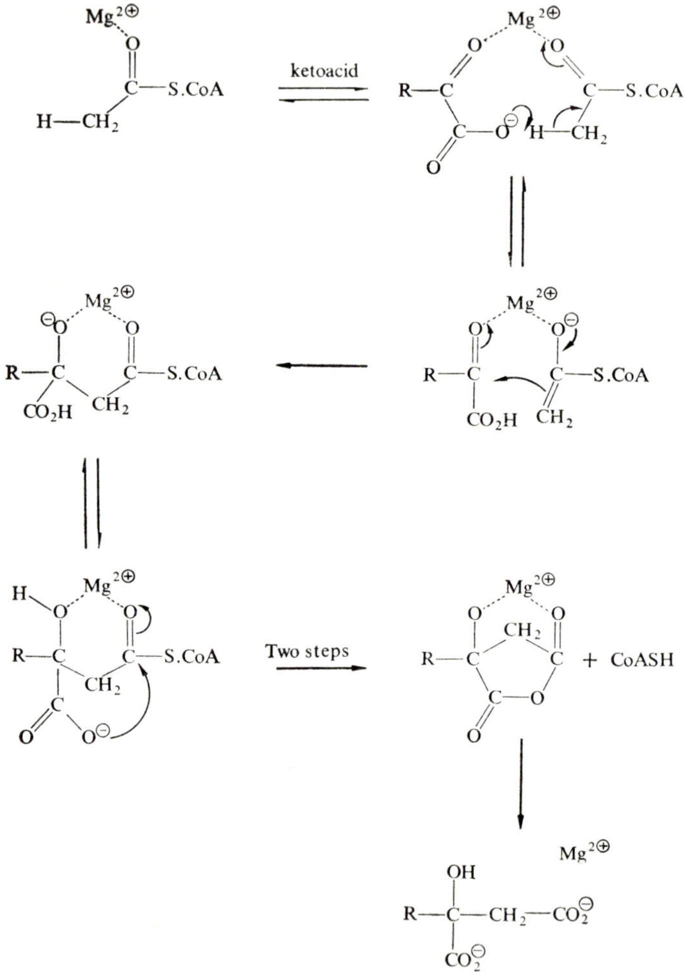


FIG. 7.14. Proposal by Eggerer and Klette [102] for the mechanism of action of malate synthase (R = H) and citrate synthase (R = $CH_2CO_2^{\ominus}$)

The next stage is the aldol condensation itself in which the methylene group attacks the carbonyl of the glyoxylate. The metal-ion may be imagined as assisting this process also. The result of this reaction is then an enzyme-bound malyl-CoA. That such a species is an intermediate in the reaction is indicated by the observation that the hydrolysis of malyl-CoA is catalysed by the enzyme. Eggerer and Klette propose that this hydrolysis is brought about by the assistance of the metal-ion and also of the carboxylate group present, in such a way that an intermediate cyclic anhydride is formed.

When the reaction is carried out in tritiated water the malate obtained contains no tritium. This observation, together with the fact that the rate of enolization in the presence of pyruvate (which is very similar to glyoxylate) is almost identical to the rate of malate synthesis, indicates that the rate-determining step for the whole process is the enolization of the acetyl-CoA.

Many enzymes are thought to function by mechanisms which involve a concerted acid–base catalysis. This particular mechanism is of interest in that the basic catalyst required for the concerted process is not provided by the enzyme, but by the second substrate.

An essentially similar mechanism is proposed for the enzyme citrate synthase (citrate oxaloacetate-lyase CoA-acetylating, 4.1.3.7) [103]. In this case, experiments carried out with $^{18}O$-labelled water [104] do not eliminate the possibility of a cyclic anhydride intermediate.

## REFERENCES


1. SNELL, E. E., *Chemical and Biological Aspects of Pyridoxal Catalysis*, Eds. E. E. Snell, F. M. Fasella, A. E. Braunstein and A. R. Faneli, p. 1 (Pergamon Press, London, 1963)
2. MANDELES, S., KOPPELMAN, R. and HANKE, M. E., *J. biol. Chem.* **209,** 327 (1954)
3. GUIRAD, R. M. and SNELL, E. E., *Comprehensive Biochemistry* Vol. 15, p. 138 (Elsevier, 1964)
4. KALYANKAR, G. D. and SNELL, E. E., *Biochemistry* **1,** 594 (1962)
5. FLAVIN, M., *Chemical and Biological Aspects of Pyridoxal Catalysis*, p. 377 (Pergamon Press, London, 1963)
6. SHUKUYA, R. and SCHWERT, G. W., *J. biol. Chem.* **235,** 1649, 1653 (1960)
7. MANDELES, S. and HANKE, M. E., *Abst. 124th Meeting A.C.S.* 55C (1953)
8. BELLEAU, B. and BURBA, J., *J. Am. chem. Soc.* **82,** 5751 (1960)
9. CATTANEO, J. and SENEZ, J. C., *Chemical and Biological Aspects of Pyridoxa Catalysis* p. 217 (1963); MEISTER, A., NISHIMURA, J. S. and NOVOGRODSKY, A., *Chemical and Biological Aspects of Pyridoxal Catalysis* p. 229 (Pergamon Press, London, 1963)
10. NISHIMURA, J. S., MANNING, J. M. and MEISTER, A., *Biochemistry* **1,** 442 (1962); NOVOGRODSKY, A., NISHIMURA, J. S. and MEISTER, A., *J. biol. Chem.* **238,** PC1903 (1963)
11. WILSON, E. M., *Biochim. biophys. Acta* **67,** 345 (1963)
12. WILSON, E. M. and KORNBERG, H. L., *Biochem. J.* **88,** 578 (1963)

13. Newton, W. A. and Snell, E. E., *Proc. natn. Acad. Sci. U.S.A.* **51,** 382 (1964); Newton, W. A., Morino, Y. and Snell, E. E., *J. biol. Chem.* **240,** 1211 (1965); Morino, Y. and Snell, E. E., *J. biol. Chem.* **242,** 2793 (1967)
14. Snell, E. E., *Vitamins and Hormones* **16,** 77 (1958); Braunstein, A. E., *The Enzymes* Vol. 2, p. 113 (Academic Press, New York, 1960)
15. Fischer, E. H., Kent, A. B., Snyder, E. R. and Krebs, E. G., *J. Am. chem. Soc.* **80,** 2906 (1958)
16. Metzler, D. E., *J. Am. chem. Soc.* **79,** 485 (1957)
17. Schirch, L. and Jenkins, W. T., *J. biol. Chem.* **239,** 3801 (1964); Jenkins, W. T., *J. biol. Chem.* **239,** 1742 (1964); Jenkins, W. T. and D'Ari, L., *J. biol. Chem.* **241,** 2845 (1966); Schirch, L. and Slotter, R. A., *Biochemistry* **5,** 3175 (1966)
18. Smith, T. A., Cordelle, F. H. and Abeles, R. H., *Archs. biochem. Biophys.* **120,** 724 (1967)
19. Kallan, K. G. and Jencks, W. P., *J. biol. Chem.* **241,** 5864 (1966)
20. Lohman, K. and Shuster, P., *Biochem. Z.* **294,** 188 (1943)
21. Ukai, T., Tanaka, S. and Dokawa, S., *J. pharm. Soc. Japan* **63,** 269 (1943)
22. Mizuhara, S., Tamura, R. and Arata, H., *Proc. Japan. Acad.* **27,** 302 (1951); Mizuhara, S. and Arata, H., *Proc. Japan. Acad.* **27,** 700 (1951); Mizuhara, S. and Oono, K., *Proc. Japan. Acad.* **27,** 704 (1951); Mizuhara, S. and Handler, P., *J. Am. chem. Soc.* **76,** 571 (1954); Takeuchi, Y., Okuda, T., Hayakawa, S. and Mizuhara, S., *J. Biochem.* (*Tokyo*) **42,** 93 (1955)
23. Koffler, H. and Krampitz, L. O., *Bacteriol. Proc.* 7 (1955)
24. Metzler, D. E., *The Enzymes* Vol. 2, p. 215 (Academic Press, New York, 1960)
25. Yatco-Manzo, E., Roddy, F., Yount, R. G. and Metzler, D. E., *J. biol. Chem.* **234,** 733 (1959)
26. Breslow, R., *J. Am. chem. Soc.* **79,** 1762 (1957)
27. Breslow, R., *Chemy Ind.* 893 (1957)
28. Mieyal, J. J., Votaw, R. G., Krampitz, L. O. and Sable, H. Z., *Biochim. biophys. Acta* **141,** 205 (1967)
29. Bruice, T. C. and Benkovic, S. J., *Bioorganic Mechanisms* Vol. 2, p. 181 (Benjamin, 1966)
30. Krampitz, L. O., Suzuki, I. and Greull, G., *Fedn Proc.* **20,** 971 (1961)
31. Krampitz, L. O., Greull, G. and Suzuki, I., *Fedn Proc.* **18,** 266 (1959)
32. Breslow, R. and McNelis, E., *J. Am. chem. Soc.* **81,** 3080 (1959)
33. Holzer, H. and Beaucamp, K., *Angew. Chem.* **71,** 776 (1959); *Biochim. biophys. Acta* **46,** 225 (1961)
34. Juni, E., *J. biol. Chem.* **236,** 2302 (1961)
35. Schellenberger, A. and Hubner, G., *Z. Physiol. Chem.* **348,** 491 (1967); Schellenberger, A., Wendler, K., Creutzburg, P. and Hubner, G., *Z. Physiol. Chem.* **348,** 501 (1967); Schellenberger, A., Heinroth, I. and Hubner, G., *Z. Physiol. Chem.* **348,** 506 (1967); Schellenberger, A. and Winter, K., *Z. Physiol. Chem.* **344,** 16 (1966); Schellenberger, A., Muller, V., Winter, K. and Hubner, G., *Z. Physiol. Chem.* **344,** 244 (1966); Schellenberger, A., Rodel, W. and Rodel, H., *Z. Physiol. Chem.* **339,** 122 (1964)

36. Schellenberger, A., Hubner, G., Schwaiberg, R., Helbig, D., Schumacher, S., Thiene, R., Bouillon, G. and Radler, K.-P., *Z. Physiol. Chem.* **346,** 123 (1966)
37. Holzer, H. and Kohlhow, G., *Biochem. biophys. Res. Commun.* **5,** 452 (1961)
38. White, F. G. and Ingraham, L. L., *J. Am. chem. Soc.* **82,** 4114 (1960); **84,** 3109 (1962)
39. Horecker, B. L., *Comprehensive Biochemistry* Vol. 15, p. 48 (Elsevier, 1964)
40. Horecker, B. L., Smyrniotis, P. Z. and Klenow, H., *J. biol. Chem.* **205,** 661 (1953)
41. Racker, E., *Adv. Enzymol.* **15,** 141 (1954)
42. Datta, A. G. and Racker, E., *J. biol. Chem.* **236,** 624 (1961)
43. Fonseca-Wollheim, F., Bock, K. W. and Holzer, H., *Biochem. biophys. Res. Commun.* **9,** 466 (1962)
44. Bock, K. W., Jaenicke, L. and Holzer, H., *Biochem. biophys. Res. Commun.* **9,** 472 (1962)
45. Votaw, R., Williamson, W. T., Krampitz, L. O. and Wood, W. A., *Biochem. Z.* **338,** 756 (1963)
46. Heath, E. C., Hurwitz, J., Horecker, B. L. and Ginsburg, A., *J. biol. Chem.* **231,** 1009 (1958)
47. Westheimer, F. H., *Proc. chem. Soc.* 253 (1963)
48. Hamilton, G. E. and Westheimer, F. H., *J. Am. chem. Soc.* **81,** 6332 (1959)
49. Fridovich, I. and Westheimer, F. H., *J. Am. Chem. Soc.* **84,** 3208 (1962)
50. Warren, S., Zerner, B. and Westheimer, F. H., *Biochemistry* **5,** 817 (1966)
51. Tagaki, W. and Westheimer, F. H., *Biochemistry* **7,** 891 (1968)
52. Rutter, W. J., *Fedn Proc.* **23,** 1248 (1964); Blostein, R. E. and Rutter, W. J., *J. biol. Chem.* **238,** 3280 (1963); Rose, I. A., O'Connell, E. L. and Mehler, A. H., *J. biol. Chem.* **240,** 1758 (1965)
53. Venkataraman, R. and Racker, E., *J. biol. Chem.* **236,** 1883 (1961); Pontremoli, S., Prandini, B. D., Bonsignore, A. and Horecker, B. L. *Proc. natn. Acad. Sci. U.S.A.* **47,** 1942 (1961); Horecker, B. L., Pontremoli, S., Ricci, C. and Cheng, T., *Proc. natn. Acad. Sci. U.S.A.* **47,** 1947 (1961)
54. Grazi, E. Rowley, P. T., Cheng, T., Tchola, O. and Horecker, B. L., *Biochem. biophys. Res. Commun.* **9,** 38 (1962)
55. Speck, J. C., Rowley, P. T. and Horecker, B. L., *J. Am. Chem. Soc.* **85,** 1012 (1963)
56. Horecker, B. L., Rowley, P. T., Grazi, E., Cheng, T. and Tchola, O., *Biochem. Z.* **338,** 36 (1963)
57. Grazi, T., Cheng, T. and Horecker, B. L., *Biochem. biophys. Res. Commun.* **7,** 250 (1962)
58. Rose, I. A., *J. Am. chem. Soc.* **80,** 5835 (1958)
59. Hoffee, P., Lai, C. Y., Pugh, E. L. and Horecker, B. L., *Proc. natn. Acad. Sci. U.S.A.* **57,** 107 (1967)
60. Westhead, E. W., *Biochemistry* **4,** 2139 (1965)

61. Ingram, J. M. and Hochster, R. M., *Can. J. Biochem.* **45,** 929, 1505 (1967)
62. Kobashi, K., Lai, C. Y. and Horecker, B. L., *Archs biochem. Biophys.* **117,** 437 (1966)
63. Ginsburg, A., *Archs biochem. Biophys.* **117,** 451 (1966)
64. Chan, W., Morse, D. E. and Horecker, B. L., *Proc. natn. Acad. Sci. U.S.A.* **57,** 1013 (1967)
65. Fluri, R., Ramasarma, T. and Horecker, B. L., *Eur. J. Biochem.* **1,** 117 (1967)
66. Shimizu, H. and Ozawa, H., *Biochim. biophys. Acta* **133,** 195 (1967)
67. Stellwagen, E. and Schachman, H. K., *Biochemistry* **1,** 1056 (1962)
68. Kowalsky, A. and Boyer, P. D., *J. biol. Chem.* **235,** 604 (1960)
69. Deal, W. C. and van Holde, K. E., *Fedn Proc.* **21,** 254 (1962)
70. Drechsler, E. R., Boyer, P. D. and Kowalsky, A. G., *J. biol. Chem.* **234,** 2627 (1959)
71. Rutter, W. J., Richards, O. C. and Woodfin, B. M., *J. biol. Chem.* **236,** 3193 (1961)
72. Pugh, E. L. and Horecker, B. L., *Biochem. biophys. Res. Commun.* **26,** 360 (1967)
73. Lardy, *et al.*, quoted in *Biological Chemistry*, Eds. H. R. Mahler and E. H. Cordes, p. 425 (Harper and Row, 1966)
74. Kobes, R. D. and Dekker, E. E., *Biochem. biophys. Res. Commun.* **25,** 329 (1966); **27,** 607 (1967)
75. Grazi, E., Meloche, H., Martinez, G., Wood, W. A. and Horecker, B. L., *Biochem. biophys. Res. Commun.* **10,** 4 (1963)
76. Meloche, P., *Biochemistry* **6,** 2273 (1967)
77. Portsmouth, D., Stoolmiller, A. C. and Abeles, R. H., *J. biol. Chem.* **242,** 2751 (1967)
78. Russell, G. K. and Gibbs, M., *Biochim. biophys. Acta* **132,** 145 (1967); Richards, O. C. and Rutter, W. J., *J. biol. Chem.* **236,** 3185 (1961); Jagannathan, V., Singh, K. and Damodaran, M., *Biochem. J.* **63,** 94 (1956)
79. Nyman, P.-O. and Lindskog, S., *Biochim. biophys. Acta* **85,** 141 (1964)
80. Lindskog, S. and Malmstrom, B. G., *J. biol. Chem.* **237,** 1129 (1962)
81. Lindskog, S., *Biochemistry* **5,** 2641 (1966)
82. Lindskog, S. and Ehrenberg, A., *J. molec. Biol.* **24,** 133 (1967)
83. Coleman, J. E., *Nature, Lond.* **214,** 193 (1967)
84. Malmstrom, B. G., Nyman, P.-O., Strandberg, B. and Tilander, B., *Structure and Activity of Enzymes*, Eds. T. W. Goodwin, J. I. Harris and B. S. Hartley, p. 121 (Academic Press, New York, 1964)
85. Tilander, B., Strandberg, B. and Fridborg, K., *J. molec. Biol.* **12,** 740 (1965)
86. Fridborg, K., Kannan, K. K., Liljas, A., Strandberg, B., Strandberg, R., Tilander, B. and Wiren, G., *J. molec. Biol.* **25,** 505 (1967)
87. Pocker, Y. and Stone, J. T., *Biochemistry* **6,** 668 (1967)
88. Pocker, Y. and Meany, J., *Biochemistry* **4,** 2535 (1965)
89. Kernohan, J. C., *Biochim. biophys. Acta* **96,** 304 (1965); **81,** 346 (1964)
90. Roughton, F. J. W. and Booth, V. H., *Biochem. J.* **40,** 319 (1946)

91. Kiese, M., *Biochem. Z.* **307,** 400 (1941)
92. Lindskog, S., *J. biol. Chem.* **238,** 945 (1963)
93. Davies, R. P., *J. Am. chem. Soc.* **81,** 5674 (1959)
94. Whitney, P. L., Nyman, P.-O. and Malmstrom, B. G., *J. biol. Chem.* **242,** 4212 (1967)
95. Kandel, S. I., Wong, S.-C., Kandel, M. and Gornall, A. G., *J. biol. Chem.* **243,** 2437 (1967)
96. Nilsson, A. and Lindskog, S., *Eur. J. Biochem.* **2,** 309 (1967)
97. Davies, R. P., *The Enzymes* Vol. 5, p. 545 (Academic Press, New York, 1961)
98. Kernohan, J. C., *Biochim. biophys. Acta* **81,** 346 (1964)
99. Riepe, M. E. and Wang, J. H., *J. Am. chem. Soc.*, **89,** 4229 (1967)
100. Veerpoorte, J. A., Mehta, S. and Edsall, J. T., *J. biol. Chem.* **242,** 4221 (1967)
101. Lo, K.-W. and Kaiser, E. T., *Chem. Commun.* 834 (1966)
102. Eggerer, H. and Klette, A., *Eur. J. Biochem.* **1,** 447 (1967)
103. Eggerer, H., *Biochem. Z.* **343,** 111 (1965)
104. Suelter, C. H. and Arrington, S., *Biochim. biophys. Acta* **141,** 423 (1967)

# CHAPTER 8

# ISOMERASES AND LIGASES

## ISOMERASES

Although in general less is known about the isomerases than about most other classes of enzymes, it is possible for certain conclusions to be drawn about the mechanisms by which some of them function. The isomerases bring about molecular rearrangements in their substrates and among these rearrangements are included:

(a) Inversion at an asymmetric carbon (D-*ribulose-5-phosphate 3-epimerase*, 5.1.3.1, equation 8.1).

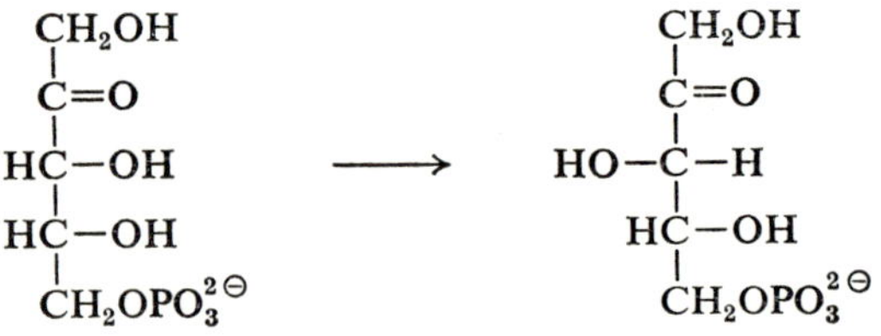


(8.1)

(b) Racemization at an asymmetric carbon (*alanine racemase*, 5.1.1.1, equation 8.2).

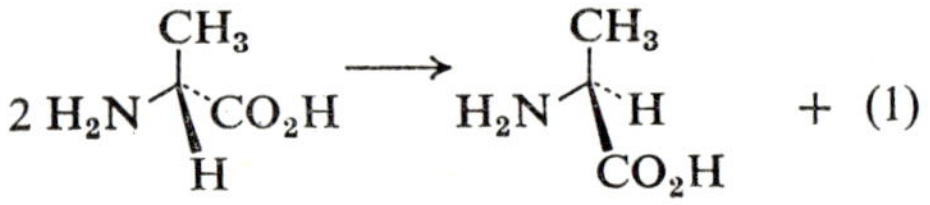


(I)

(8.2)

(c) Double bond migration (3-*ketosteroid* $\Delta^4$-$\Delta^5$-*isomerase*, 5.3.3.1, equation 8.3).

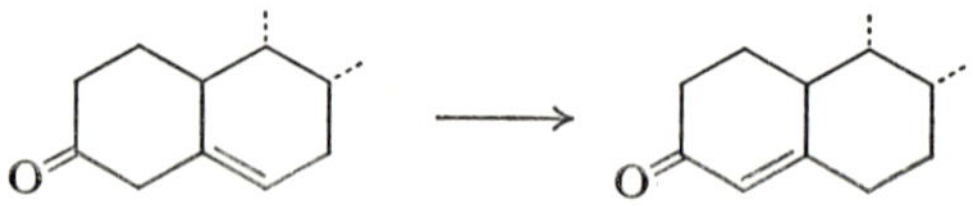


(8.3)

(d) Certain intramolecular transfer reactions (for example, L-threo-3-*methylaspartate carboxy-aminomethylmutase*, 5.4.99.1, equation 8.4).

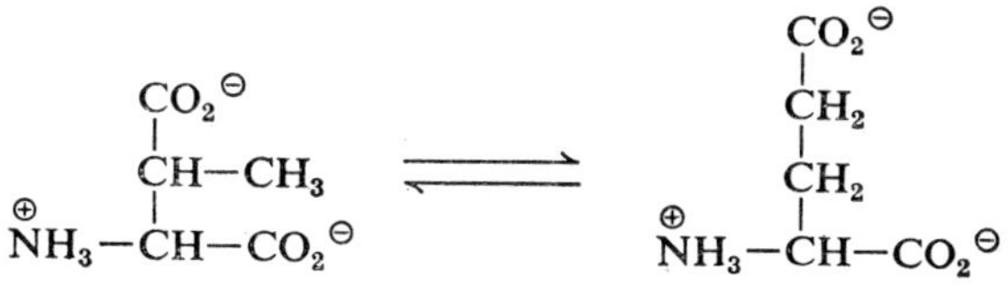

3-methyl-L-aspartate L-glutamate (8.4)

(e) Interconversions of aldoses and ketoses (for example, D-*glucose-6-phosphate isomerase*, 5.3.1.9, equation 8.5).

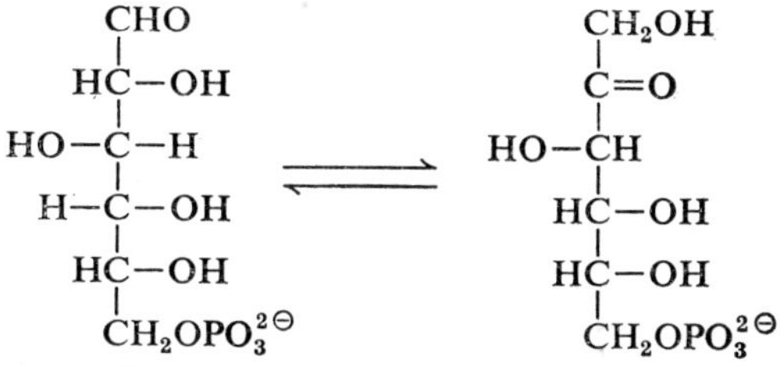

D-glucose-6-phosphate D-fructose-6-phosphate (8.5)

### Amino-acid Racemases (5.1.1.–)

These enzymes are pyridoxal phosphate-dependent and it is assumed that the mechanism of racemization in the presence of the enzymes is essentially similar to that of the non-enzymic process catalysed by this

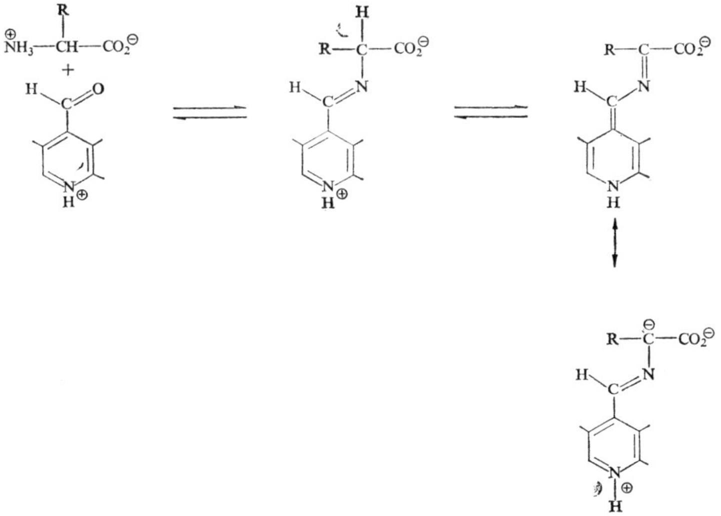

FIG. 8.1. Racemization of an amino-acid by pyridoxal. Other than the essential aldehydro-group, the substituents of the pyridoxal ring are omitted

coenzyme. The effect of forming the Schiff-base between the amino-acid and the pyridoxal is to labilize the α-hydrogen, loss of which gives a resonance stabilized carbanion (Fig. 8.1). In the case of enzymes such as the transaminases and those which catalyse β-eliminations (see Chapter 7) this type of carbanion takes up a proton from the solvent in a highly stereospecific manner so that only L-amino-acids are produced. However, in the case of the racemases, the amino-acid pyridoxal complex must be held on the surface of the enzyme in such a way that a proton can be taken from either side of the amino-group giving rise to a mixture of the D- and L-forms of the amino-acid [1].

### Enzymes Requiring the Vitamin $B_{12}$ Coenzyme

The vitamin $B_{12}$ coenzyme, 5′-deoxyadenosyl-cobalamin (see II, Chapter 6 and Fig. 6.1) is required for a variety of enzymes catalysing intramolecular rearrangements of the general type shown in equation (8.6). Thus,

$$\begin{array}{cc} | & | \\ -\overset{a}{C}- & -\overset{b}{C}- \\ | & | \\ X & H \end{array} \longrightarrow \begin{array}{cc} | & | \\ -\overset{a}{C}- & -\overset{b}{C}- \\ | & | \\ H & X \end{array} \tag{8.6}$$

X may be carboxy-aminomethyl (methylaspartate mutase), a thioester group (*methylmalonyl-CoA mutase*, 5.4.99.2) or hydroxyl (*propanediol dehydratase*, 4.2.1.28). In the last case, the transfer of hydroxyl to the b-carbon is followed by elimination of water and hence the enzyme is classified as a lyase rather than an isomerase. However, its mode of action has much in common with the general mechanism which applies to the $B_{12}$ isomerases and its inclusion in this section is appropriate.

In the reaction shown in equation (8.6) it can be seen that the hydrogen atom is lost from the b-carbon (which ultimately accepts group X) and a hydrogen is gained by the a-carbon, and therefore an intramolecular hydrogen transfer is implied. The nature of the hydrogen transfer is fundamental to the mechanisms of these and other isomerases, and studies with hydrogen isotopes have been carried out in order to clarify this point. For all the $B_{12}$ isomerases the hydrogen taken up by the a-carbon is *not* derived from the solvent medium [2, 3, 4, 5]. Reactions were carried out in deuterium oxide without incorporation of deuterium into the molecule.

However, the transfer of hydrogen from C—a to C—b is not necessarily direct, as shown by the 'cross-over' experiment of Abeles and Frey [2]. In this experiment a mixture of 1,2-propanediol (labelled with tritium in the 1-position) and ethylene glycol (unlabelled) were incubated with the enzyme propanediol dehydrase, which functions with both of these substrates. If a direct intramolecular transfer of hydrogen occurred then only the propionaldehyde would have contained tritium. However, the acetaldehyde isolated also contained tritium (equation 8.7). This result shows clearly that the reaction with the diol must take place in two distinct

$$\begin{array}{ccc} \overset{(Tr)|}{CH_3 \cdot CHOH \cdot CH_2OH} & & CH_3 \cdot CH_2 \cdot CHO \text{ (Labelled)} \\ + & \xrightarrow{\text{enzyme}} & + \\ CH_2OH \cdot CH_2OH & & CH_3 \cdot CHO \text{ (Labelled)} \end{array} \quad (8.7)$$

stages, in the first of which a hydrogen is transferred from the diol to some acceptor intermediate and in the second the hydrogen is transferred from this intermediate.

The role of the vitamin $B_{12}$ coenzyme as the hydrogen transferring agent has now been established for most of these enzymes. Thus, Abeles and Frey [2, 6] showed that, when propanediol dehydrase–coenzyme $B_{12}$ was incubated with propanediol (1-$^3$H), the coenzyme became labelled in the 5′-position. Incubation of synthetically-prepared coenzyme labelled with tritium in the 5′-position with the enzyme and unlabelled propanediol gave rise to tritiated propionaldehyde. Similarly, labelled coenzyme obtained as above from the propanediol dehydrase reaction transferred its label to succinyl-CoA in the presence of methylmalonyl-CoA mutase [7]. Methylmalonyl-CoA mutase catalyses the reaction shown in equation (8.8).

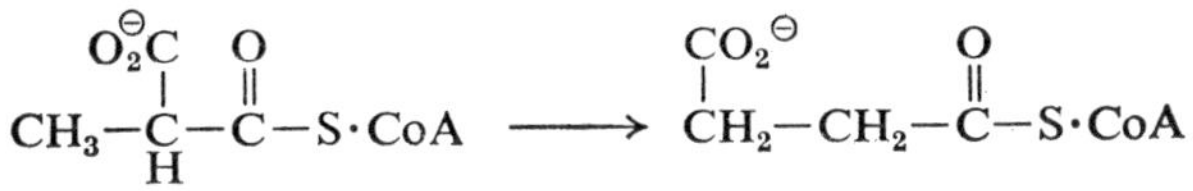

methylmalonyl-CoA succinyl-CoA

(8.8)

Quantitative experiments indicate that both hydrogens at the C-5′ position of the coenzyme participate in the reactions, and therefore the reactions are not stereospecific with respect to this site of the coenzyme. This probably means that the cobalt–carbon atom breaks during the process allowing free rotation at the C-5′ position.

In the presence of glycolaldehyde the holo-enzyme was inactivated and glyoxal (II) was produced [8]. The protein–coenzyme complex was de-

$$\begin{array}{c} CHO \\ | \\ CH_2OH \end{array} + \text{coenzyme } B_{12} \xrightarrow{\text{enzyme}} \begin{array}{c} CHO \\ | \\ CHO \end{array} + \begin{array}{c} \text{hydroxocobalamin} \\ \\ \text{5′-deoxyadenosine} \end{array}$$

(II)

natured and hydroxocobalamin and 5′-deoxyadenosine isolated. One hydrogen from the glycolaldehyde had been transferred to the 5′-position.

In this case, the coenzyme had been able to accept the hydrogen but because of the nature of the substrate, had been unable to re-donate it.

Two possible mechanisms are envisaged for the action of these enzymes, one in which the hydrogen transfer involves a hydride, the other in which a radical process is suggested.

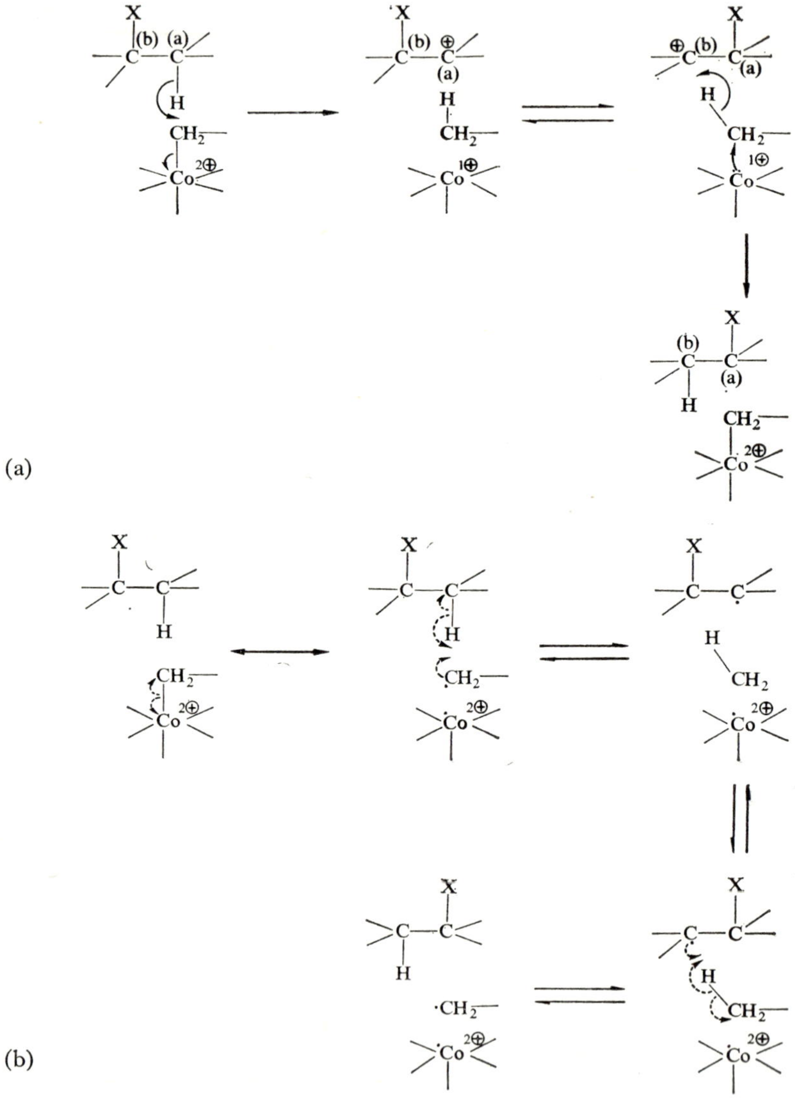

FIG. 8.2. (a) Hydride-transfer mechanism for the vitamin $B_{12}$-coenzyme isomerases. (b) Radical mechanism. Notice that in each case the hydrogen is transferred from C-(a) to C-(b) via the 5′-position of the deoxyadenosyl group

*Hydride-transfer Mechanism*

In chapter 4 it was seen that the properties of substituted cobalamins are such that in general the alkyl (or acyl) groups are susceptible to nucleophilic attack, the $Co^{2+}$ species being an effective leaving group. Similarly the $Co^{1\oplus}$ species was seen to be a good nucleophile, and these properties were responsible for the function of methyl-cobalamin as a methyl transfer agent. These ideas may be extended to incorporate the function of 5′-deoxyadenosyl-cobalamin as a hydride transfer agent. Figure 8.2(a) shows how this function may be represented. The 5′-methylene is viewed as being subject to a nucleophilic attack by a hydride species so that the cobalt–carbon bond is broken giving 5′-deoxyadenosine and the reduced $B_{12}$ compound. Loss of the hydride from the substrate gives rise to a carbonium ion and group X may migrate as shown. A hydride transfer in the reverse direction can then result in carbon-b carrying the hydrogen. The loss of hydride from carbon-a is represented as leaving a carbonium ion although this need not be the case. For example, for those compounds in which the a-carbon also carries a hydroxyl group (propanediol) loss of hydride will lead to the production of a carbonyl group. This is in fact not an uncommon process, and the oxidation of glycolaldehyde to glyoxal by the enzyme–coenzyme complex is best viewed in this way. In other cases the loss of hydride is less easy to envisage and the enzyme must clearly provide some extra assistance.

*Radical Mechanism*

The detection by Williams and his co-workers [9] of radicals in the propanediol dehydrase system (by e.s.r. and ultraviolet absorption spectroscopy) raised the possibility that homolytic rather than heterolytic bond cleavage occurred during the reaction. Figure 8.2(b) shows the probable mechanism if this is the case.

Whichever of these mechanisms is correct, it is probable that the 5′-deoxyadenosine, which is an intermediate in each case, remains closely bound to the remainder of the coenzyme; even if it were to dissociate away the hydrogen isotopes at the 5′-position would be unlikely to exchange with water.

The enzyme *ribonucleotide reductase*, which catalyses the reduction of ribonucleotides to deoxyribonucleotides is also a $B_{12}$-enzyme. It has been

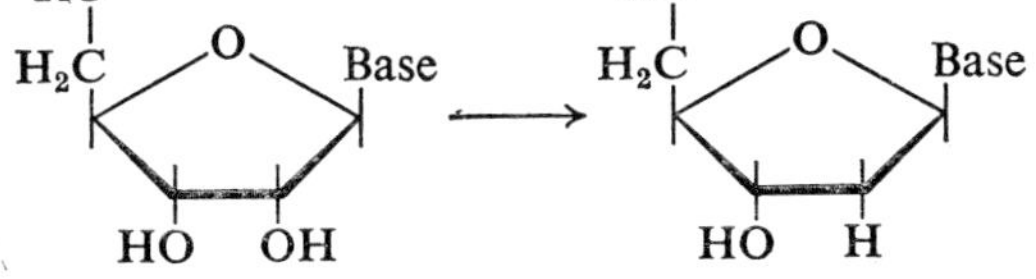

established that in this case also the coenzyme functions as the hydrogen-transfer intermediate. Thus, tritium is transferred from the 5′-position of the coenzyme to the 2′-carbon of the deoxyribose species [10]. The reduction of a ribonucleotide requires the presence of a reducing agent such as a dithiol which functions as the ultimate hydrogen donor. The function of the $B_{12}$-coenzyme in this enzyme is therefore similar to that of FAD in other oxidoreductases.

### 3-Ketosteroid $\Delta^4$-$\Delta^5$-Isomerase (5.3.1.1)

This type of enzyme catalyses the migration of a double bond. Thus androst-5-ene-3,17-dione (III) is converted into its isomer (IV). In

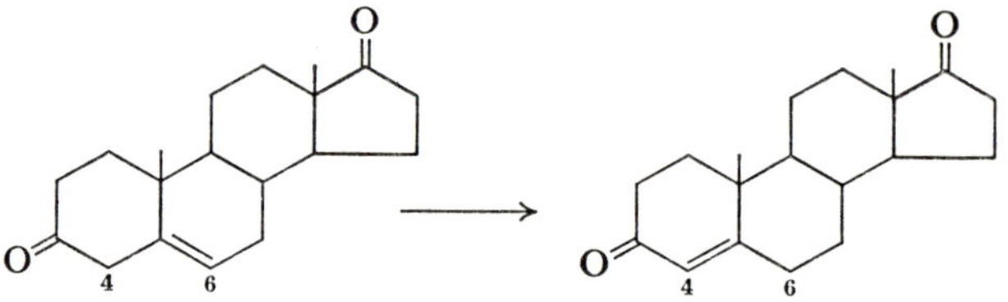


(III) (IV)

practical terms the reaction is essentially irreversible because, due to the conjugation between the olefinic and carbonyl double bonds, compound (IV) has greater thermodynamic stability than (III) and hence the equilibrium strongly favours (IV).

Interestingly, isotope substitution experiments have given different results with the bacterial and mammalian enzymes. When the isomerization catalysed by a bacterial enzyme is carried out in deuterium oxide no deuterium is incorporated into the product [11]. This is in contrast to the acid- and base-catalysed reactions and suggests that in the enzymic process the hydrogen atom acquired by the 6-position is the same atom as that lost by the 4-position. The direct nature of the transfer was confirmed when Malhotra and Ringold [12] examined the product obtained from the isomerization of steroid labelled in the 4-position. Quite clearly the hydrogen which appeared at position-6 was derived from position-4. Furthermore the transfer was quite stereospecific, in that the 4-$\beta$ hydrogen was transferred to the 6-$\beta$ position. The loss of the 4-$\beta$ hydrogen was found to be the rate-determining step.

In their mechanistic scheme, Malhotra and Ringold suggest that a proton-transfer agent was involved. The reaction thus proceeds (see Fig. 8.3) via the removal of a proton from the 4-$\beta$ position by a basic group B in the enzyme. This process might be assisted by the presence of some acidic group at the oxygen of the carbonyl. The product is therefore a conjugated enol (V) which takes up a proton from the conjugate acid of B, giving the isomeric product. That the same group B functions as proton

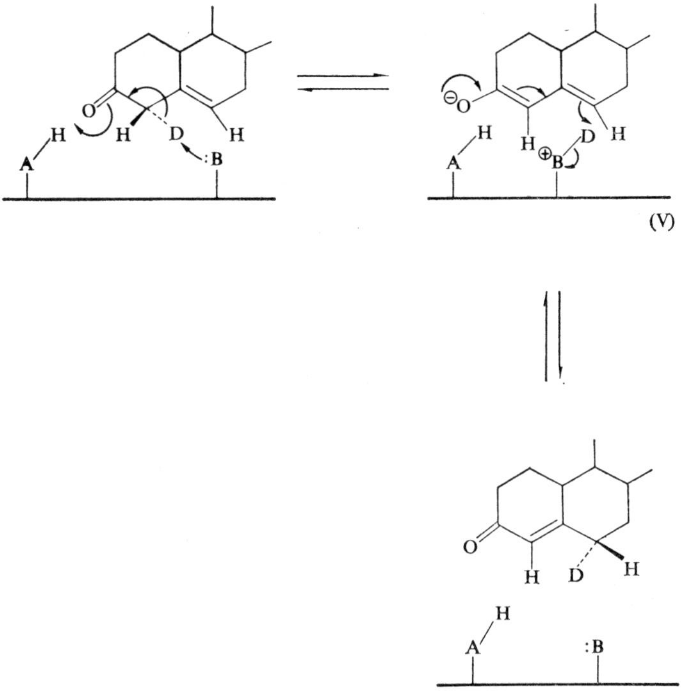


(V)

FIG. 8.3. Mechanism by which a deuterium atom may be transferred from the 4-$\beta$-position of a 3-keto-$\Delta^5$-steroid to the 6-$\beta$-position in the presence of 3-keto-$\Delta^4$-$\Delta^5$-steroid isomerase. The B-group is envisaged as being below the plane of the two six-membered rings

acceptor and then as proton donor is shown by the fact that the reactions take place on the same side of the molecule, the $\beta$-hydrogen being involved in both cases, and also that deuterium from position-4 appears at position-6.

The reaction takes place without exchange with the solvent and this is compatible with the scheme if the proton abstraction from position-4 is far slower than the proton donation to position-6. This is the case since proton removal from position-4 is rate-determining. In addition the positioning of the enol (V) on the surface of the enzyme may render the B group unavailable to solvent molecules.

These workers also comment on the rapidity of the enolization and protonation and suggest that the reactions take place in a hydrophobic region of the enzyme.

In the case of the mammalian enzymes, however, a slightly different picture emerges. A tritium label at the 4-$\beta$ position is lost to the solvent [13]. Furthermore, Oleinick and Koritz [14] showed that in the presence of a rat enzyme in deuterium oxide, deuterium is incorporated into the 6-$\beta$ position and in addition into the C-4 site. It is possible therefore that the group which is envisaged as abstracting the proton from C-4 is not the same group which protonates C-6. However, the mechanism shown in Fig. 8.3 may be compatible with the results obtained for the mammalian enzymes if (a) abstraction of the proton from C-4 is much more rapid than protonation of C-6, and (b) the protonated form of B is able to exchange freely with solvent. Oleinick and Koritz emphasize the similarity between the base-catalysed isomerization and that brought about by the rat enzyme. In the base-catalysed process the protonation of the 6-position is rate-determining. Because of the similarity noted, these authors have proposed that the enolate anion (VI) plays a critical role in the enzymic process.

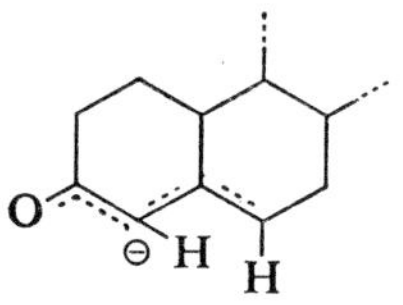


(VI)

It has been suggested [12] that an intact histidine residue is required for enzymic activity; B may therefore be an imidazole group.

**D-Glucose-6-Phosphate Ketol-Isomerase (5.3.1.9)**

The reaction catalysed by this enzyme, the interconversion of D-glucose-6-phosphate and D-fructose-6-phosphate, is another in which a hydrogen is lost from one carbon atom and gained by another. Equation (8.9) shows such an aldose–ketose interconversion for the open-chain forms of the carbohydrates. For glucose-6-phosphate and fructose-6-phosphate the equilibrium ratio is 3 : 1 in favour of the glucose derivative.

$$-\underset{\underset{O}{\|}}{C}-\underset{\underset{OH}{|}}{CH_2} \rightleftharpoons -\overset{H}{\underset{\underset{OH}{|}}{C}}-CHO \qquad (8.9)$$

As in the cases of all the other isomerases which we have considered, the nature of the hydrogen transfer is critical to the mechanism. Two points in particular are of importance, (a) whether the transfer is direct, and (b) whether the hydrogen is transferred as a proton, a hydrogen radical or a hydride ion. In general, transfer of a hydride ion must either be direct or via some hydride transfer agent usually present as a coenzyme. No coenzyme or added co-factors are necessary for glucosephosphate isomerase.

Early experiments suggested that a direct transfer was not involved. Thus, when fructose-6-phosphate was converted into the glucose derivative in deuterium oxide, deuterium was incorporated at C-2 [15, 16]. However, later studies by Rose *et al.* [17] showed that some kind of direct transfer did occur. When fructose-6-phosphate labelled with tritium in the

C-1 position was isomerized, the glucose derivative produced carried tritium on the C-2 position. However, the reaction was not quantitative in that some of the tritium did exchange with the solvent.

Perhaps the best interpretation of these results is provided by a proton-transfer mechanism with an enediol intermediate. If the same basic group B is involved in the proton loss from C-1 and the proton uptake by C-2 then the mechanism shown (Fig. 8.4) explains the apparent direct transfer.

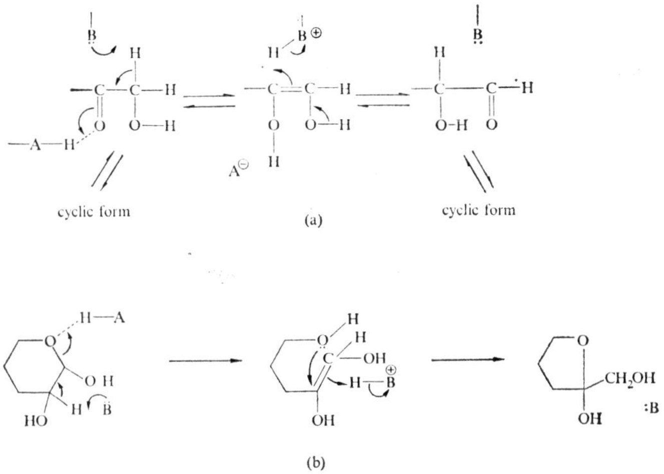


FIG. 8.4. Proton-transfer mechanisms for glucose-6-phosphate isomerase involving an enediol intermediate

The partial nature of the tritium transfer arises if the rates of the individual processes are such that the conjugate acid of B partially exchanges before donating the proton to C-2.

This mechanism is therefore very similar to that proposed for the 3-keto-$\Delta^5$-steroid isomerase. Indeed it appears to represent a situation intermediate between those of the bacterial and the mammalian enzymes.

It is not yet known whether the isomerization takes place with the sugars in the open-chain or the cyclic hemiacetal forms. Figure 8.4(b) shows how an enediol mechanism can apply to the cyclic form. In this case the enediol form is not cyclic, in fact it is identical to the enediol intermediate for the reaction of the open-chain form. Hines *et al.* [18] favour a mechanism involving the cyclic form because they have calculated that if the open-chain forms were involved, $K_m$ values of the order of $10^{-9}$ would be required. This is considerably lower than usual values for $K_m$.

These workers also showed by kinetic and pH-dependence studies that the reaction required an unprotonated group with $pK$ 6–7 and a protonated species with $pK$ 9–10. On this basis, B may be a histidine residue and AH the $\varepsilon$-ammonium group of a lysine. More recently, photo-oxidation studies have suggested that a histidine residue may be essential for the catalytic activity [19].

An enediol intermediate has been suggested for the reaction of ribulose-5-phosphate 3-epimerase, in which inversion takes place at a carbon adjacent to a carbonyl group [15]. During the reaction the substrate takes up deuterium or tritium from water [20]. Either epimer may be formed according to the way in which the enediol takes up the proton.

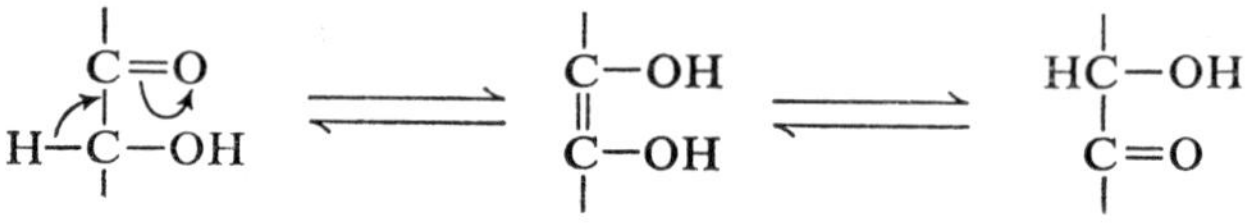

## LIGASES

Each of these enzymes brings about a synthetic reaction, that is, the combination of two molecules to give a third. The reaction is invariably accompanied by the hydrolysis of ATP, and this decomposition of ATP satisfies the energy requirement for the synthetic process. Equations (8.10 and 8.11) give examples of ligase-catalysed reactions, reaction (8.10) being catalysed by $\gamma$-glutamylcysteine synthetase (L-glutamate: L-cysteine $\gamma$-ligase (ADP), 6.3.2.2), the other by acetyl-CoA synthetase (acetate: CoA ligase (AMP), 6.2.1.1).

$$\text{Glutamate} + \text{cysteine} + \text{ATP} \rightleftharpoons \gamma\text{-glutamylcysteine} + \text{ADP} + \text{P} \quad (8.10)$$

$$\text{Acetate} + \text{CoA.SH} + \text{ATP} \rightleftharpoons \text{acetyl-CoA} + \text{AMP} + \text{PP}_i \quad (8.11)$$

In one case we see that the $\beta$–$\gamma$ pyrophosphate link of the ATP is cleaved yielding ADP and $P_i$, while in the other the $\alpha$–$\beta$ link is broken giving rise to AMP and pyrophosphate.

With the exception of the carbon dioxide ligases (6.4.–.–) which are biotin-containing enzymes and are discussed in the chapter on transferases, the reactions catalysed by the ligases are those which, in organic chemistry, are viewed as being brought about by the 'activation' and subsequent reaction of a carboxyl or similar species. In this context 'activation' means the introduction of a good leaving group into the molecule producing what is, essentially, a high-energy species which is susceptible to nucleophilic substitution. For example, a peptide synthesis such as that in equation (8.10) if performed *in vitro* non-enzymically, would require the activation of the carboxyl component as a mixed anhydride or an 'activated ester' prior to the reaction with the amino component [21, 22].

The tripeptide glutathione ($\gamma$-glutamylcysteinylglycine) is synthesized enzymically from $\gamma$-glutamylcysteine and glycine by the enzyme *glutathione synthetase* ($\gamma$-glutamylcysteine : glycine ligase (ADP), 6.3.2.3). The specificity of the enzyme is not very strict with respect to the cysteinyl residue: the synthesis of $\gamma$-glutamyl-$\alpha$-aminobutyrylglycine from $\gamma$-glutamyl-$\alpha$-aminobutyrate and glycine is also catalysed [23, 24, 25]. Early experiments

[26] indicated that glutathione synthetase catalysed exchange between ADP and ATP, suggesting the first step in the reaction to be as shown in equation (8.12), where E—P represents a phosphorylated enzyme species.

$$\text{ATP} + \text{E} \rightleftharpoons \text{ADP} + \text{E—P} \tag{8.12}$$

However, Meister and co-workers have shown that a highly-purified preparation possesses a greatly-reduced ability to catalyse this exchange [25] and have provided further evidence which indicates that reaction (8.12) is not the first stage in the overall reaction.

Although a phosphorylated enzyme is probably not an intermediate in the reaction it is now well established that an activated dipeptide derivative functions in an intermediate capacity. 'Pulse-labelling' experiments were carried out in which large quantities of the enzyme were incubated briefly with $^{14}$C-labelled $\gamma$-glutamyl-$\alpha$-aminobutyrate and ATP, and then a mixture of hydroxylamine and an excess amount of unlabelled dipeptide added. Under these conditions $^{14}$C-labelled dipeptide hydroxamate was formed, in larger quantities than would be expected if equilibration with the free dipeptide had taken place. Meister *et al.* [23, 24, 25] deduced that an activated $\gamma$-glutamyl-$\alpha$-aminobutyryl derivative was formed during the reaction, the amount of labelled dipeptide hydroxamate detected being a measure of the quantity of activated dipeptide formed. The formation of the activated species was rapid at first, eventually slowing down so that the quantity produced reached a limiting value which was dependent on the amount of enzyme present (i.e. an initial burst took place). In other experiments [24] a compound with the properties of $\gamma$-glutamyl-$\alpha$-aminobutyryl phosphate was isolated from reaction mixtures containing $\gamma$-glutamyl-$\alpha$-aminobutyrate, ATP, $Mg^{2\oplus}$ and the enzyme, and the identification of the activated dipeptide derivative as this dipeptide phosphate was strengthened by the observation that the enzyme could utilize chemically-synthesized $\gamma$-glutamyl-$\alpha$-aminobutyryl phosphate for the syntheses of the tripeptide product with glycine, and of ATP from ADP.

The activated dipeptide exists in a complex with the enzyme. This complex exchanges rapidly with free dipeptide, and the rate of this exchange is similar to that of the overall reaction, demonstrating that the activated dipeptide complex is kinetically able to function as a reaction intermediate.

On the basis of these results the overall reaction scheme appears to be:

$$\text{E} + \text{ATP} + \gamma\text{-Glu-Bu} \rightleftharpoons \text{E}.\gamma\text{-Glu-Bu-P.ADP} \tag{8.13}$$

$$\text{E}.\gamma\text{-Glu-Bu-P.ADP} + \text{Gly} \rightleftharpoons \gamma\text{-Glu-Bu-Gly} + \text{ADP} + P_i + \text{E} \tag{8.14}$$

In these equations $\gamma$-Glu-Bu-P represents the species in which a mixed anhydride link exists between the terminal carboxyl group of the dipeptide and the phosphate group.

The ADP remains firmly bound in the dipeptide phosphate–enzyme complex, and therefore in this mechanism ADP–ATP exchange would not be expected, except in the presence of dipeptide and glycine. There is nevertheless a slow release of ADP from the complex as the enzyme-bound acyl phosphate decomposes spontaneously. This would lead to a very low level of ADP–ATP exchange [25].

Equations (8.13) and (8.14) show the minimum number of stages compatible with the data. Although no phosphorylation of the enzyme could be detected [25] it is not possible to exclude entirely the possibility of a prior reaction as shown in (8.15). However the above discussion makes it

$$E + ATP \rightleftharpoons E\text{—}P.ADP \qquad (8.15)$$

clear that if such a reaction does take place, then the ADP cannot dissociate away from the E—P.ADP complex.

In overall mechanistic terms the ATP phosphorylates the carboxyl of the dipeptide, activating it so that peptide synthesis can take place. The phosphorylation takes place by a nucleophilic attack on the terminal phosphate group of the ATP, ADP being the leaving group (see Fig. 8.5). The

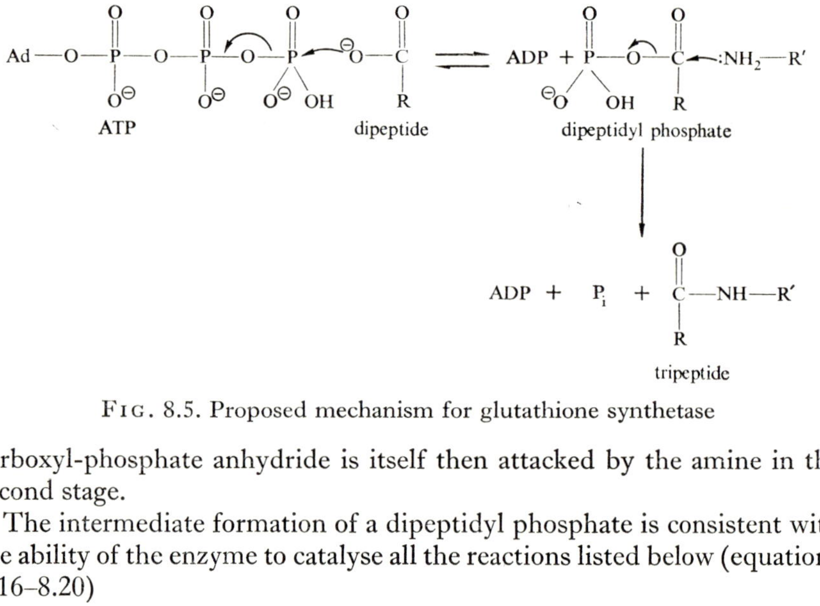


FIG. 8.5. Proposed mechanism for glutathione synthetase

carboxyl-phosphate anhydride is itself then attacked by the amine in the second stage.

The intermediate formation of a dipeptidyl phosphate is consistent with the ability of the enzyme to catalyse all the reactions listed below (equations 8.16–8.20)

$$\text{Dipeptide} + \text{ATP} + \text{Gly} \rightleftharpoons \text{Tripeptide} + \text{ADP} + P_i \qquad (8.16)$$

$$\text{Dipeptide} + \text{ATP} + NH_2OH \rightleftharpoons \text{Dipeptide-NHOH} + \text{ADP} + P_i \qquad (8.17)$$

$$\text{Tripeptide} + NH_2OH \rightleftharpoons \text{Dipeptide-NHOH} + \text{Gly} \qquad (8.18)$$

$$\text{Dipeptidylglycine}^* + \text{Gly} \rightleftharpoons \text{Dipeptidylglycine} + \text{Gly}^* \qquad (8.19)$$

$$\text{Tripeptide} + H_2O \overset{\text{arsenate}}{\rightleftharpoons} \text{Dipeptide} + \text{Gly} \qquad (8.20)$$

Reaction (8.16) is the normal synthetic reaction catalysed by the enzyme, and as in reaction (8.17) the glycine may be replaced [26] by hydroxylamine (the pulse-labelling experiment). Reactions (8.18 and 8.19) involve the transfer of the dipeptidyl group from one amino compound to another. In the case of peptidases, etc., such transfer reactions are often regarded as diagnostic of the intermediate production of a covalent complex. Reaction (8.20) in which hydrolysis of the tripeptide takes place in the presence of arsenate indicates that arsenate may replace phosphate, and in this case a carboxyl-arsenate anhydride link is probably involved. The three reactions (8.18–8.20) all require the presence of phosphate or arsenate, and also of catalytic quantities of nucleotide, ATPo r ADP. The requirements for the nucleotide in these reactions provide more evidence that ADP is necessary for the formation of the dipeptidyl-phosphate–enzyme complex [25].

An essentially similar mechanism is envisaged for the enzyme *γ-glutamylcysteine synthetase* which brings about the production of the dipeptide from glutamate and cysteine (8.10). This enzyme also is relatively non-specific with respect to the cysteine residue and can catalyse the synthesis of several γ-glutamyl peptides [27]. Enzyme-bound γ-glutamyl phosphate (VII) appears to the intermediate formed in this reaction. One difference, however, between the mode of action of this enzyme and glutathione synthetase is that the ADP is not essential for the formation of the intermediate complex. The enzyme-catalysed exchange of ADP and ATP does not require glutamate or cysteine [28]. In view of this, equation (8.12) may represent the first stage in the reaction. Phosphate is not liberated until the addition of cysteine [28] and therefore the reaction sequence after equation (8.12) proceeds as shown.

$$\text{E—P} + \text{Glu} \rightleftharpoons \text{E}.\gamma\text{-Glu-P}$$
$$\text{E}.\gamma\text{-Glu-P} + \text{Cys} \rightleftharpoons \text{E} + P_i + \gamma\text{-Glu-Cys}$$

The reaction only differs from that of glutathione synthetase in the extra step in which the enzyme is phosphorylated. Figurc 8.6 shows the hypothetical reaction mechanisms. When glutamate containing $^{18}O$ in the carboxyl was used in the reaction, the phosphate liberated contained $^{18}O$ [27]. This is in agreement with the function of γ-glutamyl phosphate as an intermediate (see Fig. 8.6).

The same intermediate, *γ-glutamyl phosphate*, is involved in the enzyme-catalysed production of glutamine from glutamate. The enzyme for this reaction is *glutamine synthetase* (L-glutamate : ammonia ligase (ADP), 6.3.1.2).

$$\text{Glutamate} + \text{ATP} + NH_3 \rightleftharpoons \text{Glutamine} + \text{ADP} + P_i \qquad (8.21)$$

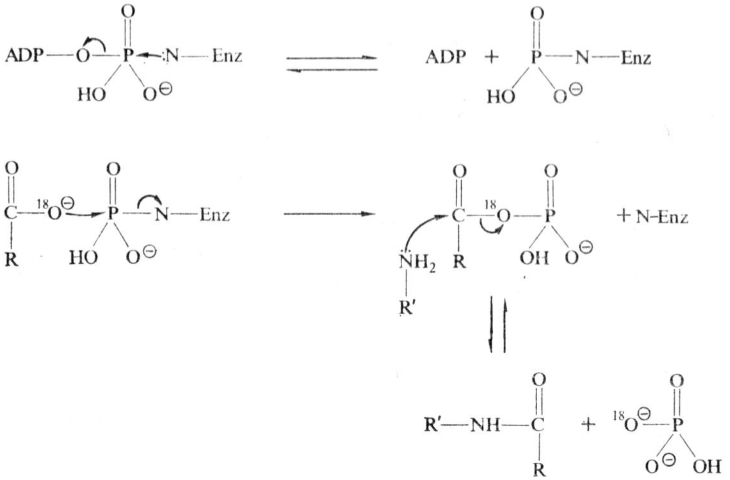

FIG. 8.6. Proposed mechanism for $\gamma$-glutamylcysteine synthetase. N represents a nucleophilic species at the active site of the enzyme

The stepwise nature of this reaction was first suggested by Levintow and Meister [29] and was confirmed by further studies by Meister and his co-workers [30, 31, 32, 33]. In the presence of ATP and $Mg^{2\oplus}$ and in the absence of ammonia, glutamate combined with the enzyme to give an enzyme-activated glutamate complex which also contained ADP and $P_i$. This complex slowly decomposed liberating ADP and producing pyrrolidone-2-carboxylate (VIII). (Compare the slow release of ADP from the activated complex of gluthathione synthetase.) With ammonia the complex gave glutamine. On denaturation of the complex, glutamate, ADP and phosphate were obtained. Again, as in the case of glutathione synthetase, the enzyme–glutamyl phosphate complex requires the presence of the nucleotide. Thus, glutamate is not bound in the absence of ATP and in

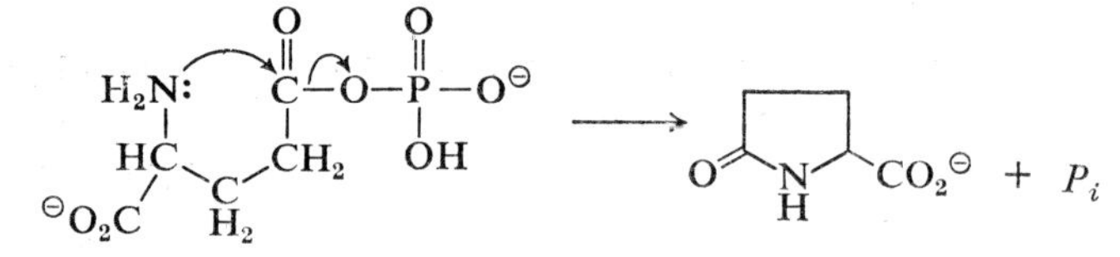

(VII) (VIII)

the reverse direction glutamine is not bound in the absence of ADP. Meister *et al.* [30] conclude that the binding site for the amino-acid does not exist until the nucleotide has combined with the enzyme, and that the ADP itself may be an important binding group.

When glutamate labelled with $^{18}O$ was used with the enzyme, the labelled oxygen appeared in the phosphate liberated [34]. This helps to confirm that glutamyl phosphate is an intermediate in the reaction. On the whole, therefore, the reaction mechanism for glutamine synthetase is closely similar to that for glutathione synthetase (Fig. 8.5).

The production of pyrrolidone-2-carboxylate by the glutamyl phosphate intermediate is due to an intramolecular nucleophilic attack by the $\alpha$-amino-group on to the carboxyl-phosphate anhydride.

The two enzymes $\gamma$-glutamylcysteine synthetase and glutathione synthetase bring about the synthesis of peptides. However, as we have seen their specificities are not absolute and other amino-acids may be substituted, especially for cysteine. For the synthesis of small peptides this type of enzymic activity may not be inconvenient for living systems, but for protein synthesis enormous disadvantages would exist. An individual enzyme to bring about the formation of each peptide bond in each protein would require an enormous number of enzymes, each one being a protein needing to be synthesized in the same way, an obviously impossible situation. In addition the lack of absolute specificity of this type of enzyme would lead to too many mistakes being made in the protein sequences, with disastrous effects on the life of the organism.

The synthesis of proteins is carried out in the cell by another mechanism entirely. An amino-acid is coupled to a specific RNA molecule (soluble or transfer RNA, t-RNA), each type of t-RNA molecule being able to accept only one particular amino-acid. Peptide bond formation takes place in the ribosomes in the presence of another type of RNA (messenger RNA, m-RNA) which determines the sequence of amino-acids in the protein. The m-RNA and t-RNA possess recognition sites by virtue of their nucleotide sequences. For each t-RNA specific for one particular amino-acid the m-RNA possess a 'codon' of three nucleotide residues, a triplet, and the t-RNA possesses an 'anticodon', a complementary triplet of nucleotide residues. Since the recognition between the m-RNA and the t-RNA, vital to correct synthesis of proteins, is entirely dependent on the nature of the two nucleic acid molecules and not on the amino-acid carried by the t-RNA, it is clearly essential that the correct amino-acid be attached to the correct t-RNA molecule. Therefore, the enzymes which bring about this attachment, the *amino-acyl–t-RNA synthetases*, must in their overall reactions demonstrate a specificity far greater than is usually apparent in enzymes. This high degree of specificity is due mainly to the interaction between the t-RNA and the protein molecule. An interaction between two large molecules such as RNA and a protein is more complex, subject to more sophisticated structural requirements and therefore more specific, than an interaction between a large and a small molecule, such as that between enzyme and amino-acid.

All the t-RNA molecules are very similar, containing 70–80 nucleotide residues per molecule, and each one ends in a free adenosine residue. An amino-acyl-t-RNA synthetase (6.1.1.–) brings about the esterification of the amino-acid by the 2′- or 3′-hydroxyl of the terminal adenosine residue of the t-RNA specific to that amino-acid. Equation (8.22) shows the pro-

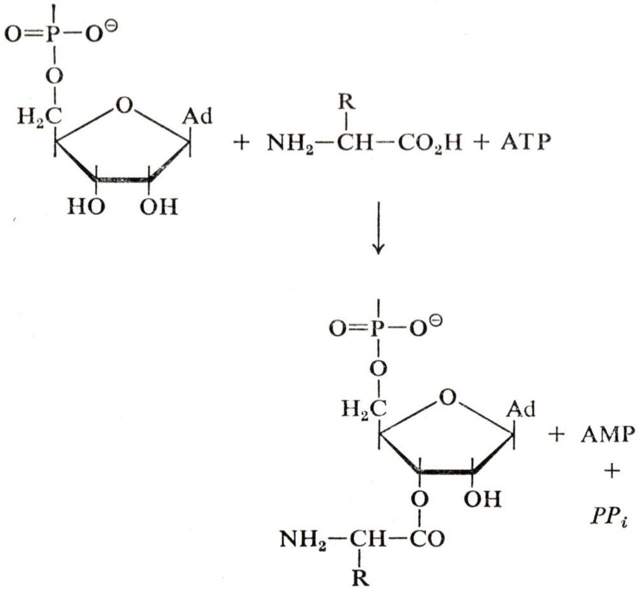

(8.22)

duct if a 3′-ester is produced. As indicated in this equation the phosphate–phosphate bond of the ATP which is cleaved is not the same as the one cleaved by the ligases previously discussed in this chapter. The bond cleaved is the $\alpha$–$\beta$ link and AMP and pyrophosphate ($PP_i$) are produced.

It is now well established that for each of these enzymes the reaction is a two-step process in which an enzyme-bound amino-acyl adenylate

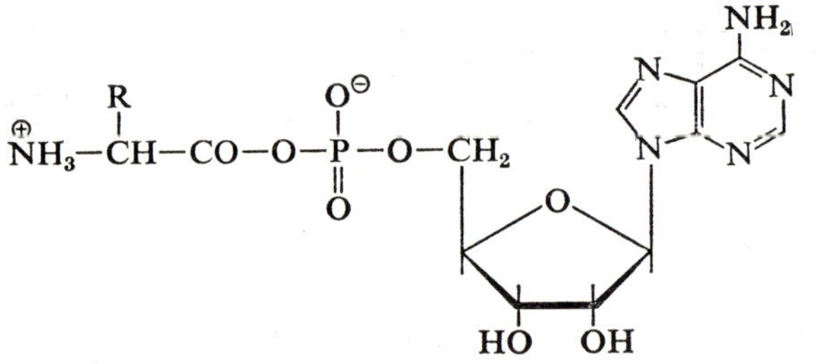

(IX)

(IX) is formed as intermediate. Thus, the amino-acid is first converted into an enzyme-bound activated state, and subsequently transferred to the terminal residue of t-RNA [35].

$$aa + E + ATP \rightleftharpoons aa\text{-}AMP.E + PP \qquad (8.23)$$

$$aa\text{-}AMP.E + t\text{-}RNA \rightleftharpoons aa\text{-}t\text{-}RNA + AMP + E \qquad (8.24)$$

The identity of the activated amino-acid intermediate has been demonstrated for many of the enzymes. Thus, incubation of the tryptophanyl–t-RNA synthetase with tryptophan and ATP gave a complex which contained amino-acid and AMP [36]. Similar complexes containing other amino-acids, AMP and the corresponding enzymes have also been isolated [37, 38, 39, 40]. These complexes undergo the reactions expected of them as intermediates; for example, threonine is transferred from the threonyl–AMP–enzyme complex to threonine–specific t-RNA [37], and isoleucine is transferred from the isoleucyl–AMP–enzyme complex to its specific t-RNA [38].

These complexes may also be formed by the interaction of chemically synthesized amino-acyl adenylates with the enzymes concerned. Whether prepared from the free amino-acid and ATP or from the amino-acyl adenylate, the complexes are able to synthesize ATP from pyrophosphate [41].

Cassio *et al.* [42] prepared some amino-alkyl adenylates (X) and tested them as inhibitors to various enzymes. These compounds, which are closely similar to the amino-acyl adenylates, were potent inhibitors, the

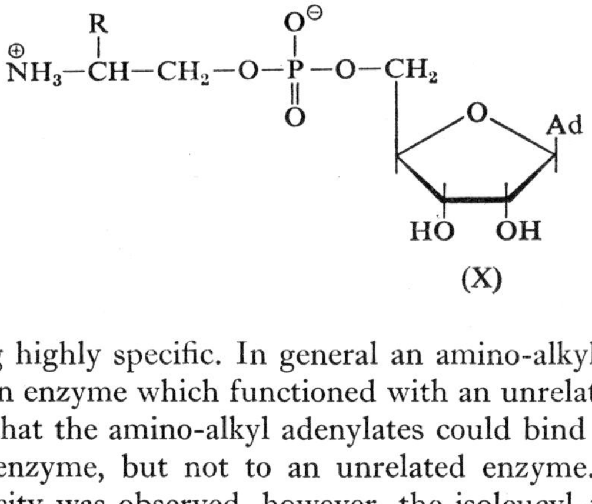


(X)

inhibition being highly specific. In general an amino-alkyl adenylate had little effect on an enzyme which functioned with an unrelated amino-acid. This indicates that the amino-alkyl adenylates could bind strongly to the corresponding enzyme, but not to an unrelated enzyme. Some lack of absolute specificity was observed, however, the isoleucyl–t-RNA synthetase being inhibited by the valinyl derivative, and valyl–t-RNA synthetase being inhibited by the derivatives corresponding to isoleucine and alanine.

This non-specificity demonstrates a general feature of these enzymes, which is that the first step in the reaction, the formation of the intermediate complex, is not absolutely specific. Most synthetic amino-acyl adenylate–enzyme systems are able to synthesise ATP from pyrophosphate in the presence of tyrosyl–t-RNA synthetase and of methionyl–t-RNA synthetase

[41]. In addition isoleucyl–t-RNA synthetase can form the adenylate–enzyme complexes with both isoleucine and valine [39]. The great specificity which is essential to the correct synthesis of proteins appears in the second stage of the reaction, equation (8.24) in which the amino-acid is transferred to the t-RNA. The specificity of this step arises from two factors. The first of these is the nature of the binding between the RNA and the enzyme. Only the corresponding t-RNA can accept an amino-acid residue from the enzyme. Even the correct RNA from another species may not always function successfully. For example, whereas the valyl-, phenylalanyl- and leucyl-enzymes from yeast are able to sterify the corresponding t-RNA molecules from *E. coli*, of the *E. coli* enzymes only valyl–t-RNA synthetase esterified yeast t-RNA [43, 44]. The second factor is perhaps more subtle. Even if the enzyme and the t-RNA are complementary, the 'wrong' amino-acid will still not be transferred. Baldwin and Berg [39] produced the valyl–AMP–enzyme complex using an enzyme specific for isoleucine. In the presence of the t-RNA specific for isoleucine, the complex broke down giving valine and AMP. When isoleucine was used the reaction proceeded normally.

Binding between the RNA and the enzyme is therefore not all that is involved in the reaction. The interaction of RNA and protein is thought to produce a conformational change in the enzyme [45, 46]. Circular dichroism measurements showed a conformational change in tyrosyl–RNA synthetase on treatment with tyrosine-specific t-RNA [46]. The hydrolysis of the 'wrong' complex above in the presence of RNA may be the result of such a change.

In some cases, when the ribose unit of the terminal adenosine residue of the t-RNA is destroyed by periodate oxidation and the RNA is then no longer able to accept the amino-acid residue, the remaining RNA is still able to bind and inhibit [47, 48]. However, this is not invariably the case [48] and there appear to be great variations in the ways that modified RNA's influence the enzymic activity. At present no really consistent pattern of evidence appears to be available concerning the interactions between these enzymes and their corresponding t-RNA molecules, and it is possible that each enzyme–t-RNA pair interacts in a highly individual way.

The activation of the amino-acid in this type of reaction therefore proceeds by an attack by the carboxylate on to the $\alpha$-phosphate group of the ATP, pyrophosphate being the leaving group (Fig. 8.7). The carboxylate–AMP anhydride group then reacts with the acceptor hydroxyl group of the t-RNA yielding the required ester.

The synthesis of acetyl-CoA from acetate and coenzyme A, catalysed by *acetyl-CoA synthetase*, also proceeds via the intermediate formation of an acyl-adenylate complex. Both acetate and CoA are required to bring about

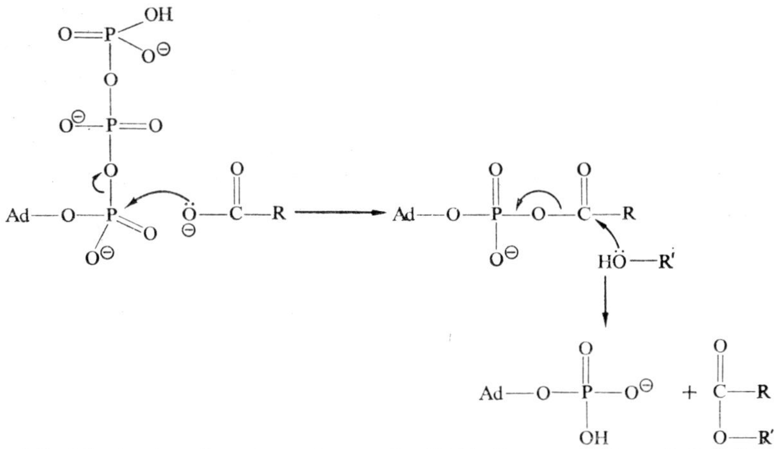

FIG. 8.7. Mechanism for the amino-acyl-t-RNA ligase reaction. $R.CO_2^{\ominus}$ represents the amino-acid which becomes 'activated', and R′-OH represents the t-RNA

the enzyme-catalysed exchange of AMP and ATP, and both AMP and pyrophosphate are needed for the acetate–acetyl-CoA exchange [49].

$$\text{ATP} + \text{acetate} + \text{E} \rightleftharpoons \text{E.acetyl-AMP} + \text{PP}_i$$
$$\text{E.acetyl-AMP} + \text{CoASH} \rightleftharpoons \text{E} + \text{acetyl-CoA} + \text{AMP}$$

Furthermore the enzyme catalysed the synthesis of ATP from synthetic acetyl adenylate and pyrophosphate at a rate similar to that of the pyrophosphate–ATP exchange. Acetyl adenylate could also be recovered from the acid-denatured intermediate complex [50].

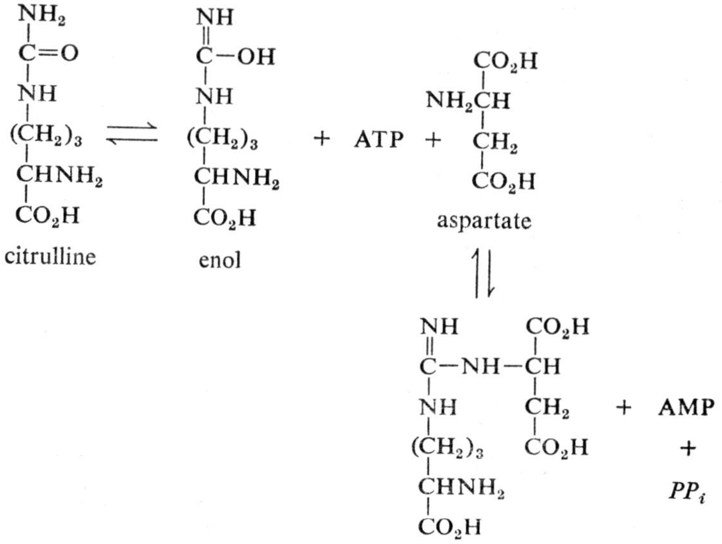


(8.25)

A similar enzyme-bound adenylate species is postulated as an intermediate in the reaction of *arginosuccinate synthetase* (citrulline : L-aspartate ligase (AMP), 6.3.4.5) which catalyses the reaction shown in equation (8.25). The similarity of the 'enol' form of the ureido group of citrulline to a carboxylic acid group is obvious, and the reaction therefore resembles the other reactions discussed in this chapter. Figure 8.8 shows the proposed mechanism for the reaction involving the adenylate intermediate,

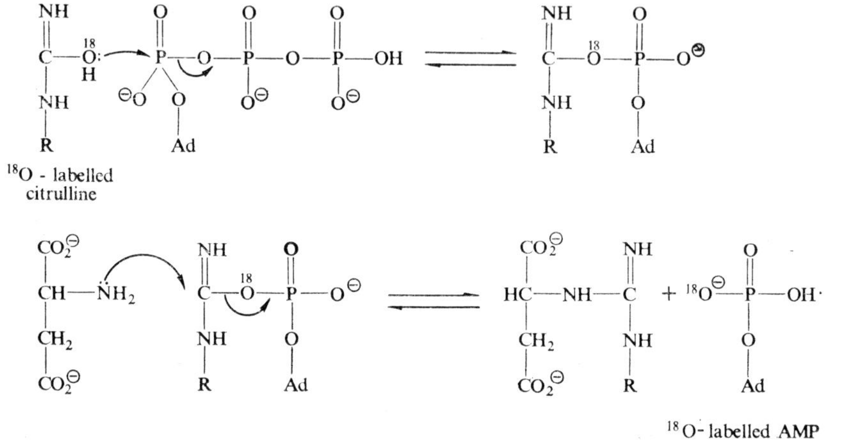


FIG. 8.8. Mechanism for the arginosuccinate synthetase reaction, showing the fate of the $^{18}$O label

and it indicates how $^{18}$O from the ureido oxygen of citrulline appears in the product AMP [51, 52].

The formation of succinyl-CoA, catalysed by *succinyl-CoA synthetase* succinate: CoA ligase (ADP), 6.2.1.5) involves an entirely different mechanism from that which relates to the synthesis of acetyl-CoA. A phosphoenzyme intermediate was suggested by Kaufmann [53] on the basis of an ADP–ATP exchange, and further evidence that such an intermediate might be involved was obtained when a phosphorylated protein was produced by incubation of the enzyme with succinyl-CoA and phosphate, or with ATP [54]. In addition, phosphate could be liberated when the phosphorylated protein was incubated with coenzyme A, coenzyme A and succinate or ADP [55]. The phosphate group is bound to the enzyme at a histidine residue [55].

Another intermediate is formed during the reaction. This is an enzyme–coenzyme A complex, which could be produced by exposure of the enzyme to succinyl-CoA [55]. This complex could also be produced from ATP, and coenzyme A in the presence of the enzyme. It, too, underwent reactions consistent with its being an intermediate. Thus it reacted with phosphate to give the phosphorylated enzyme, and with ADP plus phosphate to give ATP. The free enzyme, the phosphorylated form and the enzyme–coenzyme A complex all possessed equal catalytic activities.

On the basis of these results the reaction scheme for succinyl-CoA synthetase may be formulated as follows:

$$\begin{aligned} \mathrm{E} + \mathrm{ATP} &\rightleftharpoons \mathrm{E{-}P} + \mathrm{ADP} \\ \mathrm{E{-}P} + \mathrm{CoASH} &\rightleftharpoons \mathrm{E\text{-}CoA} + \mathrm{P}_i \\ \mathrm{E\text{-}CoA} + \text{succinate} &\rightleftharpoons \text{succinyl-CoA} + \mathrm{E} \end{aligned}$$

The nature of the link between the enzyme and coenzyme in the intermediate complex is not known. Until this is established the mechanism of action of succinyl-CoA synthetase remains in doubt.

## REFERENCES


1. OLIVARD, J., METZLER, D. E. and SNELL, E. E., *J. biol. Chem.* **199,** 669 (1952)
2. ABELES, R. H. and FREY, P. A., *Fedn Proc.* **25,** 1639 (1966)
3. ABELES, R. H. and ZAGALAK, B., *J. biol. Chem.* **241,** 1245 (1966)
4. BROWNSTEIN, A. M. and ABELES, R. H., *J. biol. Chem.* **236,** 1199 (1961)
5. ERFLE, J. D., CLARK, J. M., JR., NYSTROM, R. F. and JOHNSON, B. C., *J. biol. Chem.* **239,** 1920 (1964); IODICE, A. A. and BARKER, H. A., *J. biol. Chem.* **238,** 2094 (1963)
6. FREY, P. A. and ABELES, R. H., *J. biol. Chem.* **241,** 2732 (1966)
7. RETEY, J. and ARIGONI, D., *Experientia* **22,** 783 (1966)

7a. EGGERER, H., OVERATH, P., LYNEN, F. and STADTMAN, E. R., *J. Am. chem. Soc.* **82,** 2643 (1960)

8. WAGNER, O. W., LEE, H. A., JR., FREY, P. A. and ABELES, R. H., *J. biol. Chem.* **241,** 1751 (1966)
9. COCKLE, S., DAVIES, S., FOSTER, M., HILL, H. A. O. and WILLIAMS, R. J. P., Symposium on *Chemical Reactivity and Biological Role of Functional Groups in Enzymes* 501st meeting, Biochem. Soc. (Jan. 1970)
10. ABELES, R. H. and BECK, W. S., *J. biol. Chem.* **242,** 3589 (1966); VITOS, E., BROWNSON, C., GARDINIER, W. and BLAKLEY, R. L., *J. biol. Chem.* **242,** 3035 (1967)
11. KAWAHARA, F. S. and TALALAY, P., *J. biol. Chem.* **235,** PC1 (1960); TALALAY, P. and WANG, V. S., *Biochim. biophys. Acta* **18,** 300 (1955)
12. MALHOTRA, S. K. and RINGOLD, H. J., *J. Am. chem. Soc.* **87,** 3228 (1965)
13. WERBIN, H. and CHAIKOFF, I. L., *Biochim. biophys. Acta* **71,** 471 (1963)
14. OLEINICK, N. L. and KORITZ, S. M., *Biochemistry* **5,** 3400 (1966)
15. TOPPER, Y., *The Enzymes* Vol. 5, p. 413 (Academic Press, New York, 1961)
16. TOPPER, Y., *J. biol. Chem.* **225,** 419 (1957)
17. ROSE, J. A. and O'CONNELL, E. L., *J. biol. Chem.* **236,** 3086 (1961)
18. HINES, M. C. and WOLFE, R. G., *Biochemistry* **2,** 770 (1963)
19. CHATTERJEE, G. C. and NOLTMANN, E. A., *Eur. J. Biochem.* **2,** 9 (1967)
20. WOOD, W. A. and MCDONOUGH, M. W., *Fedn Proc.* **18,** 354 (1959)

21. SCHRODER, E. and LUBKE, K., *The Peptides* Vol. 1 (Academic Press, New York, 1965)
22. HOFMANN, K. and KATSOYANNIS, P. G., *The Proteins*, Ed. H. Neurath, 2nd edn, Vol. 1, p. 53 (Academic Press, New York, 1963)
23. NISHIMURA, J. S., DODD, E. A. and MEISTER, A., *J. biol. Chem.* **239,** 2553 (1964)
24. NISHIMURA, J. S., DODD, E. A. and MEISTER, A., *J. biol. Chem.* **238,** PC1179 (1963)
25. MOOZ, E. D. and MEISTER, A., *Biochemistry* **6,** 1722 (1967)
26. SNOKE, J. E., *J. biol. Chem.* **213,** 813, 825 (1955); SNOKE, J. E., YANARI, S. and BLOCH, K., *J. biol. Chem.* **201,** 573 (1953)
27. RATHBUN, W. B., *Arch. biochem. Biophys.* **122,** 62, 73 (1967)
28. STRUMEYER, D. H. and BLOCH, K., *J. biol. Chem.* **235,** PC27 (1960)
29. LEVINTOW, L. and MEISTER, A., *J. Am. chem. Soc.* **75,** 3039 (1953)
30. KRISHNASWAMI, P. R., PAMILJANS, V. and MEISTER, A., *J. biol. Chem.* **237,** 2923 (1962)
31. KRISHNASWAMI, P. R., PAMILJANS, V. and MEISTER, A., *Fedn Proc.* **19,** 349 (1966)
32. KRISHNASWAMI, P. R., PAMILJANS, V. and MEISTER, A., *J. biol. Chem.* **235,** PC39 (1960)
33. MEISTER, A., KRISHNASWAMI, P. R., PAMILJANS, V. and DUMVILLE, G., *Fedn Proc.* **20,** 229 (1961)
34. BOYER, P. D., KOEPPE, O. J. and LUCHSINGER, W. W., *J. Am. chem. Soc.* **78,** 356 (1956)
35. NOVELLI, G. D., *Ann. rev. Biochem.* **36,** part 2, 449 (1967)
36. WONG, K. and MOLDAVE, K., *J. biol. Chem.* **235,** 694 (1960)
37. ALLENDE, J. E., ALLENDE, C. C., GATICA, M. and MATAMALA, M., *Biochem. biophys. Res. Commun.* **16,** 342 (1964)
38. NORRIS, A. T. and BERG, P., *Proc. natn. Acad. Sci. U.S.A.* **52,** 330 (1964)
39. BALDWIN, A. N. and BERG, P., *J. biol. Chem.* **241,** 833 (1966)
40. LAGERKVIST, U. and WALDENSTROM, J., *J. biol. Chem.* **240,** 2264 (1965)
41. MOLDAVE, K., CASTELFRANCO, P. and MEISTER, A., *J. biol. Chem.* **234,** 841 (1959)
42. CASSIO, D., LEMOINE, F., WALLER, J.-P., SANDRIN, E. and BOISSONNAS, R. A., *Biochemistry* **6,** 827 (1967)
43. LAGERKVIST, U. and WALDENSTROM, J., *J. molec. Biol.* **8,** 28 (1964)
44. HAYASHI, H., *J. molec. Biol.* **19,** 161 (1966)
45. DEUTSCHER, M. P., *J. biol. Chem.* **242,** 3601 (1967)
46. OHTA, T., SHIMADA, I. and IMAHORI, K., *J. molec. Biol.* **26,** 519 (1967)
47. PREISS, J., BERG, P., OBENGAND, E. J., BERGMANN, F. H. and DIECKMANN, M., *Proc. natn. Acad. Sci. U.S.A.* **45,** 319 (1959)
48. ROY, K. L. and TENER, G. M., *Biochemistry* **6,** 2847 (1967)
49. BERG, P., *J. biol. Chem.* **222,** 991, 1015 (1956)
50. WEBSTER, L. T., *J. biol. Chem.* **242,** 1232 (1967)
51. ROCHOVANSKY, O. and RATNER, S., *J. biol. Chem.* **236,** 2254 (1961)
52. ROCHOVANSKY, O. and RATNER, S., *J. biol. Chem.* **242,** 3839 (1967)
53. KAUFMANN, S. *J. biol. Chem.* **216,** 153 (1955)

54. Mitchell, R. A., Butler, L. G. and Boyer, P. D., *Biochem. biophys. Res. Commun.* **16,** 545 (1964); Kreil, G. and Boyer, P. D., *Biochem. biophys. Res. Commun.* **16,** 551 (1964)
55. Ramaley, R. F., Bridger, W. A., Moyer, R. W. and Boyer, P. D., *J. biol. Chem.* **242,** 4287 (1967)

# CHAPTER 9
# THE NATURE OF CATALYSIS BY ENZYMES

It is always tempting for the scientist to search for a single unified theory which will embrace a whole field or range of phenomena. As the previous chapters have shown, there appears to be no single all-embracing theory to account for catalysis by enzymes, just as there is no simple mechanism for all the reactions in organic chemistry. Even within groups of enzymes which catalyse similar reactions the mechanisms by which the catalysis is achieved may be sometimes slightly, sometimes extensively, different. For example, the hydrolysis of a peptide bond by chymotrypsin is brought about in a manner which is in many ways very different from that in which the same peptide bond may be hydrolysed by pepsin.

In chapter 1 the catalysis of some organic reactions by relatively simple compounds was discussed, and certain modes or mechanisms of catalysis were distinguished. These may be summarized as (a) catalysis by acids and bases (in the most general sense) and (b) alternative pathway catalysis (sometimes called nucleophilic catalysis because the catalyst usually functions as a nucleophile). In the former the reactivity of certain groups in the substrate molecule is enhanced by association with an acid (which improves an electron terminus or sink) or a base (which increases nucleophilic character or assists an electron source). As for (b), it should be noted that the term alternative pathway catalysis covers a wide range of possible reactions and mechanisms, the essential point being that each step in an alternative pathway must be more rapid than the slowest in the uncatalysed reaction.

It was also noted that a change in the environment of the reacting molecules, for example a change in the dielectric constant, would influence the rate of a reaction. If the rate of a reaction is increased in this way by the addition of a new solvent to the existing one then the added solvent could be regarded as a catalyst.

Other factors were mentioned which, although not in themselves mechanisms for catalysis, could have profound effects on reaction rates. In essence they consist of special situations in particular molecules which result in higher reactivities as compared with similar molecules without these special factors. They include the proximity effect whereby two reacting groups, if suitably orientated with each other in the same molecule, react more rapidly than they would if they existed in two separate independent molecules. The concerted acid–base catalysis may be regarded as an extension of the proximity effect. Finally, the possession of strain by a molecule will result in a higher rate for any reaction which releases that strain.

It would be remarkable if, in seeking to explain enzymic catalysis, various workers had not at some time or another invoked these possible factors. Indeed, it is generally assumed at the present time that catalysis by any individual enzyme can be explained largely in terms of them.

The chemistry of any molecule, including its ability to catalyse a chemical reaction, must be explicable in terms of, and only in terms of, its components and the way in which those components are arranged within the molecule. Any theory for the catalysis of a reaction by an enzyme which does not comply with this condition must be regarded as unsatisfactory or at best incomplete. In other words it is not acceptable for a theory to suggest some special factor in the operation of an enzyme unless that factor is either established experimentally or justified theoretically.

The first, and immediately obvious, feature of the structure of an enzyme which is likely to be important in explaining its action is its possession of a wide variety of functional groups. In Chapter 2 it was pointed out that the side-chains of the amino-acids provide many of the functional groups of organic chemistry. In addition the ability of an enzyme to bind a coenzyme or a metal-ion increases the range of functional groups available.

In all, the functional groups which can be provided by an enzyme are suitable for either acid–base or alternative pathway catalysis, or both, and throughout this book many examples may be found. For example, in the case of lysozyme the carboxylic group of residue-35 protonates the glycoside oxygen of the substrate; a tyrosine hydroxyl is suggested as an acidic catalyst in the action of carboxypeptidase A and a protonated imidazole is suggested in the mechanism or action of ribonuclease A. In particular, this last-named group, the imidazole of histidine, has taken on great importance and has been suggested as an essential proton-transfer agent for many enzymes including the transaminases, certain glycosidases, fructosediphosphate aldolase and ribonuclease A (in two roles). The active serine in chymotrypsin, the active lysine of acetoacetate decarboxylase and coenzymes such as pyridoxal or thiamine pyrophosphate clearly take part in alternative pathway mechanisms.

At this point it must be emphasized that the establishment of an alternative pathway mechanism does not automatically explain the catalysis, indeed it *explains* very little. Again, the mere possession of functional groups does not of itself confer on a protein the remarkable properties associated with an enzyme. Albumin and other proteins, for example, possess all the functional groups of chymotrypsin or trypsin and yet show virtually no catalytic properties. Although these groups are clearly essential for the catalysis they are more effective in the active site than elsewhere, and it is this special reactivity which must be explained.

One very important factor in the production of this special reactivity must be related to the proximity effect. Two features of the proximity effect as applied to enzyme-catalysed reactions must be considered. In the first place the protein molecule is so folded that certain functional groups are close enough to each other to influence each other's activity. This will often be by acidic or basic catalysis, but may be by influencing the polarity of the environment—probably an extremely important factor in all enzymes. Probably the best example of this effect, in acid–base catalysis, is that of chymotrypsin where the nucleophilicity of the serine hydroxyl is influenced strongly by the presence of an imidazole and a carboxylate group in the correct attitudes in the three-dimensional structure. In the second place the relevant group or groups in the substrate are correctly positioned relative to the appropriate groups of the enzyme. This of course is achieved by the binding of the substrate to the enzyme in the enzyme–substrate complex. The enzyme–substrate binding is especially important in those cases where there is, in the native state of the enzyme, no pre-existing active site and the introduction of substrate causes a conformational change in the enzyme molecule. Carboxypeptidase A provides a good example of this 'induced-fit' type of mechanism.

The bringing together of the substrate and the reactive groups in the active site into close and suitable proximity has been assumed, if not stated, by most workers to be a major contributing factor in the function of an enzyme. Whenever attempts have been made to synthesize molecules which will mimic the effect of enzymes, almost invariably the approach has been to design a molecule which would bind the substrate and bring it into close proximity with the catalytic groups. Several examples are based on the known ability of imidazole to catalyse the hydrolysis of esters. Wagner, Hou and Pratt [1] have shown that N-stearylhistidine catalyses the hydrolysis of ester (I) more effectively than does N-acetylhistidine.

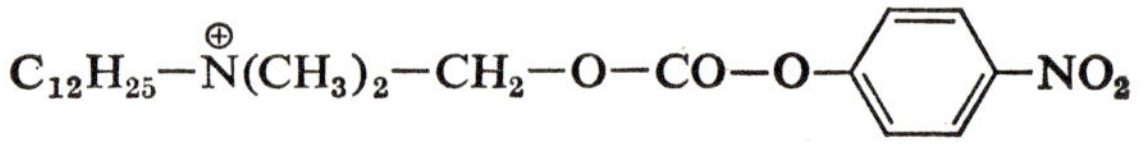


(I)

These workers attribute the enhancement to the formation of an ester–stearylhistidine complex (the large non-polar portions of the molecules binding to each other) in which the imidazole is very close to the ester group. Similarly, Schneider *et al.* [2] report that aromatic histidine derivatives (N-benzoyl-, N-α-naphthoyl-) are better catalysts for the hydrolysis of p-nitrophenyl acetate than are aliphatic derivatives. In a different field,

Bender *et al.* [3] have shown that cycloamyloses catalyse the hydrolysis of certain phenyl esters. This catalysis has properties similar to those characteristic of enzyme catalysis, such as saturation of catalyst by substrate and competitive inhibition. In this case too a catalyst–substrate complex was envisaged in which a hydroxyl group in the cycloamylose is close enough to the carbonyl of the ester to engage in a nucleophilic attack.

Certain polymers have been synthesized to bring about catalysis in the same way. Poly-ethyleneimine catalyses the release of p-nitrophenol from p-nitrophenyl acetate and p-nitrophenyl laurate [4]. When a small proportion of polar side-chains (lauroyl groups) were introduced into the polymer it became a far more effective catalyst for the reaction of p-nitrophenyl laurate than for that of the acetate. Again the enhancement of activity was attributed to the formation of a polymer–substrate complex by the non-polar interaction of the lauroyl groups, bringing together the ester group and the nucleophile. A copolymer of vinylimidazole and acrylic acid is reported to catalyse the hydrolysis of cationic esters with a degree of specificity [5]; in this case the polymer is anionic and binds the cationic esters by electrostatic bonding.

Bruice and Benkovic [6], in discussing the theoretical background to the proximity effect, indicate that they attribute it mainly to entropy considerations, assessing the problem from the point of view of the order of the reaction. Thus the enzyme–substrate complex can be regarded as one single molecule and therefore the reaction between the functional groups of the enzyme and substrate is essentially first order. Such a process has a more favourable entropy of activation than a reaction in which two independent molecules interact in a second-order process.

Very little quantitative work has been done on the thermodynamic parameters of the stages in an enzyme-catalysed reaction which could support this concept. However, one study worth mentioning is that by Bender *et al.* referred to in chapter 4. These workers studied the deacylation step in the hydrolysis of substrates by chymotrypsin. Their clear conclusion was that the variation in values for the rate constants corresponding to different substrates was mainly due to the variation in entropies of activation. The entropy of activation was also shown to be more favourable for better substrates, emphasizing the importance of enzyme–substrate binding to the catalytic effect.

The relatively rigid organization of the large protein molecule is of great importance in connection with the entropy of activation. Because of this organization the enzyme–substrate complex is already in a situation of very low entropy and the change in proceeding to the transition state of the reaction does not involve any further great drop in the entropy factor (i.e. the entropy of activation is not a large negative term).

By contrast, in the uncatalysed case we envisage the reaction as being between two independent species which in forming the transition state must associate into a well-defined structure. Although this transition state may be no more well defined than that of the enzyme-catalysed process, the *change* in entropy in the uncatalysed case must be large and negative. Therefore the overall free energy of activation will be higher and the rate lower.

Of course this argument for the higher rate of reaction in the enzyme–substrate complex is only relevant if the formation of the complex is sufficiently rapid. This stage of the process must have a very unfavourable entropy of activation since it involves the association of two molecules and the taking up of rigidly-defined positions by the reacting groups. This unfavourable factor is, however, likely to be offset in the free energy of activation by the energy released by the formation of bonds, partially or completely.

The proximity or entropy effect probably contributes to the activity of all enzymes to a certain extent. Thoma [7], however, has calculated that it cannot totally account for the high efficiencies of all enzymes. Another factor is certainly involved in some cases; this is the production of strain in the substrate [8] when it is incorporated into the enzyme–substrate complex. The X-ray crystallographic studies carried out on lysozyme [9] indicated that in the enzyme–substrate complex the glycosidic ring at the catalytic site is distorted in such a way that its configuration is more nearly like the transition state for the reaction. In this way the cleavage of the inter-saccharide link is assisted. A non-enzymic example in which the process of binding has a similar effect has been reported by Lovrien and Linn [10]. The *cis–trans* isomerizations of certain azo-dyes were found to be catalysed by some proteins not normally regarded as enzymes. This catalysed process is assumed to occur via the formation of a protein–azo-dye complex in which the substrate molecule is distorted. The distorted configuration of the molecule is similar to the transition state for the *cis–trans* isomerization.

Again we note the importance of the binding of the substrate to the enzyme, since it is this binding process which releases the energy required to bring about the distortion or strain in the substrate.

It has already been stated that the polarity of the environment can influence the rate at which a reaction takes place. Considerable emphasis is now placed on the nature of the microenvironments at the active site as an important feature in the explanation of the catalytic activity of an enzyme. In the case of chymotrypsin several environmental factors can be considered. The first is the non-polar region which accommodates the aromatic side-chain of the preferred substrate. The second is the non-polar region which surrounds the asp-102 carboxylic group and hence favours the localization of negative charge on the ser-195 hydroxyl, resulting in its enhanced nucleophilicity. Thirdly, there is the highly polar region at the active site which may assist the transfer of the negative charge even further, on to the substrate itself. Similarly the non-polar environment of glu-35 in lysozyme decreases the stability of its carboxylate anion form and in effect raises the $pK_a$. This results in its being able to function as a protonating agent at higher pH values. In another region of the active site of lysozyme a microenvironment of high polarity (in the vicinity of asp-52) serves to stabilize the carbonium ion which is postulated as an essential intermediate.

The presence of microenvironments, both polar and non-polar, can contribute therefore to the binding of the substrate and to the enhancement of the reaction rate.

In the first chapter of this book it was made clear that the critical factor which determines the rate of a reaction is the free energy of activation, and hence that a catalyst can only function as such by effectively reducing that free energy of activation. Therefore, one must explain the mechanism by which an enzyme achieves its catalysis by considering how it reduces the activation energy of its particular reaction.

In the most general of terms then we may summarize present views on enzymic catalysis by considering the energy terms of the situations which arise. Let us consider first the uncatalysed process as a single-step reaction. The free energy of activation can be considered to be made up of several factors, (a) the energy required to break partially the bonds which must be broken during the reaction, (b) the energy required to overcome any repulsive forces which may tend to prevent the transition state being achieved, (c) the entropy of activation term, which will be high if the transition state has fewer degrees of freedom than the starting materials. Serving to offset these terms are other factors, (d) the energy obtained by the partial formation of bonds, (e) any energy obtained by the release of strain in proceeding to the transition state. Now we can compare this situation with that in the enzyme-catalysed process. Let us consider a simple case involving three stages. In the first stage the enzyme–substrate complex is formed, in the second this complex is converted into the enzyme–product complex and in the third stage the enzyme–product complex dissociates. The first stage has almost always been found to be very rapid. Presumably the energy of activation is low because a large amount of energy is released in the (at least partial) formation of hydrogen-, ionic-, and non-polar bonds between substrate and enzyme. This term offsets the unfavourable entropy of activation term. It may also offset any unfavourable factor involved in the production of strain in the substrate or in inducing the protein to undergo any conformational change which might be necessary. The energy of activation for the second stage is much lower than that for the equivalent uncatalysed process for the following reasons: (i) the reaction may now be taking place in a medium of different dielectric constant and this may tend to reduce the energy required for bond cleavage and increase the energy released on bond formation, (ii) the formation of the enzyme–substrate complex may have included the formation of hydrogen bonds between certain substrate groups and the acidic and basic groups of the enzyme, i.e. a concerted acid–base catalysis can now take place. The effect of this may be again to reduce the free energy of activation by reducing the energy required for bond cleavage, (iii) the entropy of activation term, which would be very unfavourable in the uncatalysed process because it involves a highly ordered transition state, is now more favourable because most of it has already been taken care of during the first stage, (iv) in some cases the substrate is now in a situation of great strain and the release of the strain means that the energy required for the bond cleavage term is much reduced. The third stage remains. This is very similar to the reverse of the first stage. The transition for the dissociation of the enzyme–product complex is also the transition state for the first stage of the enzyme-catalysed reaction in the reverse direction. In the forward direction the entropy of activation term for the third stage is probably favourable, helping to offset the energy requirement for bond cleavage.

Even in this very brief and qualitative summary we have referred to a number of factors which are assumed to contribute to the enzymic catalysis. Only adequate quantitative information about each step of an enzyme-catalysed reaction can enable the chemist and biochemist to determine which of these factors are most important and hence to understand better the nature of catalysis by enzymes.

## REFERENCES


1. Wagner, T. E., Hou, C.-J. and Pratt, C. S., *J. Am. chem. Soc.* **89,** 6366 (1967)
2. Schneider, V. F. and Wenck, H., *Z. Physiol. Chem.* **348,** 1221 (1967)
3. Van Etten, R. L., Sebastian, J. F., Clowes, G. A. and Bender, M. L., *J. Am. chem. Soc.* **89,** 3242 (1967)
4. Klotz, I. M. and Stryker, V. H., *J. Am. chem. Soc.* **68,** 1486 (1946)
5. Morewetz, H., Overberger, C. G., Salamone, J. C. and Yaroslavsky, S., *J. Am. chem. Soc.* **90,** 651 (1968)
6. Bruice, T. C. and Benkovic, S. J., *Bioorganic Mechanisms* (Benjamin, London, 1966)
7. Thoma, J. A., *J. Theoret. Biol.* **19,** 297 (1968)
8. Jencks, W. P., *Ann. rev. Biochem.* **32,** 658 (1963); *Current Aspects Biochem. Energetics*, Eds. N. O. Kaplan and E. P. Kennedy, p. 273 (Academic Press, New York, 1966)
9. Blake, C. C. F., Mair, G. A., North, A. C. T., Phillips, D. C. and Sarma, V. R., *Proc. R. Soc.* **167B,** 365 (1967)
10. Lovrien, R. and Linn, T., *Biochemistry* **6,** 2281 (1967)

# INDEX

Enzymes are shown in italics. Numbers in bold print indicate major references